WOMEN
OF
SCIENCE
100 Inspirational Lives

About the Author

John Croucher is a Professor of Actuarial Studies and Business Analytics, Macquarie University, Sydney. He has published over 130 research papers and 30 books, and for eight years was a television presenter on football. John holds 4 PhDs and in 2013 won the prestigious Prime Minister's Award for Australian University Teacher of the Year. A fellow of both the Royal Society of Arts and the Australian Mathematical Society, in 2015 John was made a Member of the Order of Australia for 'significant service to mathematical science in the field of statistics, as an academic, author and mentor and to professional organisations'.

WOMEN
OF
SCIENCE
100 Inspirational Lives

John S. Croucher AM
BA (Hons) (Macq) MSc PhD (Minn) PhD (Macq)
PhD (Hon) (DWU) PhD (UTS) FRSA FAustMS

Introduction by Rosalind F. Croucher AM

AMBERLEY

Dedicated to all those bold pioneering women
who have made this world a better place

This edition published 2024

Amberley Publishing
The Hill, Stroud
Gloucestershire, GL5 4EP

www.amberley-books.com

British Library Cataloguing in Publication Data.
A catalogue record for this book is available from the British Library.

ISBN 978 1 3981 1958 1 (paperback)
ISBN 978 1 4456 8472 7 (ebook)

Typeset in 10pt on 12pt Sabon.
Typesetting by Aura Technology and Software Services, India.
Printed in India.

Contents

Foreword by Kate Jenkins   9

Preface   11

Introduction by Emeritus Professor Rosalind F. Croucher AM   13

Frances 'Fran' Elizabeth Allen   21

Virginia Apgar   24

Phoebe Sarah 'Hertha' Marks Ayrton   27

Massimilla 'Milla' Baldo-Ceolin   30

Alice Augusta Ball   32

Nora Stanton Blatch Barney   34

Florence Bascom   36

Laura Bassi   39

Ulrike Beisiegel   41

Elizabeth Helen Blackburn   43

Elizabeth Blackwell   46

Mary Adela Blagg   49

Katharine 'Katie' Burr Blodgett   52

Rachel Littler Bodley   54

Alice Middleton Boring   56

Elizabeth Brown   58

Linda Brown Buck   61

(Susan) Jocelyn Bell Burnell   64

Nina Byers   67

Annie Jump Cannon   69

Rachel Louise Carson   72

Mary Lucy Cartwright   75

Yvette Cauchois   78

Edith Clarke 80
Anna Botsford Comstock 83
Esther Marly Conwell 86
Gerty Theresa Radnitz Cori 88
Marie Skłodowska Curie 90
Ingrid Daubechies 93
Olive Wetzel Dennis 97
Mildred Spiewak Dresselhaus 100
(Gabrielle) Émilie du Châtelet 103
Alice Eastwood 106
Elsie Eaves 109
Gertrude Belle Elion 111
Thelma Austern Estrin 114
Margaret Clay Ferguson 118
Lydia Folger Fowler 120
Rosalind Elsie Franklin 123
(Marie) Sophie Germain 126
Catherine 'Kate' Anselm Gleason 129
Gertrude 'Trude' Scharff Goldhaber 132
Anna Jane Harrison 135
Caroline Lucretia Herschel 138
Beatrice Alice Hicks 141
Dorothy Crowfoot Hodgkin 144
Erna Schneider Hoover 149
Grace Brewster Murray Hopper 152
Margaret Lindsay Murray Huggins 155
Frances Betty Sarnat Hugle 158
Shirley Ann Jackson 160
Sophia Louisa Jex-Blake 163
Nalini Joshi 166
Helen Dean King 169
Maria Margaretha Winckelman Kirch 172
Margaret Galland Kivelson 175
Sofia Vasilyevna Korvin-Krukovskaya Kovalevskaya 178
Doris Kuhlmann-Wilsdorf 181
Stephanie Louise Kwolek 183
Hedwig Eva Maria Kiesler 'Hedy Lamarr' 186
Inge Lehmann 189
Rita Levi-Montalcini 192
Barbara Jane Huberman Liskov 195

Contents

(Augusta) Ada Byron Lovelace	198
Elizabeth 'Elsie' Muriel Gregory MacGill	201
Margaret Eliza Maltby	204
Lynn Petra Alexander Margulis	207
Mileva Marić-Einstein	210
Antonia Caetana de Paiva Pereira Maury	212
Maria Göppert Mayer	215
(Eleanor) Barbara McClintock	218
Elise 'Lise' Meitner	221
Maud Leonora Menten	224
Maria Sibylla Merian	227
Maryam Mirzakhani	230
Maria Mitchell	233
Maria Marcia Neugebauer	236
Ida Tacke Noddack	239
Amalie 'Emmy' Noether	242
Christiane 'Janni' Nüsslein-Volhard	245
Muriel Wheldale Onslow	248
Ruby Violet Payne-Scott	251
Rózsa Péter	255
Agnes Luise Wilhelmine Pockels	258
Helen Rhoda Arnold Quinn	261
Mina Spiegel Rees	264
Ellen Henrietta Swallow Richards	267
Mary Ellen Estill Rudin	271
Hazel Marguerite Schmoll	274
Beatrice ('Tilly') Shilling	277
Michelle Yvonne Simmons	280
Mary Fairfax Greig Somerville	283
'Janet' Jane Ann Ionn Taylor	287
Ruth Lichterman Teitelbaum	290
Valentina Vladimirovna Tereshkova	292
Mildred Trotter	295
Anna Johnson Pell Wheeler	298
Sheila Marie Evans Widnall	301
Rosalyn Sussman Yalow	304
Ada E. Lifshitz Yonath	308
Bibliography	312
Photograph Credits	348

Foreword

So often I hear in my work advancing gender equality in Australia, 'You can't be what you can't see.' This is raised as a lament on how hard it is for pioneering women to forge a path into male domains. But the error too quickly made by many is that there were *no* women in the past working in male-dominated fields like science, technology, engineering and mathematics. I am grateful that Professor John Croucher has written this fascinating book to bring to public prominence the stories of 100 remarkable women of science across several centuries and several countries. These stories both inspire us and remind us that women have for many years played an important role in solving some of our most complex problems.

This book is a page turner. Through short eloquent vignettes, it tells the story of achievement in the face of discouragement, discrimination and denied recognition. Professor Croucher honours these women through sharing their personal stories as well as their professional achievements – reminding us that for women, talent was not sufficient: resilience, persistence and determination were also essential to overcome the barriers they faced. It also shows the importance of supportive men in their lives, men who valued the talents of women as well as men – men like John Croucher.

In reading this book, I encourage you to remember that women today continue to face barriers in accessing the important disciplines now described as 'STEM', wrapping together science, technology, engineering and mathematics. In Australia, women make up only around 20 per cent of the enrolments in tertiary STEM education, make up only 17 per cent of the STEM qualified careers and earn less than their male counterparts (Australian Government, *Advancing Women in STEM Report*, 2019, p. 7). But research indicates that 75 per cent of the fastest growing occupations now require STEM skills, and over 70 per cent of Australian employers identify STEM employees as being among the most innovative (PWC, *A Smart Move: Future Proofing Australia's Workforce by Growing Skills in STEM*, April 2015, pp. 4, 14).

The causes of poor attraction and retention of women and girls in STEM begin from an early age and continue in the workplace and at senior levels. If women and girls continue to be underrepresented in STEM education and careers, Australia will not reach its full potential (*Advancing Women in STEM Report*, p. 8). This book reminds us of the significant opportunities the world stands to gain by increasing the representation of women in STEM.

This volume adds to a process of highlighting, and honouring, the contributions of women in fields traditionally the province of men. Their stories, like those of the men, are driven by the same essential energy and passion. But their stories – the women's – need considerably more amplification to overcome the historic invisibility of women in STEM. There is so much more to be written and if this book, written so accessibly, leaves you 'hungry' for more, then I think it has achieved its purpose.

Kate Jenkins
Sex Discrimination Commissioner
Australian Human Rights Commission

Preface

This book was inspired by the success of a previous work, *Mistress of Science: The Story of the Remarkable Janet Taylor, Pioneer of Sea Navigation*, which I co-authored with Rosalind Croucher. Janet was an outstandingly gifted mathematician, navigator, author, teacher and compass adjuster in 19th-century England. And yet, until the biography on her life, she remained largely unknown.

Across the centuries there have been many other tales of remarkable women in science and their incredible achievements serve as an encouragement to others who have a yearning to follow in their footsteps. One of the hardest tasks here was to limit the number of women included in this volume to just one hundred. To be honest, this number could have easily been doubled, tripled and even more, but space did not permit such a luxury. And so I had to make a selection of a cross-section from different time periods, countries and disciplines. It is not as comprehensive as an encyclopaedia, but provides a valuable reference tool and starting point for those who wish to further expand their knowledge of those women who are included.

The women embraced by this work are largely from the USA, Britain, Europe and Australia, although other countries are also included. This was mainly as a result of historical information being difficult to obtain from other parts of the world, although this leaves scope for a further volume, or volumes. The structure of the book is similar for each entrant, beginning with their families, upbringing as a child and what motivated them to pursue a career in science.

The research involved spanned several years and numerous sources were called upon to create the final product. In some case there were discrepancies in the backgrounds of the women, but all efforts were made to include the one that seemed correct. There are over 500 references provided at the end of the volume, and special thanks must go to the exceptional Macquarie University researcher Stephnie Hon who worked tirelessly to create and format a comprehensive list that the reader can easily access. This was no easy assignment and she performed it with great distinction.

Much appreciation goes to the historian Dr Annette Salt, who took on the challenge of not only finding appropriate images of the women included, but to obtain copyright permission where it was required. This was a daunting assignment that involved liaising with numerous people worldwide, sometimes frustratingly so, as in some cases nobody knew who owned the copyright or those that did refused to respond or wanted a very large sum of money for its use. In the end, Annette was able to secure photographs or portraits of ninety-six women, an incredible achievement that warrants much well-deserved praise. For the reader, being able to see a likeness of the woman portrayed brings her story to life and arouses the imagination. The acknowledgement of each one is shown in a separate section.

Special thanks also go to Alex Bennett at Amberley Publishing for believing in this volume that honours the lives of so many women and allowing me to bring it to life. It has enabled me to breathe life into the historical remnants to create a tribute to these women of science who paved the way for others to follow in their footsteps. Of course none of this would have been possible without the wonderful assistance of my loving wife Rosalind, who wrote the Introduction and whose laughter, patience and guidance made it such an enjoyable project.

John S. Croucher AM

Introduction

Women of Science: 100 Inspirational Lives recognises extraordinary achievements of women across five centuries in a wide range of scientific disciplines. In the condensed cameos presented in this volume, we get a glimpse into women's lives that are indeed inspirational. There are common themes across the lives of these women – often an early passion for the subject that became their life's work, and an abiding curiosity. This hunger to learn, and to solve problems, was essential for scientific pursuit, which requires patience, observation and application, the essence of empirical work, often repeated many, many times to test sometimes fragile, and bold, hypotheses. For every one of the women we meet in this volume, we see a story of incredible, persistent, even stubborn, determination and courage.

The example of Italian chemist Agnes Pockels, born in 1862 in Venice, is illustrative. Agnes had a passionate interest in natural science, especially physics, but in her day, women were not allowed to enter universities. Her younger brother was, and he passed on what he could, giving access to his textbooks, but Agnes lacked the equipment for experiments or a laboratory. Undeterred, and while looking after the household of her ageing parents, she used her daily household activities for experiments on the surface tension of liquids while doing the washing-up in her kitchen. Through curiosity and application, Agnes measured the surface tension of water by devising an apparatus known as the 'slide trough', a key instrument in the emerging discipline of 'surface science'. Through her brother, she learned of an American chemist's work in this field. She wrote to him, telling of the results of her own experiments, which he was welcome to utilise in his own work. Lord Rayleigh (John William Strutt) was so impressed by Agnes's exposition of her work that he submitted her letter to the prestigious journal *Nature*, under her name, together with an article of his own. With such a sponsor, her letter was indeed published, which heralded the commencement of Agnes's career and reputation in studying surface films.

Not all the women in this volume were credited in this way for their achievements. Consider the example of the story of Austrian physicist Lise

Meitner. With the chemist Otto Hahn, she discovered several new isotopes and, in 1918, while studying radioactivity, they discovered the element 'protactinium'. In 1926, Lise became the first woman in Germany to assume a post of full professor in physics, at the University of Berlin. But, in 1938, Austria was annexed by Germany, and Lise, as a Jew, was forced to flee. In 1944 it was Hahn who was awarded the Nobel Prize for Chemistry for his research into fission, partly because he downplayed Lise's role after she left Germany. Her omission was never acknowledged, but was partly rectified in 1966, when Hahn and Meitner were recognised in the Enrico Fermi Award, an American award honouring lifetime achievements of scientists of international stature in the development, use, or production of energy.

For American chemist Alice Augusta Ball, her pioneering work towards a treatment for leprosy was essentially taken by another chemist after her death in 1916 at the age of twenty-four, without crediting Alice at all and labelling the method with his name. Some restoration of her standing was made in 2000, when 29 February was named as 'Alice Ball Day' in Hawaii, where she had undertaken her research.

Whereas British astronomer Caroline Herschel, born in 1750, was credited with discoveries of eight comets and locating a number of new nebulae and star clusters, German astronomer Maria Kirch was not so lucky. On 21 April 1702, Maria discovered the so-called 'Comet of 1702', but her husband took the credit for it. It was not until 1710, the year he died, that he finally admitted the truth, and Maria was then deemed to be the first woman to have discovered a comet.

British chemist Rosalind Franklin should have been credited for her role in discovering the structure of DNA, but she died in 1958. In 1962, James Watson, Francis Crick and Maurice Wilkins were awarded the Nobel Prize in Physiology or Medicine for solving the problem. Watson had suggested that Rosalind should be recognised, but the Nobel Committee cannot make posthumous nominations. In his 1968 book *The Double Helix*, Watson outlined how the two had become friends while working together and that he would never have won a Nobel Prize or published a famous paper if it wasn't for Rosalind's work.

The Serbian mathematician Mileva Marić-Einstein's contribution to the early work of her husband, the theoretical physicist Albert Einstein, is a matter for conjecture. There appears to be at least some evidence that she helped him in his early research. She told a friend in 1905 that she and Albert 'finished some important work that will make my husband world famous'. Mileva's evident brilliance and the possibility of such contribution, and its non-recognition, merits her inclusion in this volume.

The author, Professor John S. Croucher, is one who has lived and taught his whole life in science disciplines; and, having written a biography of a nineteenth century woman scientist (co-authored with me) and a relative of his, lovingly and conscientiously written, has embarked on the wider project of honouring peers of his ancestor across the centuries in the one hundred women selected here.

The author uses the first names of the women he has included. For women, surnames are always an issue. In countries with English-speaking traditions, women's surnames are customarily their father's; and, in times past, women usually adopted their husband's surnames. In other countries the tradition may be different. But in every place, the only thing that a woman can truly say is her own, is her given name. So the adoption of given names is respectful of the individual identity of the women represented here.

Of the women in this volume, half did not marry. While being unmarried did not necessarily lead to regret, it was an issue that many felt they had to address. Italian neurologist and 1986 Nobel Prize winner Rita Levi-Montalcini, for example, never married or had children. In an interview in 2006 she said, 'I never had any hesitation or regrets in this sense. My life has been enriched by excellent human relations, work and interests. I have never felt lonely.' On 22 April 2009, Rita became the first Nobel laureate ever to reach the age of 100.

Some of those who married lost their jobs as a result. Australian physicist Ruby Payne-Scott kept her marriage in September 1944 secret for as long as she could. During the Second World War, Ruby had been involved in top-secret work in a government agency and she had become an expert on the radar detection of aircraft. When her marriage was discovered, after the war, she lost her permanent position as a result of a government rule against employment of married women. In her private life, however, Ruby was also a member of the Communist Party, an atheist and an avid advocate for women's rights, which meant that she had come to the attention of the Australian Security Intelligence Organisation (ASIO) as a 'person of interest' from 1948 to 1959. Regardless of such associations, as a result of Ruby's work with others, Australia became a world leader in radio astronomy.

For those who did marry, many found partners who were their intellectual equals in similar fields, or who were supportive of their endeavours. As eighteenth-century Italian physicist Laura Bassi wrote, 'I have chosen a person who walks the same path of learning, and who from long experience, I was certain would not dissuade me from it'. For some of the women, this kind of partnership was found in their second marriages, like American engineer Nora Stanton Barney. But if they sought appointments together in academic institutions, this was often prevented by the policies of the times. This was the case for biochemist Gerty Cori. Arriving in the USA from Vienna in the 1920s, she and her husband Carl were discouraged from working together, Gerty even being threatened with dismissal. In the 1930s, while Carl had a number of university job offers, almost all universities refused to hire Gerty, one even declaring that it was 'un-American' for a married couple to work together. Washington University in St Louis, Missouri, offered them both positions, but Gerty's salary as a research associate was only ten per cent of that of her husband, who was an associate professor. It would be thirteen years before she attained that rank. Yet, in 1947, they won a Nobel Prize together.

Similarly for Maria Göppert Mayer, physicist and 1963 Nobel Prize winner, she and husband Joseph moved from Germany to the United States in the 1930s, where he had been offered a position as associate professor of chemistry at Johns Hopkins University. But Maria, as his wife, could not also be hired as a faculty member, but instead was given a role as an assistant in the Physics Department working with German correspondence. For this she received only a very small salary, a place to work and access to the facilities. In 1937, Joseph accepted a position as chairman of the Physics Department at Columbia University, part of the arrangement being that Maria would have an office there, but receive no salary. In December 1941, Maria took up her first paid professional position as a part-time science teacher at Sarah Lawrence College and, in the spring of 1942, with the United States heavily involved in the Second World War, she joined the 'Manhattan Project' – a research and development programme that produced the first nuclear weapons.

There is no doubt that wartime provided an opportunity for a number of the women included in this volume to realise their potential – as it had for Ruby Payne-Scott. For some, such as US engineer Thelma Estrin, talent in mathematics took her into an entirely different career path, engineering rather than accountancy. For others it provided an opportunity to shine. British aeronautical engineer and daredevil motorcycle racer Beatrice 'Tilly' Shilling, for example, is credited by her peers as helping the Allies to win the Second World War. In March 1941, she solved a problem that had jeopardised the lives of pilots during the Battle of France and the Battle of Britain. 'As a child,' she said, 'I played with Meccano. I spent my pocket money on penknives, an adjustable spanner, a glue pot and other simple hand tools.' In 1940, Royal Air Force pilots discovered a serious problem of stalling in the dive in fighter planes powered by Rolls-Royce's famous Merlin engine. Tilly led a small team that designed a simple device to solve this problem – a brass thimble with a hole in the middle, which could be fitted easily into the engine's carburettor. Pilots affectionately named her device 'Miss Shilling's orifice', or simply the 'Tilly orifice'. It remained in use as a stop-gap to help prevent engine stall for a number of crucial wartime years.

Readers may be surprised to encounter Hedy Lamarr, the Hollywood movie star of the 1930s and 1940s, among the women included in this volume. But her other life was as an inventor, obtaining a patent in 1942 in the area of radio-controlled systems, which proved to be a forerunner to Bluetooth technology and the digital communications era.

The years preceding the Second World War witnessed the oppressiveness of the Nazi influence on Jewish scientists in Europe, as had been the case for Austrian physicist Lise Meitner. German mathematician and physicist Amalie 'Emmy' Noether's experience also testifies to this lamentable chapter. She was at the University of Göttingen in Germany when, in April 1933, she and other Jewish faculty members were given a directive that read 'I hereby withdraw from you the right to teach at the University of Göttingen'. After continuing to teach in secret for a time, she escaped to the United States.

This was also the case for Hungarian mathematician Rózsa Péter who, as a result of the laws passed in 1939, was forbidden to teach, due to Jewish origin, and was briefly confined to the Jewish ghetto in Budapest. In her distaste for all things German, Rózsa changed her German-style surname 'Politzer' to a Hungarian one, 'Péter'.

Introduction

Twelve women in this volume won Nobel Prizes, Marie Curie was awarded two. Erna Schneider Hoover was one of the first women in the USA to receive a software patent for a computerised telephone switching system, and US pharmacologist and biochemist Gertrude Elion developed forty-five patents in medicine. Two of the women won A. M. Turing awards, the computer science community's equivalent of the Nobel Prize. Dorothy Hodgkin was the first British woman scientist to have been awarded a Nobel Prize in any of the three sciences it recognises and in 1965 was the second woman, after Florence Nightingale, to be appointed to the Order of Merit by a British monarch.

Computer science provided the pathway for a number of the women to shine. US Naval Commander Grace Hopper became one of the first programmers in computing history, with the aspiration that programming languages should be as easily understood as English. Consequently, she was highly influential in the development of one of the first programming languages called COBOL (COmmon Business-Oriented Language) which became one of the most widespread languages in the business of its time. In 1964, Grace was awarded the Society of Women Engineers Achievement Award, the Society's highest honour, 'In recognition of her significant contributions to the burgeoning computer industry as an engineering manager and originator of automatic programming systems'. In 1996 the USS *Hopper* (DDG-70) was launched, nicknamed '*Amazing Grace*', one of the very few US military vessels named after women. Encouraging young people to push themselves intellectually, she said 'A ship in port is safe; but that is not what ships are built for. Sail out to sea and do new things.'

American engineer Frances Hugle was a significant pioneer in the development of Silicon Valley, with one of her most notable solo patents being her 1966 process for Automated Packaging of Semiconductors (granted in 1969 after her death) and the development of TAB (tape-automated bonding) for the first time, allowing the miniaturisation that is used today in thousands of products from hearing aids to personal computers. In her husband Bill Hugle, whom she met at university, Frances found an intellectual colleague, who supported her in her engineering pursuits and, together, they were awarded many patents.

Over time, women like those included in this book found strength and fellowship in their own associations when in earlier times they were denied membership of societies and academies only open to men. For example, Marie Curie, despite being the first person to win two Nobel Prizes, was rejected for membership of the French Academy of Sciences in 1911, both as a woman and for being Polish. In the United States, for example, associations of scientific women included the Society of Women Engineers, established in 1950, with engineer Beatrice Hicks as its first president, the Association for Women in Science, established in 1971, and established in the same year the Association for Women in Mathematics. The Association for Women in Computing (AWC), founded in Washington DC in 1978, was one of the first professional organisations for women in computing, and is dedicated to promoting the advancement of women in the computing professions. The

AWC established the 'Augusta Ada Lovelace Award', in honour of the first computer programmer, Augusta Ada Byron Lovelace, who is included in this volume. Lovelace's writings in the first part of the nineteenth century developed the idea of programming and explained the operation and theory of Charles Babbage's Analytical Engine. The Award honours outstanding scientific and technical achievement and extraordinary service to the computing community through accomplishments and contributions on behalf of women in computing. Award winners included in this volume are Frances 'Fran' Elizabeth Allen, Thelma Estrin and Grace Hopper.

There are many other awards recognising the distinct contributions of a number of the women included. The contributions of physicist Maria Göppert Mayer, the second woman to receive a Nobel Prize in physics after Marie Curie, for example, are honoured in an annual prize presented by the American Physical Society. Established in 1987, the award recognises outstanding contributions to physics research by women physicists in the early years of their careers.

In honour of the US astronomer Annie Jump Cannon, the American Astronomical Society presents the annual Annie Jump Cannon Award for distinguished work in astronomy. US astronomer Antonia Caetana de Paiva Pereira Maury won this award in 1943. Each year, the American Medical Women's Association honours Elizabeth Blackwell, through the presentation of a medal that is named in recognition of her having been the first woman to receive a medical degree in the USA. The US anatomist and anthropologist Mildred Trotter is similarly recognised by the American Association of Physical Anthropologists, in the Mildred Trotter Prize for exceptional work in the field of physical anthropology, while German physicist and mathematician Emmy Noether is recognised in an annual lecture in her honour, held by the Association for Women in Mathematics.

On the other side of the Atlantic Ocean, the Royal Society awards the Dorothy Hodgkin Fellowship in honour of British biochemist Dorothy Hodgkin, for outstanding scientists at an early stage of their research; and, established in Norway in 1991, the Rachel Carson Prize is awarded in honour of marine biologist Rachel Carson, for women who have made a contribution in the field of environmental protection. French mathematician and physicist Sophie Germain is recognised in a prize awarded by the French Academy of Sciences since 2003 to honour a French mathematician for research in the foundations of mathematics. The Russian mathematician and physicist Sofia Vasilyevna Kovalevskaya is recognised in substantial awards to promising young researchers by the Alexander von Humboldt Foundation of Germany.

In honour of all women mathematicians, Women in Mathematics Day has been celebrated since 2018, on 12 May, marking the birthday of Iranian mathematician Maryam Mirzakhani on 12 May 1977. Ada Lovelace Day, the second Tuesday in October, has, since 2009, celebrated achievements of women in science, technology, engineering and mathematics. Some women scientists have local days in their honour, such as Alice Ball Day, 29 February, in Hawaii; and Katharine Blodgett Day, 13 June, in Schenectady, USA.

Another way of honouring many of the women in this volume is through the naming of astronomical objects after them – such as craters, comets and asteroids. In honour of Caroline Herschel, for example, is a comet and a crater on the Moon; and, for British astronomer and mathematician Mary Fairfax Somerville, there is an asteroid and a Moon crater, as there is for US astronomer Annie Jump Cannon. Egyptian-born British biochemist Dorothy Hodgkin is honoured by an asteroid. Russian engineer and astronaut Valentina Tereshkova has a crater on the Moon named after her, as does Russian mathematician Sofia Kovalevskaya and US astronomer Maria Mitchell, while US geologist Florence Bascom has a crater on Venus and an asteroid named after her, as well as a lake.

Other ways of honouring the women included in this volume include their being selected to appear on stamps and banknotes. Maria Göppert Mayer and Virginia Apgar both had US postage stamps in their honour, while Dorothy Hodgkin was honoured in a series of British postage stamps, and astronomer Mary Somerville had a Scottish banknote in her honour.

The cameos also include a few quotes that may be seen, perhaps, to represent the lived experience of the women. A particular issue was how, if at all, to identify their gender and their professional pathways. While engineer Olive Dennis, one of the most remarkable women in US railroad industry history, said that, 'No matter how successful a business may seem to be, it can gain even greater success if it gives consideration to the woman's viewpoint'. US engineer Betty Hugle resisted the elision of her professional identity with her gender, stating, 'I am a woman and an engineer; I am not a woman engineer', and resented the notion that her gender defined the type of engineer she was. While gender may have been resisted as a categorisation of the women scientists in this volume, for some of them marriage and motherhood had to be managed, somehow. Speaking in 1996, when she was in her early fifties, Jocelyn Bell Burnell spoke about the differences of life experiences for women from those of men, 'Although we are now much more conscious about equal opportunities I think there are still a number of inbuilt structural disadvantages for women. I am very conscious that having worked part-time, having had a rather disrupted career, my research record is a good deal patchier than any man's of comparable age … The life experience of a female is rather different from that of the male.'

Reflecting on the larger driving force in her intellectual quests, Italian neurologist Rita Levi-Montalcini wrote in her autobiography, 'It is imperfection – not perfection – that is the end result of the program written into that formidably complex engine that is the human brain and of the influences exerted upon us by the environment, and whoever takes care of us during the long years of our physical, psychological and intellectual development.' In a letter she wrote in 1890, Russian mathematician Sofia Kovalevskaya observed that, 'It seems to me that the poet must see what others do not see, must see more deeply than other people. And the mathematician must do the same.' US geologist Florence Bascom spoke of the motivating force in scientific inquiry, that is seen in the stories of all the women in this volume, 'The fascination of any search after truth lies not in the attainment ... but in the pursuit, where all the powers of the mind are absorbed in the task. One feels oneself in contact with something that is infinite and one finds a joy that is beyond expression in "sounding the abyss of science" and the secrets of the infinite mind.'

The book is limited to one hundred women. It is a 'sampler' of a wide array of talent and achievement, in concise biographical cameos. 'Sampler' is a sewing term, but perhaps appropriate to the feminine subjects contained here; and, as a sampler, it can be used in just that way. Flip through the pages – randomly, or sequentially, as you please. But, for every one hundred examples there are likely to be one hundred, or even one thousand more. And for each woman included here the cameo is but a short snippet of a rich, lived life, chasing a hunger from their youth in pursuit of answers. Let it be an example of, and an inspiration for, the thousands upon thousands of women in science yet to have their stories shared.

Emeritus Professor Rosalind F. Croucher AM

Frances 'Fran' Elizabeth Allen

Born: 4 August 1932 in Peru,
New York, USA
Died: 4 August 2020 (aged 88)
Field: Technology
Awards: US Academy of
Science and Engineering (1987);
IEEE Computer Society Charles
Babbage Award (1997); Augusta
Ada Lovelace Award (2002); Turing
Award (2006)

The eldest of six children, Frances ('Fran') Elizabeth Allen was born on 4 August 1932 in Peru, New York, USA. She was raised on a farm near Lake Champlain, her father a farmer and mother an elementary-school teacher. Inspired by her high school mathematics teacher, she wanted to follow the same profession, so she attended New York State College for Teachers (now the State University of New York, Albany) for four years where she earned a Bachelor of Arts in mathematics with a minor in physics. For two years she then taught mathematics at the same high school she had attended, covering topics ranging from elementary algebra to advanced trigonometry.

Aware that she would need a master's degree to be a fully qualified teacher, she enrolled in summer courses at Columbia University, followed by the University of Michigan at Ann Arbor, graduating with a Master of Science in mathematics in 1957. As part of her studies, Fran undertook some of the first-ever-offered courses in computing, learning how to program an IBM 650 from American mathematician and computer scientist Bernard Galler, who was a co-developer of the Monitored Automatic Debugging (MAD) programming language and later president of the Association for Computing Machinery (ACM) and an ACM Fellow. After attending an on-campus interview, IBM offered Fran a research position, which she accepted. Her idea at the time was to earn enough money to repay her student loans and then return to teaching, but she ended up staying at IBM for the next forty-five years.

It was a propitious time when she joined IBM on 15 July 1957, as only two months previously the FORTRAN programming language had been released. Fran's initial task was to teach research scientists within IBM how to use it, and indirectly encourage IBM customers to adopt it as well. In reality, her understanding of the subject matter was only just ahead of her students' in reading the source code for the FORTRAN compiler that had been developed by the American computer scientist John Backus (later a Turing Award winner) and his team. Fran stated: 'It set my interest in compiling, and it also set the way I thought about compilers, because it was organised in a way that has a direct heritage to modern compilers.'

Most of her career was spent developing programming language compilers for IBM Research, initially working on IBM 704's MAD operating system, with her first major project involving the *Stretch/Harvest* computer. *Stretch* was one of the first supercomputers, and *Harvest* was a co-processor for *Stretch* that had been designed for the US National Security Agency (NSA) to break the codes in secret messages. Fran and her colleagues designed a single compiler framework to handle three programming languages: FORTRAN, Autocoder (similar to COBOL), and a new language, Alpha. It was arduous and difficult work, but she and her team were successful, and she acted as the liaison between IBM and NSA, coordinating the design of the Alpha language and its acceptance tests. In 1962, Frances spent a year at NSA where she oversaw the installation and testing of the system. It remained in operation for fourteen years before it was retired in 1976.

As it involved extensive travel, she declined the opportunity to work on a programming language for the IBM System/360, and instead returned to IBM Research at the new T. J. Watson Research building in Yorktown, New York. She joined Project Y, later called the Advanced Computing Systems project (ACS), that included another set of cutting-edge advances in computer system design. This allowed her to again collaborate with the US computer scientist John Cocke, a future Turing Award winner, who had worked on the hardware for *Stretch*.

Fran took a sabbatical from IBM to teach graduate courses on compilers at the Courant Institute for Mathematical Sciences at New York University at the invitation of Jacob 'Jack' T. Schwartz, creator of the SETL programming language and the NYU Ultracomputer. Schwartz had previously visited IBM to collaborate with her and John Cocke on ACS. Fran and Schwartz were married in 1972, and divorced ten years later. They had no children.

Her next major project for IBM was the Experimental Compiler Systems (ECS) project. This system, like the earlier compiler framework for *Stretch/Harvest*, was designed to support multiple programming languages. In this case the main focus was on a new language called PL/I, which presented far more difficult problems for an optimising compiler. Her last big project for IBM was the Parallel Translator (PTRAN), a system for compiling FORTRAN programs, as part of which she consulted with a professor of computer science, David Kuck, at the University of Illinois. PTRAN introduced the concept of the program dependence graph, a concept used today by many parallelising compilers.

In 1989, Fran was named the first female IBM Fellow, in 1991, an Institute of Electrical and Electronics Engineers (IEEE) Fellow and in 1994, an ACM Fellow. She retired from IBM in 2002, the same year she won the Augusta Ada Lovelace (*see entry*) Award from the Association of Women in Computing, an award that recognises outstanding scientific technical achievement and/or extraordinary service to the computing community through accomplishments and contributions on behalf of women in computing. In 2006, she won the prestigious A. M. Turing Award, presented by the Association for Computing Machinery (ACM), which is generally regarded as the 'Nobel Prize of computing' and the highest distinction in computer science. The citation for Fran's award read, in part, 'Fran Allen's work has had an enormous impact on compiler research and practice. Both alone and in joint work with John Cocke, she introduced many of the abstractions, algorithms, and implementations that laid the groundwork for automatic program optimization technology.'

In 2007, the Fran Allen PhD Fellowship Award was created in her honour. She continued to advise IBM as a Fellow Emerita on a number of projects including the Blue Gene supercomputer, encouraging the involvement of other women in computer-related fields. Throughout her illustrious career she held many positions, including serving on the US National Science Foundation (1972–78), the IEEE, the American Academy of Arts and Sciences and the ACM.

Fran passed away on her eighty-eighth birthday, 4 August 2020, from complications with Alzheimer's disease.

Virginia Apgar

Born: 7 June 1909 in Westfield, New Jersey, USA
Died: 7 August 1974 (aged 65)
Field: Medicine
Awards: Distinguished Service Award (American Society of Anesthesiologists, 1966); Alumni Gold Medal for Distinguished Achievement (Columbia University College of Physicians and Surgeons, 1973); Ralph M. Waters Award (American Society of Anaesthesiologists, 1973); Woman of the Year in Science (*Ladies' Home Journal*, 1973)

Virginia Apgar was born on 7 June 1909 in Westfield, New Jersey, USA. She was the youngest of three children who were raised there. Virginia attended Westfield High School, finishing her secondary education in 1925. Her family shared a love of music and she played the violin during family concerts and in amateur chamber quartets. She was also an enthusiastic gardener, with other interests including fly-fishing, golf and stamp collecting. The basement of her family home contained a laboratory where her businessman father, Charles, built a telescope and conducted scientific experiments with electricity and radio waves.

After completing her high school education, Virginia went to Mount Holyoke College where she studied zoology, minoring in physiology, with the aim of becoming a doctor. She completed her bachelor's degree in 1929 after taking a number of jobs to pay her way. Still in poor circumstances, she borrowed money so she could attend Columbia University College of Physicians and Surgeons from which she graduated in 1933 with a degree in medicine, placing fourth in her class.

Now heavily in debt, in the middle of the Great Depression, she realised even male surgeons were having difficulty finding employment. To improve her chances of finding paid work, she turned her attention to the field of anaesthesiology, the study of giving patients anaesthetics. Although at the time anaesthetics were normally administered by nurses, doctors were beginning to take an interest and women were particularly encouraged to enter the field. And so she began a two-year programme in anaesthesiology at Columbia University, the University of Wisconsin, Madison, and Bellevue Hospital in New York.

Her instincts proved correct and, in 1938, Virginia was appointed as the director of the anaesthesia division at Columbia University. Although funding was always an issue, she managed to increase the number of physicians working for her and in 1949 a separate department of anaesthesia was created at the university with the aim of training medical practitioners and carrying out research. When a department head was appointed, a male colleague was preferred to her. She was, however, made a full professor in the department, being the first women to attain this level at the university.

For the next ten years she studied the effects of anaesthetic on childbirth, realising that this was a crucial time for babies as they were generally not assessed for health issues at birth as the attention was focused on the mother. As a result, many serious and life-threatening conditions were missed. To overcome this, Virginia devised a method to easily determine which babies needed special attention. The five-part test gave a score to the child's heart rate, respiration, muscle tone, colour and reflex. This became known as the Apgar Newborn Scoring System (Apgar stands for Appearance, Pulse, Grimace, Activity, and Respiration) and was initially designed to be performed 1 minute after birth. In time, this was extended to include assessment at 5 and 10 minutes after birth. The test was eventually globally adopted to become the standard for determining the survival chance and development of a child.

After spending more than twenty years at Columbia, Virginia undertook a new career. She completed a Master of Public Health degree at Johns Hopkins University, then joined the March of Dimes organisation that provided support services to children and pregnant women. In 1959, she became the head of the division on congenital birth defects and ten years later she became their head of research, steering the organisation's focus towards the prevention of defects. During this period, she not only lectured but wrote extensively in both research journals and popular magazines.

In 1964–65 there was a pandemic in the USA of rubella (German measles), a disease that can cause serious congenital disorders for children if a pregnant woman is infected. There were an estimated 12.5 million cases leading to 11,000 miscarriages and 20,000 cases of congenital rubella syndrome. It was a grave situation that saw 2,100 deaths of babies, as well as 12,000 who were left deaf, 3,500 left blind due to cataracts and 1,800 who were cognitively impaired. Virginia became a strong supporter and vocal advocate for a universal vaccination to prevent the transmission of rubella from mother to child. She was indefatigable, travelling thousands of miles each year to address audiences on the importance of early detection and to stress the need for more research in the area.

In 1972, Virginia co-authored a book, with Joan Beck, *Is My Baby All Right?*, and lectured teratology (essentially, the study of birth defects) between 1965 and 1971 when she became a clinical professor and, in doing so, becoming the first woman to hold a faculty position in the new field of paediatrics. In 1973, just one year before her death, Virginia took up a post as lecturer in medical genetics at the Johns Hopkins School of Public Health.

Virginia's research included over sixty scientific articles and many shorter pieces for newspapers and magazines. She did not participate in women's causes or organisations, although at times she expressed her frustration with the salary imbalance of women compared to men. She was a woman of the people, just as at ease speaking to teenagers as she was to eminent scientists, always with the same degree of enthusiasm and energy. She was not fazed by topics considered 'off limits' and spoke of teen pregnancy and congenital disorders when others would not.

Virginia Apgar died at age sixty-five, of cirrhosis of the liver, on 7 August 1974 at Columbia-Presbyterian Medical Center, where she had worked for many years. She is buried at Fairview Cemetery in Westfield, New Jersey. Never marrying, she continued her work until ill-health overtook her, being remembered as a warm and lively woman who had a keen sense of humour along with professional integrity and great intelligence. In 1994, twenty years after her death, she was honoured by appearing on a US commemorative postage stamp, and the following year she was inducted into the National Women's Hall of Fame. Virginia's legacy still protects babies all over the world.

Phoebe Sarah 'Hertha' Marks Ayrton

Born: 28 April 1854 in Portsea, Portsmouth,
Hampshire, England
Died: 26 August 1923 (aged 69)
Field: Physics, mathematics and electrical
engineering
Awards: Fellow of the Royal Society (1902);
Hughes Medal (1906)

Born the third child of Levi Marks, a Polish Jewish watchmaker, and Alice Moss, the daughter of a glass merchant, Phoebe Sarah Marks was so full of energy that, in her teens, her friends nicknamed her 'Hertha' after a German earth goddess, a name she adopted for the rest of her life. When her father died she was seven years old; her mother was left with seven children and another on the way. Hertha had showed early her intellectual abilities and her mother Alice was determined to enable her daughter to develop her education. In 1863, when Hertha was nine, she went to live with her maternal aunt Marion Hartog, who ran a school in London. Not only did Hertha's intellect stand out but also her inclination towards social justice, as she staged a hunger strike for two days due to the injustice of being wrongly accused of something.

At the age of sixteen, Hertha worked as a governess with the aim of helping her mother and also raising funds to enable her to undertake further study. In 1874, she passed the Cambridge University for Women examination with honours in both mathematics and English. She entered Girton College in October 1876, her application supported by the writer George Eliot. Hertha passed the Mathematical Tripos but, as a woman, she was only awarded a certificate. She did, however, pass an external examination at the University of London, which awarded her a Bachelor of Science degree in 1881. She was not solely interested in academic endeavours; she was responsible for establishing the Girton College Fire Brigade, led the Choral Society and formed a mathematical club.

After leaving Girton, Hertha returned to London and taught mathematics to ladies in preparation for them to undertake the London

University Matriculation Exam and the Cambridge entrance examinations. Her active interest in devising and solving mathematical problems resulted in being published in 'Mathematical Questions and Their Solutions', in the *Educational Times*. Hertha also came up with applications for her solutions. As a student at Girton, she constructed a sphygmograph for recording blood pressure and, by 1884, she had invented and obtained a patent for an engineering drawing instrument that could divide a line into any number of equal parts, a project that had taken her twelve months to complete. It is reported that engineers, artists and architects found it to be a useful device. This success triggered a passion to follow a career in research and invention, funded by an eccentric philanthropist Barbara Bodichon, who was a benefactor to the causes of women as well as a leader in the suffragette movement. Hertha was to obtain twenty-six patents in her lifetime.

Hertha also continued her own studies at Finsbury Technical College, attending evening classes on electricity in 1884. It was there that she met the physicist and electrical engineer Professor William Edward Ayrton FRS, a widower and known to be supportive of women's education and their legal rights. They were married in 1885.

A year after the birth of their daughter Barbara, named after Barbara Bodichon, in 1892, Hertha became intrigued with the problem of hissing and flickering electrics that were prevalent in street lighting and theatres. In fact, the term the 'flicks' for the movies originated from the flicker that came from the early arc lighting of the projector, created by two carbon rods being electrified when they were placed next to each other. She and her husband wanted to improve the technology to make them less noisy. Although all of Hertha's notes were accidentally burned in the fireplace after being mistaken for kindling by the maid, Hertha started all over again and discovered that the issue was the rods themselves. So she designed something that would be much quieter.

Hertha then published a number of invited papers on the arc in the journal *Electrician*. Her research paper, 'The Hissing of the Electric Arc', was read to the Institute of Electrical Engineers in 1899, for which she was awarded a prize of £10. The Institute also elected Hertha their first female member. She chaired the physical science section of the International Congress of Women and gave a presentation to the Congress in London. This was pioneering work, especially for a woman in those days, and it paved the way for others to follow. Her audiences were amazed to see a woman being involved in such dangerous experiments, with one newspaper reporting that 'lady visitors were astonished ... one of their own was in charge of the most dangerous-looking of all the exhibits – a fierce arc light enclosed in glass. Mrs Ayrton was not a bit afraid of it.'

Although the Royal Society did not permit a woman to read a paper, they did agree to allow an associate of her husband's, John Perry, to read her work, 'The Mechanism of the Electric Arc', in 1901. In 1902, her only book, *The Electric Arc*, was published. This 450-page volume was based on twelve papers she wrote for the *Electrician* and it quickly became the standard textbook on the subject. A year later, she met Marie Curie (*see entry*) and the two became friends, later proving that they were outstanding scientists on their own merit and not merely collaborators with their husbands.

In 1901, when her husband's illness required long periods by the seashore, Hertha began to study sand ripples. In 1904 the Royal Society allowed Hertha to read a paper, the first woman to do so, and 'The Origin and Growth of Ripple Marks' was well received. The next year, she became involved in a project, assigned by the Admiralty, to supply specifications for carbon that would successfully burn in searchlight projectors. As a result, Hertha wrote four reports between 1904 and 1908 for which her husband officially received the credit. However, in a small gesture, he insisted that she be listed as a co-author on the fourth one.

Hertha's work was now so impressive that, in 1906, the Royal Society awarded her their Hughes Medal for 'an original discovery in the physical sciences, particularly as applied to the generation, storage and use of energy'. However, as a married woman she could not be a Fellow. It was now that she became involved with the suffragette movement, being appalled at the discrimination felt by women, especially in science. Following the death of her husband in 1908, Hertha continued her scientific research, but also became more outspoken on women's rights. In this regard she provided assistance to women protestors on hunger strikes, as well as being part of a boycott of the 1911 English census, demanding that women be given the right to vote. In 1912, Hertha provided refuge for Marie Curie (*see entry*) from the tragedy surrounding her husband's death, writing in Marie's defence that, 'An error that ascribes to a man what was actually the work of a woman has more lives than a cat.' In 1920 Hertha became a founding member of the National Union of Scientific Workers.

During the First World War, Hertha invented a device that would drive away poisonous chlorine gasses that were reportedly being used on British soldiers. Making it her goal to find a way of protecting them, she developed a large fan that resembled a long broomstick with a sizeable rectangular paddle at one end that, when flapped manually, would drive the gasses away. However, the army wasn't convinced that a 'fan', an accessory traditionally carried by a woman, would be accepted as an instrument to protect men in wartime. However, after a field demonstration in 1917, some 100,000 of her fans were eventually shipped to the Western Front. Just two years later, she examined an automatic version that would work on more powerful winds. In fact, most of her scientific research from that point onwards involved various applications of fan principles, although she retained a keen interest in political causes. Hertha Ayrton was a pioneer in every sense of the word. She passed away in August 1923 in Sussex.

Massimilla 'Milla' Baldo-Ceolin

Born: 12 August 1924 in Legnago, Italy
Died: 25 November 2011 (aged 87)
Field: Particle physics
Awards: Feltrinelli Prize (*Accademia dei Lincei*, 1976); Gold Medal for Education and Arts (*Benemeriti della Scuola, della Cultura e dell'Arte*, 1978); Gold Medal for Science (*Benemeriti della Scienza e della Cultura* (1995); Enrico Fermi Prize of the Italian Physical Society (2007)

Massimilla 'Milla' Baldo-Ceolin was born in Legnago, Italy, near Verona, on 12 August 1924, the daughter of the owner of a small mechanical workshop. She was raised under the government of Benito Mussolini and wrote that, in her early years, she was 'faced with a world in disorder and disaster. ... My family had to leave town and take refuge in the nearby countryside ... During these long days, I read many books of popular science.'

Due to the Allied bombing in northern Italy during the Second World War, Milla could not commence university studies until the end of the conflict. In 1952, she received her *laurea* (master's degree) in physics from the University of Padua with a thesis on pion nuclear capture, under the supervision of Professor Nicolò Dallaporta. The 1950s also witnessed the emergence of the nuclear photographic emulsion method, when kaons and hyperons were being discovered and their properties examined. At first, the only sources of those new particles were cosmic-ray collisions but, by 1955, a new generation of accelerators was producing beams of particles with energies in the GeV range. Milla initially worked on charged and neutral kaons, sub-atomic particles, with William F. 'Jack' Fry in Padua. Their work confirmed the model proposed by the American physicist Murray Gell-Mann, who later won the Nobel Prize for physics in 1969, and Dutch-born American physicist and science historian Abraham Pais.

After the discovery of the proton and neutron antiparticles, Milla collaborated with the physicist Derek Prowse, to extend this work to the realm of hyperons (particles made of quarks). After Prowse exposed nuclear emulsions to a pion beam, they detected the first antihyperon, the antilambda. In the early 1960s, Milla began using bubble chambers at the Argonne National Laboratory, CERN (*Conseil Européen pour la Recherche*

Nucléaire – European Organisation for Nuclear Research), and the Institute for Theoretical and Experimental Physics in Moscow, to explore selection rules and conservation laws in the kaon system. After the discovery of neutral currents, her experimental programme was directed to neutrino physics, working with a group led by Helmut Faissner, one of the pioneers of neutrino experiments at CERN, at Rheinisch-Westfälische Technische Hochschule Aachen (RWTH Aachen). They used spark chambers to measure scattering cross-sections of neutrinos off electrons.

Between 1965 and 1968, Milla was the head of the local section of the *Istituto Nazionale di Fisica Nucleare* (National Institute of Nuclear Physics) (INFN) in Padua and, between 1973 and 1978, the head of the Physics Department. By the mid-1970s, she was a member of a large international collaboration using a liquid deuterium bubble chamber at CERN's Super Proton Synchrotron, exploring neutrino interactions with protons and neutrons. She also had an interest in solar neutrinos, which she studied as part of her research on the ICARUS detector in Italy's Gran Sasso Laboratory, the largest underground research centre in the world.

In 1988, Milla started the Workshops on Neutrino Telescopes at the *Istituto Veneto di Scienze, Lettere ed Art* (Venetian Institute for Sciences, Humanities, and Arts) and, with others, coordinated European Networks of neutrino oscillators. As well as an active research career, she had a passionate interest in Galileo Galilei, managing to combine celebrations of the 400th anniversary of Galileo's telescopic discoveries with a neutrino workshop. A special commemorative medallion, showing Galileo presenting a telescope to the Doge, an elected Italian Head of State, was produced for the occasion. Its reverse side carried an inscription in Italian of one of her favourite quotes by Galileo, translated as: 'I deem it of more value to find out some truth about however light a matter than to engage in long disputes about the greatest questions without achieving any truth'. This was also the motto of the Padua Institute of Physics, which Milla used in her 2002 paper in the *Annual Review of Nuclear and Particle Science* on the nuclear emulsion era of the 1950s, each volume of the *Annual Reviews* starting with an historical review by a distinguished scientist.

Milla Baldo-Ceolin received many honours and awards, including the prestigious Antonio Feltrinelli Prize for Physics from the Lincean Academy in 1976. With a significant monetary grant, the five-yearly prize is awarded for achievement in the arts, music, literature, history, philosophy, medicine, and physical and mathematical sciences. In each of 1978 and 1993, she was awarded gold medals from the Italian Ministry for Universities and Scientific and Technological Research. She was a respected leader on the world stage with an outstanding record in research and experimental expertise. There was one event of which she was most proud, declaring, 'A national competition for a full professorship in physics opened up. I competed and I won. And it was my own department that offered me the professorship at the University of Padova, where I became the first woman to obtain a chair since the University's foundation in 1222.' This was in 1963.

In 1998, Milla was made a Professor Emeritus of the University of Padua. She never married. She died on 25 November 2011, aged eighty-seven.

Alice Augusta Ball

Born: 24 July 1892 in Seattle, Washington, USA
Died: 31 December 1916 (aged 24)
Field: Chemistry
Award: University of Hawaii's Regents' Medal of Distinction (post., 2007)

One of four children, Alice Ball was born on 24 July 1892 in Seattle, Washington, to Laura Louise (Howard) Ball and James Presley Ball, a newspaper editor, photographer and lawyer. Her paternal grandfather, James Ball Sr, was a well-known photographer who suffered from arthritis and, as a result, during Alice's early childhood, in 1903 the family moved to the warmer climate of Honolulu, Hawaii. But after twelve months her grandfather passed away and the family returned to Seattle.

Alice graduated from Seattle High School with top grades in science in 1910. She then attended the University of Washington, earning a bachelor's degree in pharmaceutical chemistry and, two years later, a further degree in pharmacy. With her pharmacy instructor as a co-author, she published a ten-page research article, 'Benzoylations in Ether Solution' in the prestigious *Journal of the American Chemical Society*, a major feat for a woman of that era, especially for one who was African American.

Despite several offers of scholarships to study at prominent universities, including the University of California Berkeley, Alice returned to Hawaii where she completed a master's degree in which she investigated the herb chaulmoogra and its chemical properties. Chaulmoogra oil had been used topically and orally since the 1300s, with intermittent success, to lessen the effects of leprosy (Hansen's disease). It can affect the nerves, skin, eyes and lining of the nerves, which are attacked by bacteria and become swollen under the skin. This in turn causes the affected areas to lose the ability to feel pain or sense touch and injuries from cuts and burns can result. For centuries, leprosy carried a significant social stigma and, due to its contagious nature, many societies created leper colonies for the sufferers.

In practice, the chaulmoogra oil was originally too sticky to be used effectively topically and it was extremely painful to be used as an injection. This did not stop some hospitals attempting to use it as such, even though the thick consistency of the oil caused it to clump under the skin and form blisters. Drinking the oil was also ineffective because it had such an acrid taste that patients usually vomited it up again.

At the age of twenty-three, Alice developed a remarkable new technique by which a dosage could be injected directly and absorbed by the body into the dermis and, in doing so, greatly increased its effectiveness. Her method involved isolating ethyl ester compounds from the fatty acids of the chaulmoogra oil. This became known as 'the Ball Method'. It was the only treatment for leprosy that was effective and, moreover, left no abscesses or had a bitter taste.

Soon after her discovery, Alice become ill during her research and returned to Seattle for treatment. She passed away a few months later, on 31 December 1916, at the young age of twenty-four. The cause of her death is unknown, as her original death certificate was later altered to declare it as being due to tuberculosis. Her illness meant that she was unable to formally publish her revolutionary findings, although Arthur L. Dean, a chemist and the president of the University of Hawaii, continued her work and published the results, producing large quantities of the injectable chaulmoogra extract. However, his publications did not give any credit to Alice and he renamed her technique 'the Dean Method', although in 1922, Harry T. Hollmann, a doctor during the time at Kalihi Hospital in Hawaii, claimed that Dean's method offered no improvement over that found by Alice.

In 1918, it was reported in the *Journal of the American Medical Association*, that a total of seventy-eight patients were released from Kalihi Hospital by the board of health examiners, after treatment with injections using Alice's method. The isolated ethyl ester remained the preferred treatment for leprosy until sulphonamide drugs were developed and utilised in 1942, when a slight further improvement took place.

In 1977, University of Hawaii professor Kathryn Takara, undertook research on Alice and concluded that she did not receive the accolades she was due and it was the result of her efforts that many leprosy sufferers worldwide were either cured or had their symptoms considerably lessened using her technique. It is said that her findings directly affected the 8,000 people diagnosed with leprosy who were removed from their homes and exiled to the Kalaupapa community on the island of Moloka'i. Now they could be treated in their own homes.

It was not until many years later that Alice Augusta Ball's contributions to medical science were duly recognised. These included on 29 February 2000, when Mazie Hirono, the former Lieutenant Governor of Hawaii, named 29 February as 'Alice Ball Day', which only occurs in leap years. The University of Hawaii honoured her memory with a plaque by the school's lone chaulmoogra tree, which still stands on the campus behind Bachman Hall.

In 2007, Alice was posthumously awarded the University of Hawaii's Regents' Medal of Distinction. She was not only the first black student to receive a master's degree from the institution, but she was also the first black woman to graduate with that distinction.

Nora Stanton Blatch Barney

Born: 30 September 1883 in Basingstoke, Hampshire, England
Died: 18 January 1971 (aged 87)
Field: Civil engineering and architecture
Awards: Fellow of the American Society of Civil Engineers (ASCE) (post., 2015)

Nora Stanton Blatch was born on 30 September 1883, in Basingstoke, Hampshire, England, to William Blatch, a brewery manager, and Harriot Stanton, the daughter of Elizabeth Cady Stanton. Harriot and her mother were both leaders of the suffragette movement in the USA. In 1902, Nora's family moved to the USA. At fourteen, Nora commenced studies at the Horace Mann School in New York in Latin and mathematics, returning to England during the summer months. She enrolled in the civil engineering programme at Cornell University in New York, graduating with a bachelor's degree in 1905 and becoming among the first women in the country to graduate with a civil engineering degree. While at the university, Nora was actively involved as the manager of the Women's Fencing Club and a member of the *Kappa Kappa Gamma*, a women's 'fraternity', being founded before the term 'sorority' was used.

In her civil engineering master's studies at the Beebe Lake hydraulics laboratory, Nora solved an important problem in hydrodynamics, set out in her thesis, 'An Experimental Study of the Flow of Sand and Water in Pipes under Pressure'. This assisted her election to *Sigma Xi*, an honorary scientific society founded at Cornell in 1886. In 1905 Nora became the first woman to be accepted as a junior member of the American Society of Civil Engineers (ASCE). She found employment with the New York City Board of Water Supply and, in 1905–06, also worked for the American Bridge Company.

Nora enrolled in courses in electricity and mathematics at Columbia University where she worked as a laboratory assistant to Lee De Forest, inventor of the radio vacuum tube. In 1908, he and Nora were married. Their honeymoon was spent in Europe, marketing radio equipment he had developed involving the new technology of wireless radio. However, De Forest insisted that Nora leave her profession and assume home duties, an attitude that led to their separation after only twelve months of marriage and while she was pregnant. Not long after, in June 1909, she gave birth to their daughter, Harriet, and began working as an assistant engineer and chief draftsman for the Radley Steel Construction Company until 1912. Her divorce was finalised in 1911 and she continued her engineering career with the New York Public Service Commission.

In 1915, Nora became the president of the Women's Political Union, a role in which she succeeded her mother, and edited the organisation's magazine, *Women's Political World*. She subsequently participated in the National Woman's Party's efforts for a federal Equal Rights Amendment and gained some notoriety when she applied to be made a full member of the ASCE. She was refused, based solely on her gender, as at the time women could only be associate members. She then took her cause to the New York State Supreme Court on the grounds that she met all requirements. She lost her case and it was a further ten years before women could receive full membership status.

In 1919, she remarried, this time to a marine architect, Morgan Barney. Their daughter, Rhoda, was born on 12 July 1920 in New York. The family moved to Greenwich, Connecticut, in 1923, and Nora worked as a real estate developer. In 1944 she published *World Peace Through a People's Parliament* and continued to be politically active for the rest of her life, in 1946 writing a pamphlet entitled 'Women as Human Beings'.

She was posthumously awarded the status of ASCE Fellow in 2015 and, in 2017, the New York City Department of Environmental Protection dedicated a tunnel boring machine 'Nora' in her honour. It was used as part of tunnelling for the $1 billion USD project to repair the Delaware Aqueduct.

Nora Stanton Barney died at her home in Connecticut on 18 January 1971, aged eighty-seven. She is buried in Woodlawn Cemetery and Conservancy, Bronx, Bronx County, New York, USA with Memorial ID 92785151. She was survived by two daughters, Harriet De Forest and Rhoda Barney Jenkins, as well as six grandchildren.

Florence Bascom

Born: 14 July 1862 in Williamstown, Massachusetts, USA
Died: 18 June 1945 (aged 82)
Field: Geology

The youngest of five children, Florence was born on 14 July 1862 in Williamstown, Massachusetts, USA, into a family that encouraged her to obtain a college education. Her father, John Bascom, was a professor at Williams College and later became president of the University of Wisconsin-Madison. Her mother, Emma (Curtiss) Bascom, was an activist for women's rights and was involved with the suffrage movement. However, it was her father who was the driving force in his daughter's career, introducing her to the field of geology.

Florence attended the University of Wisconsin and graduated with a bachelor's degree in arts and letters in 1882, followed by a Bachelor of Science in 1884 and a Master of Science in 1887. She then started a college teaching career in Virginia at the 'Hampton School of Negroes and American Indians', now known as Hampton University. After a year, she returned to Wisconsin to undertake a master's degree in geology, which she completed in 1887, although as a woman she had limited access to resources, including the library and gymnasium, and her admission to classrooms was restricted if there were men present. From there she began her career as a teacher at Rockford Female Seminary (now Rockford University) in Illinois, from 1887 until 1889.

In 1893, she obtained her PhD from Johns Hopkins University, the first woman to do so, and the second in her field with a thesis on petrology (the origin, structure, and composition of rocks). The college president, opposed to co-education, required her to sit behind a screen during lectures so that she would not disturb or distract the men. At the time, geology was viewed as exclusively the domain of men. Because of these and other challenges, she earned the title of 'Pioneer of Women Geologists'.

Florence then took up a position as an instructor and associate professor in geology and petrography at Ohio State University for two years. Then she left to work at Bryn Mawr College to teach advanced geology and undertake research. It was there that she was allocated an office that was essentially a storage space in a building designed exclusively for chemistry and biology. Despite these trying conditions, in 1901 she set up the department of geology where she lectured and trained classes of young women. That same year, she became the first female geologist to present a paper before the Geological Survey of America. From 1896 to 1905, she was the Associate Editor of *The American Geologist* and the first female officer of the Geological Society of America, subsequently elected as vice president in 1930.

Florence's programme at Bryn Mawr College was considered to be among the finest in the country and she set high standards both for herself and her students, emphasising the importance of laboratory and field work, acquiring a significant collection of minerals, fossils and rocks. The majority of female geologists in the country came from her programme and her students went on to achieve enormously successful careers in their own right. This was one of Florence's most outstanding achievements – the education of women.

Florence's work was so highly regarded that, in 1906, the first edition of the journal *American Men of Science* ranked her among the top 100 leading geologists in the USA. Later editions also included some of her female students, but it was not until 1971 that they changed the title to *American Men and Women of Science*.

Florence undertook intensive work in geomorphology, examining how the earth's geography changes over many thousands of years. Her work also focused on an area of the Appalachian Mountains, creating valuable geographic maps of New Jersey and Pennsylvania that are still used today.

Although Florence retired from teaching in 1928, she continued to work at the US Geological Survey until 1936. On her retirement she wrote,

The fascination of any search after truth lies not in the attainment ... but in the pursuit, where all the powers of the mind are absorbed in the task. One feels oneself in contact with something that is infinite and one finds a joy that is beyond expression in 'sounding the abyss of science' and the secrets of the infinite mind.

Florence was well aware of the tremendous influence she had on her field, particularly when it came to training women, declaring in 1932 that

I have always claimed there was no merit in being the only one of a kind. ... I have considerable pride in the fact that some of the best work done in geology today by women, ranking with that done by men, has been done by my students.

In 1937, a remarkable eight of the eleven women who were Fellows of the Geological Society of America were graduates of her programme.

Florence Bascom never married. On 18 June 1945, at age eighty-two, she died of a stroke in Northampton, Massachusetts. She is buried in a Williams College cemetery in Williamstown. Her legacy has been memorialised in the Bascom Crater on Venus in 1994 and in 6084 Bascom, an asteroid discovered in 1985. Lake Bascom, also named in her honour, is a prehistoric, postglacial lake located in Berkshire Country in Massachusetts, being formed when receding glacial ice acted as a dam, preventing drainage of the Hoosic River watershed.

Laura Bassi

Born: Between 20 and 31 October 1711 in
Bologna, Papal States, Italy
Died: 20 February 1778 (aged 66)
Field: Physics

Laura Bassi was born in Bologna, Papal States, Italy, sometime between 20 and 31 October 1711, to an upper-middle-class family. Her father, a wealthy lawyer, recognising early the intellectual aptitude of his only child, ensured her private tuition in Latin, French and mathematics. From the age of thirteen, the family physician and professor of medicine and philosophy at the University of Bologna, Gaetano Tacconi, took control of her education, providing traditional training in Aristotelian and Cartesian thinking. Word of her intellectual capabilities spread around Bologna, particularly when she decided to investigate the new field of Newtonian physics. In 1731, Tacconi invited philosophers from the university to meet her. They were so impressed that her reputation soon spread. The following year, she was the focal point of a series of lectures arranged by the Archbishop of Bologna, Prospero Lambertini, who later became Pope Benedict XIV.

On 20 March 1732, Laura was admitted to the Bologna Academy of Sciences as an honorary, and its first female, member. She obtained her doctor of philosophy degree on 12 May 1732, defending her thesis in the *Palazzo Pubblico* (Town Hall), rather than the customary church of the religious orders. She became only the second woman to earn a doctor of philosophy after Elena Cornaro Piscopia in 1678, some fifty-four years previously. On 27 June 1732, Laura successfully defended another set of theses about the properties of water, the university then awarding her an honorary post as a professor in physics.

Laura had now become quite prominent in the town and her thesis revealed the influence of Sir Isaac Newton's works on optics and light. When Laura received her degree, the citizens of Bologna held a public celebration of her success, with collections of poetry published in her honour. The university recognised the occasion with miniature paintings decorating official documents, and a silver coin in her honour linking her to the Roman goddess of learning, Minerva.

In 1733, Laura was appointed as an academic lecturer in universal philosophy at the Institute and Academy for Sciences and Arts of Bologna, but was prevented from teaching in public without approval from her male colleagues. They were concerned about creating a precedent for other women who may also wish to be teachers, along with the disruption of having a young woman in their midst. In 1732, at the age of twenty-one, she was appointed Professor of Anatomy at the University of Bologna and, two years later, was given a position in philosophy. On 7 February 1738 Laura married Giuseppe Veratti, a doctor of philosophy and medicine and a fellow lecturer in physics at the University of Bologna, explaining, 'I have chosen a person who walks the same path of learning, and who from long experience, I was certain would not dissuade me from it'.

In 1745, she was appointed to an elite group of academics known as the 'Benedettini', set up by Lambertini, who was now the Pope. Laura and Giuseppe shared their domestic and professional life, setting up a laboratory in their home where they conducted experiments with an emphasis on the effectiveness of electric therapy. In addition, they operated a private school that Laura managed from 1749 to 1778, allowing her to teach experimental physics, basing her courses on Isaac Newton's *Principia*. These private courses offered students innovative experimental physics classes and she quickly became a key figure in the dissemination of Newtonian science in Italy. In the 1760s, Laura undertook experiments with her husband on possible medicinal applications of electricity, but they did not publish any papers on their work. In 1776, aged sixty-five, Laura was appointed to a professorship in experimental physics at the Institute of Sciences, with Giuseppe as her teaching assistant.

For twenty-eight years, Laura Bassi was instrumental in introducing Newtonian physics and philosophy to Italy and she paved the way for female academics. It is said that she was admired for her 'good character' and her charitable works, as well as her knowledge of Latin, logic, metaphysics, natural philosophy, algebra, geometry, Greek and French. She was an early pioneer for women in the fields of physics and anatomy. Her dissertations (one on chemistry, thirteen on physics, eleven on hydraulics, two on mathematics, one on mechanics and one on technology) are now kept at the Academy of Science in Bologna, although only four of her papers appeared in print, with an emphasis on physics and hydraulics. Laura was the first woman to hold an academic chair and, by the end of her career, she was the highest paid lecturer in the entire university.

Laura and Giuseppe had twelve children, only five of whom survived infancy. She passed away suddenly after experiencing chest pains on 20 February 1778 in Bologna, at the age of sixty-seven. Her tomb is located in the Corpus Domini Church in Via Tagliapietre, Bologna, in front of the tomb of her fellow scientist Luigi Galvani. She was survived by her husband Giuseppe who did his best to continue their laboratory. He died in 1793.

Ulrike Beisiegel

Born: 23 December 1952 in Mülheim an der Ruhr, North Rhine-Westphalia, West Germany
Field: Biochemistry, Molecular biology
Awards: Heinz Maier-Leibnitz Prize from the German Ministry of Education and Science (1983); Honorary Doctor at the Faculty of Medicine, University of Umeå (Sweden) (1996); Rudolf Schönheimer Medal of the German Atherosclerosis Society (2008); Ubbo Emmius Medal of the University of Groningen (Netherlands) (2014); Honorary Doctor of Science, University of Edinburgh (Great Britain) (2015)

Ulrike Beisiegel was born on 23 December 1952 in Mülheim an der Ruhr, North Rhine-Westphalia, West Germany. She completed her undergraduate studies in biology in Münster and Marburg, graduating in 1979 with a PhD from the Department of Medicine in Human Biology at the University of Marburg. Following this, she moved to the USA where she undertook postdoctoral research under the guidance of Nobel Prize winners Joseph L. Goldstein and Michael S. Brown in the Department of Molecular Genetics at the University of Texas in Dallas. Ulrike then returned to the University of Marburg where, for two years, she worked as a research assistant.

In 1983, Ulrike was awarded the Heinz Maier-Leibnitz Prize for her work on 'receptor defects as a cause of disease'. This prize is awarded annually to young researchers in recognition of outstanding achievement. The following year, she began serving as an academic counsellor at the University Medical Centre Hamburg-Eppendorf. In 1989, she undertook joint research on dietary fats in the blood and their effect on the liver with the Swedish scientist Gunüla O. Uvecrona of Umeå University. They examined triglycerides and enzymes in the fat particle deposits of the liver, as a means of preventing cardiovascular disease. In 1996, Umeå University awarded her an honorary doctorate for her collaboration with Uvecrona.

In 2000, Ulrike was elected to serve as a review board member for the German Research Foundation and, in 2001, also accepted the position of director of the Institute of Biochemistry and Molecular Biology at the University Medical Centre Hamburg-Eppendorf. The following year she was elected as a member of the evaluation committee of the Leibniz Association, a union of German non-university research institutes. In 2005, she was appointed speaker of the Ombudsman panel of the German Research Foundation, where she established rules for good scientific practices and monitoring scientific misconduct. The next year, she became a member of the German Council of Science and Humanities (GCSH) and, in 2008, was elected as chair of the GCSH research commission.

In 2008, Ulrike received the Rudolf Schönheimer Medal from the German Society for Arteriosclerosis Research – the highest award for achievement in this research field. The next year, she became a senator of the Leibniz Association and, twelve months later, was appointed the first woman president of the Georg-August-Universität, with a six-year term from 1 January 2011. During that year, she also became a senator of the Max Planck Society, one of the leading research organisations in Europe. In 2012, she assumed the position of vice president of the German Rectors' Conference (the association of public and government-recognised universities in Germany) and, in 2014, as a result of her research on cardiovascular diseases and her commitment to good science practice, was awarded the Ubbo-Emmius Medal.

In 2015, Ulrike was awarded an honorary doctorate of science from the University of Edinburgh for her contributions to university management and promotion of interdisciplinary and international collaboration aimed at improving the academic community. In the same year, she was confirmed for a second term as president, from 1 January 2017, a role which represents the University of Göttingen (founded in 1737) and the Foundation, internally and externally, chairs the Presidential Board as well as setting the board's guidelines. Ulrike is also responsible for the university's developmental and financial planning, along with being responsible for the premises.

Ulrike Beisiegel is determined to maintain high levels of scholarship and eradicate scientific misconduct, working tirelessly to maintain the highest standards in her field. She is an outstanding researcher and has deservedly received many awards for her work. In March 2020, she was nominated as the Chairperson of the Commission for Ethics in Research at the German Electron Synchrotron DESY in Hamburg.

Elizabeth Helen Blackburn

Born: 26 November 1948 in Hobart,
Tasmania, Australia
Field: Biology
Awards: NAS Award in Molecular Biology
(1990); Australia Prize (1998); Harvey Prize
(1999); Dickson Prize (2000); Heineken Prize
(2004); ASCB Public Service Award (2004);
Lasker Award (2006); Meyenburg Prize
(2006); Louisa Gross Horwitz Prize (2007);
L'Oréal-UNESCO Award for Women in
Science (2008); Nobel Prize in Physiology
or Medicine (2009); AIC Gold Medal (2012);
Royal Medal of the Royal Society (2015)

Elizabeth Blackburn was born on 26 November 1948, in Tasmania, Australia, the second of eight children. Both her parents were family doctors. The family lived in the small seaside town of Snug near the capital, Hobart. When she was four, her family moved to Launceston, a city some 125 miles (200 km) away. It was here that she attended Broadland House Church of England Girls' Grammar School until she turned sixteen. As a child, Elizabeth had a fascination with animals and was an avid reader of illustrated scientific, books – especially the biography of Marie Curie (*see entry*) written by Marie's daughter Evie that first appeared in 1937.

As her school did not offer physics, Elizabeth took evening classes at a nearby public school. Before she finished high school, the family moved to Melbourne, the capital of the state of Victoria on mainland Australia, where she completed her secondary education, spending her final year at University High School.

Elizabeth graduated with a Bachelor of Science (Hons) degree in 1970 and in 1972 was awarded her Master of Science degree, both in the field of biochemistry from the University of Melbourne. From there she travelled to England to the Medical Research Council Laboratory of Molecular Biology to undertake a PhD in molecular biology at the University of Cambridge, completing her degree in 1975 with a thesis 'Sequence studies on bacteriophage Phi X 174 DNA by transcription'. Her supervisor was the British biochemist Frederick Sanger, who twice won the Nobel Prize for Chemistry (1958 and 1980). The field of DNA (deoxyribonucleic acid) was now emerging and it held a great deal of fascination for her.

Elizabeth planned to undertake postdoctoral work at the University of California in San Francisco (UCSF), but fell in love with the accomplished scientist John Sedat and they were married in 1975. As John was going to Yale University in New Haven, Connecticut, Elizabeth changed her plans and was accepted into Yale as a postdoctoral fellow. There she began work on how to sequence the DNA found at terminal regions of the abundant, short, linear ribosomal gene-carrying 'minichromosomes'. In 1977, John was offered a position as assistant professor at UCSF, where Elizabeth accepted a research track role. She was provided space in the Department of Biochemistry in the Genetics unit where she continued her work on DNA. In 1978, she accepted a position of associate professor in the Molecular Biology Department at the University of California, Berkeley, where she was given her own laboratory.

Elizabeth stayed there for twelve years while she continued her work on chromosomes, at the end of which lies a cap or 'telomere' which protects them. During 1975, while undertaking a study of pond scum, she discovered that these telomeres have a certain DNA. In 1982, she collaborated with the Canadian American biologist Jack Szostak to discover that it is this DNA that prevents chromosomes from being broken down. In 1984, Elizabeth, with the molecular biologist and graduate student Carol Greider, discovered the enzyme telomerase that produces the telomere's DNA. After eight years on the faculty at Berkeley, Elizabeth was made a full professor in 1986, the same year as her son, Benjamin David, was born. Four years later, in mid-1990, Elizabeth accepted a full professorship at UCSF and continued her work on the biological systems consisting of telomeres and telomerase.

In 1998, Elizabeth became president of the American Society for Cell Biology. Three years later, she was asked to become a member of the President's Council on Bioethics, a body of the newly created US Federal Commission, a position that would see her be an adviser on national science policy. The Council, appointed by President George W. Bush's administration, was asked to consider topics that were of political significance and, at times, controversial. Quite outspoken in her views, Elizabeth was informed after two years that her services would no longer be required, due to her opinions not being in keeping with those of the White House.

Elizabeth's analysis of pond scum, with molecular biologist Carol Greider and Jack Szostak, led to their winning the 2009 Nobel Prize in Physiology or Medicine. Their research led to a field that shows it is within our capabilities to slow the cellular rate at which bodies age, in this way extending lifespan and youthfulness along with reducing the risk of contracting types of cancer, diabetes, Alzheimer's disease and cardiovascular disease.

In 2015, she was awarded a Royal Medal of the Royal Society, 'for her work on the prediction and discovery of telomerase and the role of telomeres in protecting and maintaining the genome'. That same year, Elizabeth assumed the presidency of the Salk Institute for Biological Sciences in San Diego, California, a position she described as the 'honour of my life'.

However, after two years, in December 2017, she resigned, declaring that, 'at this stage of my career and life, I've concluded that my energies will be best devoted to wider issues of science policy and ethics ... and spent advocating for measures I feel are critical to supporting ongoing scientific research and discovery worldwide'.

In addition to the Nobel Prize, Elizabeth Blackburn has been the recipient of numerous prizes and awards for her outstanding research.

Elizabeth Blackwell

Born: 3 February 1821 in Bristol, England
Died: 31 May 1910 (aged 89)
Field: Medicine

Elizabeth Blackwell was born on 3 February 1821 in Bristol, England, one of nine children who were all taught at home by private tutors. Her parents were Hannah (Lane) Blackwell and Samuel Blackwell, a Dissenter and sugar trader, and an activist for social reform. When Elizabeth was eleven, Samuel's sugar refinery was destroyed by fire and the family emigrated to the USA, spending the first six years there in New York City and Jersey City. All the family became activists for the anti-slavery movement, their home sometimes serving as a refuge for fugitives.

When she turned seventeen, the family moved to Cincinnati, Ohio. When Samuel passed away shortly afterwards, Hannah and the family of surviving children were left in dire financial circumstances. Despite this setback, the girls opened a private boarding school, the Cincinnati English and French Academy for Young Ladies, which operated for four years. Following this venture, Elizabeth accepted a position teaching at a girls' school in Henderson, Kentucky, at an annual salary of $400 USD. She was considering studying medicine, although, as she stated in her autobiography, she 'hated everything connected with the body, and could not bear the sight of a medical book'. She declared that 'the idea of winning a doctor's degree gradually assumed the aspect of a great moral struggle, and the moral fight possessed immense attraction for me'. Elizabeth said that she turned to medicine when a close friend suggested that she could have been spared suffering if she had been treated by a woman. Several family physicians indicated that, while studying medicine was an admirable idea, it was expensive and in any case, women were not permitted to do so. Undeterred, Elizabeth convinced two medical practitioners to let her study privately with them.

She then applied to every medical school in New York and Philadelphia, as well as a dozen more in the country areas of the north-east states. Elizabeth was turned down by all of them, including Harvard and Yale; one of the main reasons that were given was that she was a woman. Finally, however, she was accepted by Geneva Medical College in upstate New York in 1847, but only as the result of a prank, when the faculty assumed that an all-male body would never accept a woman. Her admission was not welcomed by most other students and faculty and, being the only female, she had a difficult time, especially with her initial experience with patients in 1848 at Philadelphia Hospital. She reported that 'the young resident physicians, unlike their chief, were not friendly. When I walked into the wards they walked out. They ceased to write the diagnosis and treatment of patients on the card at the head of each bed, which had hitherto been the custom, thus throwing me entirely on my own resources for clinical study.'

Despite this treatment, Elizabeth graduated with a medical degree in 1849, the first woman to do so in the USA. Elizabeth then did clinical work in London and Paris for two years to gain additional training, as well as studying midwifery at *La Maternité*, the lying-in hospital. It was during this time, following trauma to one eye, that she contracted purulent ophthalmia, an inflammation of both eyes, which later led to her losing the sight in one of her eyes. This put an end to Elizabeth's ambition to be a surgeon.

She travelled to New York in 1851. While her efforts to practise medicine were thwarted, she operated a dispensary in a single rented room where she saw patients on three afternoons each week. In 1854, the dispensary was incorporated and relocated to a small house that she had purchased on 15th Street. In 1856, she was joined by her sister Dr Emily Blackwell and Dr Marie Zakrzewska, and together they opened the New York Infirmary for Women and Children at 64 Bleecker Street, in 1867. It was Elizabeth's intention to expand the services of her institution to include a medical college and nursing school for women, a process that was delayed because of the Civil War. The institution remained in operation until 1899, essentially being run by Emily and Marie. In 1874, Elizabeth returned to England, establishing a medical school in London, the London School of Medicine for Women, with the English physician, teacher and feminist Sophia Jex-Blake (*see entry*). Its aim was to prepare women for the licensing examinations of Apothecaries Hall, the headquarters of the Worshipful Society of Apothecaries of London, which manufactured medicinal and pharmaceutical products.

In her later years, Elizabeth spent a great deal of time at a retreat at Kilmun on the north shore of the Holy Loch, on the Cowal peninsula in Argyll and Bute in the Scottish Highlands. It was during a holiday there in 1907 that she fell down a flight of stairs, an incident that left her with physical and mental disabilities. She never fully recovered from the accident and, as she aged, spent an increasing amount of time at the retreat.

She died of a stroke in 1910 in her home in Hastings, Sussex, aged eighty-nine. Her ashes were buried in the cemetery of St Munn's Parish Church in Kilmun. Her legacy was in making medicine more accessible and acceptable for women and in emphasising the importance of the health benefits of personal hygiene. Neither Elizabeth nor any of her five sisters married.

Elizabeth Blackwell was a pioneer in every sense of the word. When she enrolled in the Medical Register of the United Kingdom, she became Europe's first modern female doctor. She worked and was friends with Florence Nightingale, among others, and paved the way for other women to enter medicine.

Mary Adela Blagg

Born: 17 May 1858 in Cheadle,
Staffordshire, England
Died: 14 April 1944 (aged 85)
Field: Astronomy

Mary Blagg was born on 17 May 1858 in Cheadle, Staffordshire, England, the daughter of Frances Caroline (Foottit) Bragg and the solicitor John Charles Blagg. Although she received her formal education in London at a private boarding school, she trained herself in mathematics by reading her brother's textbooks, while undertaking an assortment of activities to benefit the community. Astronomy became her passion after hearing a lecture by Joseph Hardcastle, the grandson of the eminent astronomer William Herschel, as part of a series on astronomy in 1904–05. Mary began corresponding with Hardcastle and he arranged publication in 1906 of her analysis of 4,000 star observations in the course of a year.

Hardcastle suggested that there was a need to standardise lunar nomenclature as there were inconsistencies in the way in which formations were being named. To this end, in 1905 the *International Association of Academies* formed a committee on which Mary was given the task of collating the labels assigned to the formations found on existing lunar maps, two German and one English. With the assistance of Hardcastle and Herbert Hall Turner, the Savilian Professor of Astronomy and Director of the Radcliffe Observatory at Oxford University, Mary eventually published her 1913 book-length table, *Collated List of Lunar Formations*, which contained an index of 4,789 entries. In this volume, she proposed a new standard of names for various features of the moon, using telescopes and photographs to find the exact locations. Although she received assistance from her eminent colleagues, this was very much her project.

At the same time, Mary undertook a meticulous analysis of variable stars, ones whose brightness changes, either irregularly or regularly, in which she combined the raw data obtained by her predecessors. To achieve this, Turner had acquired a number of notebooks written by an astronomer who had compared, over time, the brightness of variable stars with those that were firmly lit. Her task was to identify those nearby stars and calculate each variable star's brightness. She established a system to measure a star's brightness, her findings serving as a basis for future astronomers. Part of the mathematical skills she employed involved the calculation of the length of the cycle of brightening and dimming, compared to the efforts of previous astronomers who had simply graphed the data from what they observed. Without the aid of modern computers, it was an incredible feat.

One of Mary's interesting conclusions was that in a particular star, *Beta Lyrae*, there was a slowing in the cycle of brightening and dimming. Her assertion was confirmed in 2008 when data from the Center for High Angular Resolution Astronomy (CHARA) Array Inferometer at Georgia State University confirmed that *Beta Lyrae* is a binary star, one which is actually two stars orbiting a common centre of mass. *Beta Lyrae* is located in the constellation of *Lyra*, about 960 light years from the Sun and is one of the brightest and easiest-to-find variable stars in the sky.

The senior scientist and planetary geologist Dr Chuck Wood undertook similar work for the National Aeronautics and Space Administration (NASA) in the 1960s, when NASA wanted to re-examine the naming designations. Underscoring Mary's incredible work, he declared, 'I can attest that that's very difficult. With even more modern hi-res images, it is still almost impossible sometimes to identify what the maps showed with what the photographs shows was the reality of the moon.'

By 1920, Mary's reputation had grown to the extent that she was appointed to the Lunar Commission of the newly formed International Astronomical Union. This led to her collaborating with Karl Müller, an amateur astronomer and retired government official. (Müller later had a lunar impact crater, located in the highlands near the centre of the Moon, named after him.) Their role was to prepare a definitive list of names that would later become the official authority in 1935 with the release of their joint two-volume set, *Named Lunar Formations*.

Their work remained as the benchmark until 1963–66, when it was replaced by the catalogue *System of Lunar Craters*, which gave a detailed account of all nearside lunar craters with diameters larger than 2.2 miles (3.5 km). Its creation was undertaken by the Lunar and Planetary Laboratory, established in 1961 at the University of Arizona, which made additions, deletions and changes to the contents in the Bragg-Müller volume.

Kevin Kilburn, an astronomer and astronomical historian, Fellow of the Royal Astronomical Society and founder of the Society for the History of Astronomy, is a great admirer of her work, declaring, 'Back in Mary Blagg's days, you had to have an analytical turn of mind to take that raw data and make sense of it. That's what made her so unusual at the time. I think the thing that strikes me most is that she wasn't ahead of her time; she was part and parcel of late Victorian astronomy. It was recognised that women were involved in science, they weren't just pressing flowers into books.'

Mary Blagg's charitable works extended to caring for Belgian refugee children during the First World War. One of her favourite hobbies was chess. She was honoured with a moon crater named Blagg, a lunar impact crater located on the small lunar mare *Sinus Medii*, before she passed away on 14 April 1944 at the age of eighty-five. She never married and, despite her fame, spent most of her life in Cheadle, where she communicated with her astronomical colleagues largely through her writing and correspondence.

Katharine 'Katie' Burr Blodgett

Born: 10 January 1898 in
Schenectady, New York, USA
Died: 12 October 1979 (aged 81)
Field: Physics and inventor
Awards: Achievement Award
from the American Association of
University Women (1945); Francis
Garvan Medal from the American
Chemical Society (1951); Photographic
Society of America's Progress Medal
(1972); National Inventors Hall of
Fame (post., 2007)

Born in Schenectady, New York, on 10 January 1898, Katharine or 'Katie' Burr Blodgett was the second child of Katharine (Burr) Blodgett and George Blodgett, a lawyer and head of the patent department at the General Electric Company. A month before Katie was born, George was murdered in his home by an intruder, who escaped and was never caught.

George having left his family with sufficient funds, they moved to New York City, then to France in 1901, returning to New York City in 1912. There Katie completed her schooling at the Rayson School. She displayed a talent for mathematics and completed high school at the age of fifteen, earning a scholarship to Bryn Mawr College in Pennsylvania. She also became interested in physics, deciding that a career in scientific research would allow her to follow her interest in both mathematics and physics. She graduated with a Bachelor of Arts degree in 1917.

During her vacations, Katie travelled to upstate New York in search of employment opportunities. At the Schenectady General Electric plant, some of her father's former colleagues introduced her to the research chemist Irving Langmuir. While he was showing her his laboratory, Langmuir recognised Katie's outstanding abilities and advised her to continue her scientific education. Following his advice, she undertook a master's degree in science at the University of Chicago and, on completion, she was the first woman to be hired as a scientist at General Electric. Langmuir encouraged her to participate in some of his earlier discoveries, with one of her initial tasks being to perfect tungsten filaments in electric lamps (he had received a patent for his work in 1916). Langmuir later asked her to concentrate her studies on surface chemistry. In 1919, at age twenty-one, she published an important paper on gas mask materials in the journal *Physical Review*. In 1926, she was the first woman to be awarded a PhD in physics from Cambridge University.

One of Katie's most important contributions came from her independent research on an oily substance that Langmuir had developed in the laboratory. Existing techniques for measuring this unusual substance were only accurate to a few thousandths of an inch, but Katie's method was accurate to about one millionth of an inch. In 1933, she developed a method using a colour gauge to measure the thickness of thin monomolecular films. This led to her invention of non-reflecting, 'invisible', glass in 1938, which proved to be a very effective instrument for physicists, chemists and metallurgists. While normal glass reflects a significant portion of light, by using a coating consisting of forty-four layers of liquid soap at one molecule each (four millionths of an inch), Katie was able to get 99 per cent of light to pass through the glass. This process started the entire field of optical coatings, appearing in many consumer products from picture frames to camera lenses and has also been exceptionally helpful in optics. Langmuir, who became her long-time collaborator, described Katie as a 'gifted experimenter', with a 'rare combination of theoretical and practical ability'.

During the Second World War, another important breakthrough was made when Katie developed smoke screens that saved many lives by covering the troops, thereby protecting them from the exposure of toxic smoke. She also assisted the meteorological sciences by developing a device to rapidly measure humidity as weather balloons ascended into the upper atmosphere.

Katie's efforts were acknowledged by many awards, including honorary doctorates from Elmira College in 1939, Brown University in 1942, Western College in 1942, and Russell Sage College in 1944. The 1943 edition of *American Men of Science* featured her and, in 1951, she received the prestigious Francis P. Garvan Medal (now the Garvan-Olin Medal), which recognises distinguished service to chemistry by women chemists, from the American Chemical Society for her work on monomolecular films.

In 1951, Katie was also chosen by the US Chamber of Commerce as one of fifteen 'Women of Achievement' and was honoured in Boston's First Assembly of American Women in Achievement (the only scientist in the group). The mayor of Schenectady honoured her with 'Katharine Blodgett Day' on 13 June 1951, because of the credit she had brought to her community.

She had a home on Lake George in the Adirondack Mountains and loved gardening, as well as acting in plays with the Schenectady Civic Players. Along with Langmuir, she was an active conservationist at the Research Lab and assisted other women through involvement with a club of professional women called the Zonta Club. She was a member of the American Physical Society and the Optical Society of America and won the Photographic Society of America's Progress Medal in 1972.

In 1963, at the age of sixty-five, Katie Blodgett retired from General Electric. She passed away at her home at 18 North Church Street, Schenectady on 12 October 1979, aged eighty-one. She never married. The techniques she developed have since become indispensable tools of science and technology, and she was entered into the National Inventors Hall of Fame posthumously in 2007. In 2008, an elementary school bearing her name was opened in Schenectady.

Rachel Littler Bodley

Born: 7 December 1831 in Cincinnati, Ohio, USA
Died: 15 June 1888 (aged 56)
Field: Chemistry and botany
Awards: Academy of Natural Sciences of Philadelphia (1871); Member of the New York Academy of Sciences (1876)

The third child and oldest daughter of five children, Rachel Littler Bodley was born on 7 December 1831 in Cincinnati, Ohio, USA, to Rebecca W. (Talbott) Bodley, a teacher, and Anthony R. Bodley, a Presbyterian carpenter and pattern maker who had settled in the town in 1817. Her mother, of Quaker descent, taught in a private school that Rachel attended until she was twelve years old and where the education of women was enthusiastically endorsed. After completing her time there, in 1844 she entered the Wesleyan Female Seminary from which she graduated in 1849, soon being appointed as an assistant teacher at the age of eighteen.

At the Seminary she became a preceptor in the collegiate studies department, where she became recognised for classifying the collection of local flora contained in the herbarium of specimens bequeathed to the Seminary by the heirs of Joseph Clark. (He had been born in Scotland in 1782 but was a resident of Cincinnati from 1823 until his death in 1858.) Her compilation was published in 1865 as an important guide to plants in the area.

Rachel stayed at the Seminary until 1860, leaving to move to Philadelphia to enrol as a special student (non-degree student) at the Polytechnic College, where she studied physics and chemistry. She returned to her home town two years later to teach at the Cincinnati Female Seminary. Just three years later, she was elected to the first chair of Chemistry and Toxicology at the Female Medical College of Pennsylvania (renamed the Woman's Medical College of Pennsylvania in 1867), the first chartered college for women in the world. In doing so, she also became the first woman to become a professor of chemistry at a medical school.

In 1863, Rachel had become one of the first women visitors appointed by the State Board of Public Charities to inspect local charitable institutions. The following year she was made corresponding member of the State Historical

Society of Wisconsin, and in 1867 and 1868 she delivered a series of key lectures on both land and sea plants. In 1871, she became an elected member of the Academy of Natural Sciences of Philadelphia (presently Academy of Natural Sciences of Drexel University). The same year, she had the degree of Master of Arts conferred on her by her alma mater in Cincinnati, one of the first three people to receive the honour. In 1873, she was elected corresponding member of the Cincinnati Society of Natural History.

Rachel became the dean of the faculty of the Woman's Medical College in 1874, a position she held for the rest of her life. The college's first new building being undertaken and, as the dean, she oversaw its construction. She also conducted a statistical study of the school's graduates, reportedly the first such factual study of women in the medical profession; her findings appeared in a pamphlet titled 'The College Story'. At that time, the graduates were practising in China, India, Europe, and in every state in the USA. The replies she received to her survey questions revealed a remarkable degree of success on the part of the school's graduates.

Rachel was elected first vice president of the meeting of the American Chemical Society of New York, called in August 1874 to commemorate the discovery of oxygen by the famed Joseph Priestley exactly a century before. In 1876, she was elected a corresponding member of the New York Academy of Sciences and, in 1883, a member of the Academy of Natural Sciences of Philadelphia (now known as the Academy of Natural Sciences of Drexel University). In 1879, the Woman's Medical College of Pennsylvania conferred upon her the honorary degree of MD and, in 1880, she was made a member of the Franklin Institute of Philadelphia, delivering a series of lectures on 'Household Chemistry' in the regular course of the institute.

In 1882, Rachel was chosen a member of the Educational Society of Philadelphia and was elected as a school director in Philadelphia's 29th School Section between 1882 and 1885. On 1 January 1884, she was elected to the Board of Public Education for three years; she and Mary Mumford became the first female directors on the board. In 1885, Rachel assisted in the formation of the Christian Association at the Woman's Medical College of Pennsylvania, securing for it a permanent home in Brinton Hall. In 1887, she was re-elected for another term in her role as a school director.

On 15 June 1888, at the age of fifty-six, Rachel Bodley unexpectedly died of a heart attack in Philadelphia. She never married. Her contributions were in administration, teaching and botany, and most importantly, in the establishment of women in science.

Alice Middleton Boring

Born: 22 February 1883 in Philadelphia, USA
Died: 18 September 1955 (aged 72)
Field: Biology and zoology

Alice Middleton Boring was born on 22 February 1883 in Philadelphia, USA, one of four children of Edwin (a pharmacist) and Elizabeth (Truman) Boring. The family was heavily involved in science; her younger brother Edwin Garrigues Boring headed a laboratory at Harvard University. Alice attended a Quaker school, the Friends' Central School, in Philadelphia, completing her studies there in 1900. She then went to Bryn Mawr College in Pennsylvania where she studied as an undergraduate under evolutionary biologist and geneticist Thomas Hunt Morgan (who later won a Nobel Prize in 1933). She received her Bachelor of Arts degree with a major in cytology in 1904. The following year, she enrolled in the graduate programme, where she was supervised in her doctoral thesis by early American geneticist Nettie Maria Stevens, and was awarded her PhD in 1910. Her dissertation, 'A Study of the Spermatogenesis of Twenty-two Species of the *Membracidae*, *Jassidae*, *Cercopidae* and *Fulgoridae*', investigated the behaviour of chromosomes in the formation of spermatozoa in insects.

Alice commenced her career as a cytologist and geneticist in 1910 and, until 1918, she lectured in zoology at the University of Maine. In 1920, she taught briefly at Wellesley College, but left the USA when she obtained a position as an assistant professor of biology at Peking Union Medical College. Funded by the Rockefeller Foundation and founded in 1906, it became one of the most selective medical colleges in China. Although she requested only a two-year leave of absence from Wellesley, her involvement in Chinese social and political matters soon became the focal points of her life's work. She changed the direction of her research and began the study of China's lizards and amphibians, making significant contributions to literature in the field. In her teaching, she provided Chinese students with her Western viewpoints and her contribution included creating a better scientific understanding between Eastern and Western cultures.

Between 1923 and 1950, Alice worked at Peking University (later called Yenching University), but the Second World War resulted in a marked change in her circumstances. During the 1937 Japanese invasion, with money very tight and no mail delivery, life at the university became difficult. In 1941, the English and American faculty members of the university were placed in a concentration camp in Shantung, during which time Alice's family had no contact with her. She was repatriated in 1943 and, in a letter to a friend written during the voyage back to the USA, she wrote that she and her colleagues were in good shape, declaring, 'We shall not look like physical wrecks when you see us in New York, even if our clothes may be rather dilapidated.'

Alice became prominent in the West for her knowledge of Chinese amphibians and reptiles and she was fascinated by the vast array Chinese herpetofauna that offered her a stimulating research field. She collaborated and published with American biologist Nathaniel Gist Gee, herpetologist Clifford Hillhouse Pope and Chinese biologist Hu Jingfu, as well as her students. To collect specimens, she undertook expeditions to various locations in China, including Jiangxi, Zhejiang and Anhui. Her programme in Yenching University became one of two major centres in China for the study of amphibians and reptiles. She was one of the charter members of Peking Natural History Society and had twenty-one papers and one handbook published by it between 1929 and 1950.

At the end of the war, Alice held positions at Columbia University and Mount Holyoke, but returned to China for a few years. In 1945, she published the important work *Chinese Amphibians: Living and Fossil Forms*, and presented American museums and scientists with specimens and information on China.

Alice was described as both strict and warm by her students, insisting on very high standards when in the laboratory or undertaking academic study. However, she also had a soft side and demonstrated her care of them by requiring every biology student to have dinner at her house at least once a year. To prepare her Chinese students to study abroad, she also personally taught them Western table manners.

Alice Boring spent her final years working at Smith College, a private, independent women's liberal arts college in Northampton, Massachusetts. She passed away on 18 September 1955, aged seventy-two, in Cambridge, Massachusetts. She never married.

Elizabeth Brown

Born: 6 August 1830 in Cirencester, Gloucestershire, England
Died: 5 March 1899 (aged 68)
Field: Astronomy

The elder of two daughters, Elizabeth Brown was born on 6 August 1830 in Cirencester, Gloucestershire, England, to Jemimah and Thomas Crowther Brown, a wine merchant and meteorological recorder. She received some early schooling from a governess, but was largely self-taught, reading widely in both literature and the sciences. As her mother passed away before Elizabeth's eleventh birthday, she was raised by her father. He introduced her to the telescope and encouraged her to develop her scientific interests, a most unusual pursuit for a woman of that era. She used the telescope to observe sunspots. In 1871, she recorded the daily rainfall for the Meteorological Society, taking over a task that her father had undertaken for the previous twenty-six years.

For much of her life until middle age, Elizabeth undertook domestic duties, which mainly involved caring for Thomas until he passed away in 1883, at the age of ninety-one. The then fifty-three-year-old Elizabeth was then free to follow her passion in the field of solar astronomy. Before long she joined the Liverpool Astronomical Society that had been formed the previous year, making the 124 mile (200 km) round trip from her home in Cirencester to attend its meetings in Liverpool. The society catered for amateur astronomers, including allowing female members. In December that year, she presented a paper on sunspots and, shortly afterwards, was appointed as their Solar Section director, using her position to encourage other women to enter the field.

In the beginning she had old equipment that was hardly suitable for her purposes, declaring, 'When I first took up solar work, I possessed no

Elizabeth Brown

observatory ... only an old refractor of 3 inch aperture, which had already seen a good deal of service'. Her method for observing the sun was to project images from the telescope onto a white card in a darkened room.

Demonstrating considerable enterprise, Elizabeth set about building two observatories in the grounds of her house, one for telescopes and the other for meteorological equipment. By then she had acquired an equatorially mounted refractor with a driving clock and a 6.5 inch reflector and an astronomical clock. In 1884, she attended a meeting of the British Association for the Advancement of Science in Montreal, afterwards travelling widely in Canada and the USA.

Elizabeth journeyed all over the world on scientific expeditions to observe solar eclipses. These included trips to Russia in 1887 and, in 1889, Trinidad, in the West Indies, resulting in her producing two books about these journeys, *Pursuit of a Shadow* (the title being a tribute to the Quaker meteorologist Luke Howard, who used the expression to describe his own observations of clouds) and *Caught in the Tropics*. Her trip to Russia included visiting a small community 200 miles (320 km) northeast of Moscow where, along with some of the world's leading astronomers, she observed the eclipse from a beautiful country home. However, it was a cloudy night, which interfered with their viewing. She reported, 'For a second or so we had a view of the coronal light ... and a glimpse of the rose-coloured prominences ... but it was all over before we had realised it'. It was a similar situation in Trinidad when clouds also filled the night sky but, at the last moment, she distinguished 'the silvery light of the corona, encircling the death-like blackness of the moon's orb'.

Elizabeth's daily recording and highly detailed drawings of sunspots gained her a distinguished reputation among other astronomers. When the Liverpool Astronomical Society closed its doors in 1890, she played an integral key role in the formation of the British Astronomical Association (BAA) in the same year, becoming the director of its Solar Section and a member of its governing council. The *Journal of the British Astronomical Society* published many of her observational reports and she was an enthusiast of methodical data collection, advising potential astronomers to 'look for no great or stirring discoveries, be prepared for long periods when there will be little or nothing to record but persevere'. The meticulous observer would eventually find it 'worth the labour involved, and the labourer who once begins to cultivate will rarely, if ever, leave it in disappointment or disgust'.

In 1892, Elizabeth was one of the first women to be nominated for election to Fellowship of the Royal Astronomical Society, but none of the women gained sufficient votes to be admitted to the society. It would not be until 1916 that women were finally admitted as Fellows. However, in 1893, Elizabeth was elected as a Fellow of the Royal Meteorological Society and held memberships in the astronomical societies of France, Wales and the Pacific.

In 1896, she made a third trip to observe a solar eclipse, this time to Vadsø in northern Norway. While preparing for a fourth journey to observe a solar eclipse in the western Mediterranean, she suddenly died at home of cardiac thrombosis at age sixty-eight on 5 March 1899. She is buried in the cemetery attached to the Cirencester Friends Meeting House and she bequeathed her astronomical observatory, its contents and the sum of £1,000 to the British Astronomical Association. She never married.

Elizabeth Brown was a great supporter and encourager of women astronomers, especially solar astronomers. In her words, 'The sun is always at hand. No exposure to the night air is involved and there is no need for a costly array of instruments.'

Linda Brown Buck

Born: 29 January 1947 in Seattle,
Washington, USA
Field: Physiology
Awards: Takasago Award (Japan,
1992); Lewis S. Rosenstiel Award
(1996); Unilever Science Award (1996);
Gairdner Foundation International
Award (2003); National Academy
of Sciences (2003); Nobel Prize in
Physiology or Medicine (2004);
Foreign Member of the Royal
Society (2015)

Linda Buck was born in Seattle, Washington and was the middle daughter of three born to her parents. Her mother enjoyed solving puzzles of all types and her father, an electrical engineer by profession and an enthusiastic inventor, spent a deal of time building his creations in their basement. Linda revealed that 'it may be that my parents' interest in puzzles and inventions planted the seeds for my future affinity for science, but I never imagined as a child that I would someday be a scientist.' Her life was initially relatively unstructured, learning to appreciate music and beauty from her mother and learning how to use power tools and building things with her father.

Linda enrolled at the University of Washington, just a few miles from her home, initially majoring in psychology with the aim of becoming a psychotherapist. However, she became intrigued by immunology, and completed a Bachelor of Science in psychology and microbiology in 1975. From there she attended graduate school in the Microbiology Department at the University of Texas Southwestern Medical Center at Dallas, earning her PhD in immunology in 1980 under the supervision of Professor Ellen Vitetta.

In 1980, Linda began postdoctoral research at Columbia University in New York City under Dr Benvenuto Pernis, two years later joining Columbia to study neuroscience at the Institute of Cancer Research in the laboratory headed by Dr Richard Axel. There, she became fascinated with the sense of smell and set out to determine how odour molecules are detected by olfactory sensory neurons in the nose. By inventing a novel approach to search for a family of receptor genes used in the nose, she discovered the odorant receptor family, which is responsible for odour detection in the nose. Her research showed that this family comprises up to 1,000 related receptors with slight variations, consistent with our ability to recognise a multitude of odour molecules with varied structures. Linda and Axel published this work in a landmark journal article in 1991.

Later that year, Linda moved to Harvard, becoming an assistant professor in the Neurobiology Department at Harvard Medical School. Olfactory sensory neurons in the nose send signals to the olfactory bulb of the brain and Linda's lab discovered that each neuron in a mouse's nose expresses only one of 1,000 different odorant receptor genes. Thus, the signals that each neuron sends to the brain reflect the odour specificity of a single receptor type. By isolating neurons that detect different odours and then determining what receptors they expressed, her lab made another key discovery: combinatorial receptor codes for odours. Each odour molecule can be detected by multiple receptors, and each receptor can detect multiple odours, but each odour molecule is detected by a unique combination of receptors. This explains how we can detect not only a multitude of different odours, but also distinguish them as having different scents.

Another important question was how signals from 1,000 different odorant receptors are organised in the nervous system to generate different scents. Linda's lab, and also Axel's, found that thousands of sensory neurons with the same receptor are dispersed in specific zones in the nose, but they all send signals to a few specific locations in the olfactory bulb. The result was a precise map of odorant receptor inputs that is similar among individuals.

Linda rose through the ranks at Harvard to become a full professor with tenure in 2001 and became an investigator of the Howard Hughes Medical Institute in 1994. That year she also married Roger Brent, whom she described as 'a marvellous intellect and fellow scientist'. In 2002, she returned to Seattle to join the Division of Basic Sciences at the Fred Hutchinson Cancer Research Center. She also held the position of affiliate professor of physiology and biophysics at the University of Washington. The return to her home city brought Linda closer to her family and friends and to Roger, who was located in Berkeley.

At Harvard, and then at Fred Hutchinson, Linda's lab made numerous additional contributions to sensory biology. These included the discoveries of additional chemosensory receptors in the nose and vomeronasal organ, an olfactory structure that detects pheromones, as well as candidate bitter and sweet taste receptors in the mouth. They also included definition of the highly dispersed chromosomal organisation of odorant receptor genes in human and mouse and, more recently, the description of brain neural circuits that transmit pheromone and stress signals from the olfactory system to elicit instinctive physiological responses.

The importance of Linda Buck's research has been acknowledged by numerous honours and awards. These include election to the National Academy of Sciences, National Academy of Medicine, American Academy of Arts and Sciences, and as a foreign member of the Royal Society. There are also numerous scientific prizes and awards, including the LVMH Moët Hennessy Louis Vuitton Science for Art Prize, the Unilever Science Award, the Lewis S. Rosenstiel Award for Distinguished Work in Basic Medical Research, and the Gairdner Foundation International Award. In

2004, Linda received the Nobel Prize in Physiology or Medicine, jointly with Richard Axel, 'for their discoveries of odorant receptors and the organization of the olfactory system'.

At the time she received the Nobel Prize, Linda wrote,

> Looking back over my life, I am struck by the good fortune I have had to be a scientist. Very few in this world have the opportunity to do every day what they love to do, as I have. I have had wonderful mentors, colleagues, and students with whom to explore what fascinates me and have enjoyed both challenges and discoveries. I am grateful for all of these things and look forward to learning what Nature will next reveal to us. As a woman in science, I sincerely hope that my receiving a Nobel Prize will send a message to young women everywhere that the doors are open to them and that they should follow their dreams.

(Susan) Jocelyn Bell Burnell

Born: 15 July 1943 in Lurgan, Northern Ireland
Field: Astrophysics
Awards: J. Robert Oppenheimer Memorial Prize (1978); Beatrice M. Tinsley Prize (1986); Herschel Medal (1989); FRS (2003); FRSE (2004); DBE (2007); Michael Faraday Prize (2010); Royal Medal (2015); *Grande Médaille* (2018); Special Breakthrough Prize in Fundamental Physics (2018)

(Susan) Jocelyn Bell was born on 15 July 1943 in Lurgan, Northern Ireland, to educated Quaker parents, Allison and (George) Philip Bell. Her father was an architect who worked for the Armagh Observatory and helped design their Planetarium. During her visits there, the staff encouraged Jocelyn to pursue astronomy as a career. She was raised in the town of Lurgan and, from 1948 to 1956, attended the Preparatory Department of Lurgan College. As was the case with all the girls, she was not allowed to study science until her parents, with others, protested against the school's policy. The girls' previous curriculum had included subjects such as cooking and cross-stitching rather than science.

When she failed the eleven-plus exam, her parents sent Jocelyn to The Mount School, a Quaker girls' boarding school in York, England. It was there that she was inspired by her physics teacher, Mr Tillott, remarking: 'You do not have to learn lots and lots ... of facts; you just learn a few key things, and ... then you can apply and build and develop from those ... He was a really good teacher and showed me, actually, how easy physics was.'

Jocelyn graduated with an honours degree in natural philosophy (physics) from Glasgow University, Scotland, in 1965. She then enrolled in the doctoral programme at the University of Cambridge, completing her PhD in 1969 under the supervision of British radio astronomer Antony Hewish. Her thesis was titled 'The measurement of radio source diameters using a diffraction method'. As part of her research, she built and operated an 81.5 megahertz radio telescope, as well as examining interplanetary scintillation of compact radio sources.

In 1967, while analysing data obtained from the telescope, she came across a few unusual signals, which she termed 'scruff'. They were too fast and regular to come from quasars. Both she and Hewish ruled out orbiting satellites as the cause and, several years later, they ultimately determined that these signals must have arisen from rapidly spinning, super-dense, collapsed stars that the media dubbed 'pulsars'. The BBC Horizon series later documented the discovery, crediting Jocelyn with 'one of the most significant scientific achievements of the 20th century'. While the breakthrough was recognised by the award of the 1974 Nobel Prize in Physics to Antony Hewish and Sir Martin Ryle, Jocelyn was controversially excluded from the recipients of the prize, despite the fact that she was the first to observe and precisely analyse the pulsars.

In 1968, Jocelyn married Martin Burnell, a government worker whose position took them to various parts of England. They had a son, Gavin, who, in 2019, was an Associate Professor in Condensed Matter Physics at Leeds University. For many years Jocelyn worked part-time while raising Gavin, and moving with her husband to various job locations, during which time she began studying almost every wave spectrum in astronomy and, as she said, seeking out 'whatever job [she] could get in astronomy or physics wherever he was'. In doing so, she acquired enormous expertise, maintaining a teaching fellowship from 1970 to 1973 at the University of Southampton where she developed and calibrated a 1–10 million electron volt gamma-ray telescope. At the same time, she held research and teaching positions in X-ray astronomy at the Mullard Space Science Laboratory in London, as well as examining infrared astronomy in Edinburgh, Scotland. Jocelyn and Martin separated in 1989 and divorced in 1993, Jocelyn retaining Martin's surname.

In 1991, when she was appointed Professor of Physics at the Open University in Milton Keynes, the number of women professors of physics in the United Kingdom doubled. Speaking in 1996, when she was in her early fifties, Jocelyn spoke about the differences of life experiences for women from those of men:

> Although we are now much more conscious about equal opportunities I think there are still a number of inbuilt structural disadvantages for women. I am very conscious that having worked part-time, having had a rather disrupted career, my research record is a good deal patchier than any man's of comparable age … The life experience of a woman is rather different from that of the male.

From 2002 to 2004, Jocelyn served as president of the Royal Astronomical Society, and from October 2008 until October 2010 she was president of the Institute of Physics. She also held the role of interim president of the Institute following the death of her successor, Marshall Stoneham, in early 2011. In 2018, she was awarded the Special Breakthrough Prize in Fundamental Physics, worth £2.3 million, for her discovery of radio pulsars. She donated the entire amount 'to fund women, under-represented ethnic minority and refugee students to become physics researchers', with the funds to be administered by the Institute of Physics.

Jocelyn Bell Burnell received many accolades and awards for her outstanding achievements, including, in 1999, being appointed Commander of the Order of the British Empire (CBE) for services to Astronomy. In 2007 she was promoted to Dame Commander of the Order of the British Empire (DBE), and in February 2013 she was deemed to be one of the 100 most powerful women in the United Kingdom by *Woman's Hour* on BBC Radio 4.

In February 2014, Jocelyn became the first woman to hold the elected office of president of the Royal Society of Edinburgh, holding the position from April 2014 to April 2018, when she was succeeded by Dame Anne Glover. In 2019, she was a Visiting Professor of Astrophysics at the University of Oxford and a Fellow of Mansfield College.

In 2020, Jocelyn was included by the BBC in a list of seven important but unheralded British female scientists, and in the same year was made an Honorary Fellow of Trinity College Dublin. In the following year she was awarded the Karl Schwarzschild Medal, presented by German Astronomical Society to eminent astronomers and astrophysicists, as well as becoming the second female recipient (after Dorothy Hodgkin in 1976) of the prestigious Copley Medal. In 2022 she was awarded the Prix Jules Janssen, the highest award of the Société astronomique de France (SAF), the French astronomical society, and in 2023 she won the Royal Irish Academy's Cunningham Medal.

Nina Byers

Born: 19 January 1930 in Los
Angeles, California, USA
Died: 5 June 2014 (aged 84)
Field: Physics
Award: Guggenheim
Fellowship for Natural
Sciences, USA &
Canada (1964)

Nina Byers was born on 19 January 1930 in Los Angeles, California. She attended the University of California, Berkeley, where she majored in physics, graduating with the highest honours in 1950. She then enrolled as a graduate student, under the supervision of Gregor Wenzel, a German physicist known for development of quantum mechanics. Nina completed her PhD in 1956 with a thesis on pi-mesic atoms, her work forming the basis of her first publication, 'Interactions of Low-Energy Negative Pions with Nuclei', which appeared in the journal *Physical Review* in August 1957.

Following her PhD, Nina held a postdoctoral position at the University of Birmingham, England, in a research group led by Rudolf Peierls in the Department of Theoretical Physics. In 1958, she accepted a research associate role at Stanford University in California, leading to an assistant professorship with a focus on superconductivity.

In 1961, Nina joined the Physics and Astronomy Department at the University of California, Los Angeles (UCLA) as an assistant professor, turning her attention to particle physics and collaborating on studies in CP-violation and pion-nucleon charge-exchange scattering. She was the first and only female professor in the department for the next two decades. In the late 1960s and early 1970s, she held appointments at the University of Oxford, becoming the first female physics lecturer there. She then divided her time between Los Angeles and Oxford, where she was the Jane Watson Visiting Fellow in Somerville College, as well as a period as a Science Research Council Senior Scientist.

In late 1973, Nina returned to UCLA where her physics research now concentrated on the new gauge theories of electroweak interactions. By the early 1980s, Nina's interest in quarkonium and bound state systems was rekindled, driven partly by her desire to find suitable thesis topics for her students. By now, Nina was undertaking cutting-edge research in addition to fulfilling commitments to the American Physical Society (APS). She also played vital roles in the Panel of Public Affairs and in the Forum on Physics and Society of the APS, serving as Councillor-at-large of the Society (1977–81) and then acting as chair in 1982. Her leadership in the American Association of the Advancement of Science (AAAS) was also outstanding, and she was named a Fellow of both societies. Nina was heavily involved in two very important issues: the role of women in physics and the examination of the role of physicists in the development of nuclear weapons. As a result of her efforts, she served as president of the APS Forum on History of Physics.

In 1993, Nina officially retired from UCLA, but in her role as Professor Emerita she remained actively engaged in the pursuit of increasing the role of women in academia. Soon after her retirement, she commenced work on setting up a website to detail the contributions by women to the field of physics. After overcoming a number of hurdles, she received a significant grant from the Sloan Foundation and, in 1996, was able finally able to establish her scholarly archive, *Contributions of 20th Century Women to Physics*, as an internet resource hosted by UCLA. She researched and wrote about the contributions of eighty-three women in subfields including astrophysics, atomic molecular and optical physics, crystallography and nuclear physics. In 2010, she co-edited a book (with Gary Williams), *Out of the Shadows: Contributions of Twentieth-Century Women to Physics*, published by Cambridge University Press.

On 5 June 2014, at age eighty-four, Nina Byers died as a result of a haemorrhagic stroke in her Santa Monica home. As part of her final wishes, her ashes were interred at Woodlawn Cemetery in Santa Monica in the early autumn. She left an enduring legacy, reflected in the establishment of the Nina Byers Lectureship, an endowed chair established to invite leading physicists to come to UCLA for several weeks to give a series of talks and interact with faculty and students. This programme is a prestigious honour for lecturers as well as a mechanism for bringing the excitement of frontier results from external scientists to UCLA.

Steven Moszkowski, an Emeritus Professor of the UCLA Physics and Astronomy Department, attended graduate school with Nina at the University of Chicago in 1951. He said of her, 'Nina was a warm-hearted person (who cared) passionately about things she had an interest in. She was meticulous about her work and passionate about the strength (of) scientific standards.' He considered her a leading figure in the study of elementary particle theory and superconductivity, a competitive field of physics. Nina achieved a high stature in the international physics community and paved the way for other women to follow in her footsteps.

Annie Jump Cannon

Born: 11 December 1863 in Dover, Delaware, USA
Died: 13 April 1941 (aged 77)
Field: Astronomy
Award: Henry Draper Medal (1931)

Annie Cannon was born on 11 December 1863 in Dover, Delaware, USA, the eldest daughter of shipbuilder and Delaware State Senator Wilson Lee Cannon and his second wife, Mary Elizabeth (Jump) Cannon, a mathematician and keen astronomer. The house attic had been turned into a makeshift observatory where Annie and her mother, who suggested that Annie also pursue a career in mathematics, could survey the stars. Annie attended a public school and then Wilmington Conference Academy in Dover, following which she entered Wellesley College, a private women's liberal arts college, graduating in 1884. It was there, under the guidance of the American astronomer and physicist Sarah Frances Whiting, that Annie developed an interest in spectroscopy, the study of the interaction between matter and electromagnetic radiation.

After completing her studies, Annie spent much of her time enjoying an active social life, notwithstanding that, from a young age, she was very hard of hearing. However, following the death of her mother in 1893, she decided to undertake postgraduate studies at Wellesley under the direction of Whiting, her field including advanced astronomy. As a result of her growing expertise, for the two years following 1895 she was a special student (non-degree student) of astronomy at Radcliffe, also a women's liberal arts college. Following this, she was appointed as an assistant at the Harvard College Observatory as part of the team led by the renowned astronomer and physicist Professor Edward Pickering. Annie formed part of a group known as 'Pickering's Women' that included astronomers Williamina Paton Stevens Fleming and Antonia Maury (*see entry*). Annie's hourly rate was a lowly 50 cents and her principal role was to study the bright southern hemisphere stars. It was then that she became deeply involved with the intensive work of the laboratory, specialising in the study of stellar spectra (the different colours of light coming from a star).

The Harvard Observatory had systematically photographed the sky from the North to the South Pole. Pickering's project had begun in 1885 and involved recording, classifying, and cataloguing the spectra of all stars down to those of the ninth magnitude, around sixteen times fainter than can be seen by the human eye. Fleming had initially classified stellar spectra by letters, in alphabetic sequence from A to Q, largely according to the strength of their hydrogen spectral lines. Subsequently, Maury created a new scheme, this one with twenty-two groups labelled from I to XXII and further added three subdivisions based on the sharpness of the spectral lines. She also placed Fleming's B stars before the A stars.

Not satisfied with these methods of categorisation, in 1901 Annie published her own catalogue of 1,122 stars in which she considerably simplified Fleming's classifications to the classes O, B, A, F, G, K, and M (given the mnemonic device, 'Oh, Be A Fine Girl – Kiss Me!'), while retaining P for planetary nebula and Q for unusual stars. She also added numerical divisions, further dividing each class into 10 steps from 0 to 9 (e.g., the Sun's spectral type is G2). It soon became apparent that her scheme was actually classifying stars according to their temperature, and subsequently her spectral classifications were universally adopted. Pickering was full of praise for Annie's ability to rapidly identify stars, declaring, 'Miss Cannon is the only person in the world – man or woman – who can do this work so quickly.'

Of the work of studying the spectra of stars Annie wrote,

> The results will help to unravel some of the mysteries of the great universe, visible to us, in the depths above. They will provide material for investigation of those distant suns of which we know nothing except as revealed by the rays of light, travelling for years with great velocity through space, to be made at last to tell their magical story on our photographic plates.

Annie remained at Harvard until her retirement in 1940, serving as the curator of the observatory's astronomical photographs between 1911 and 1938, after which she was appointed the William Cranch Bond Astronomer at Harvard University. During this period, she lectured, travelled and attended meetings, including those of the International Astronomical Union. She was also a strong supporter of the suffragette movement and an active member of the National Woman's Party.

Her work consisted mainly of observing, classifying and undertaking the spectroscopic analysis of stars. In this way she produced a comprehensive survey of the skies in a simple way that is still in use today. It was said of her that it was her 'keen visual memory as well as her patience and discipline' that uniquely qualified her to sort out the line patterns and to place each star in its proper category. The speed at which she worked was 'phenomenal', classifying 5,000 stars a month in the period 1911–15. For sparsely populated regions of the sky she achieved a rate of more than three stars per minute. She was meticulous in her recording of detail, noting the date and time that she began and ended each classification session. Annie was also said to be modest about her remarkable talent, although taking 'an inner pride in performing efficiently and effectively'. By 1915 she had completed classifying 225,300 stars, but it took several more years for the work to be published.

Her compilation of these spectral classifications was enormous, with her major publications *The Henry Draper Catalogue* (1918–24) and *The Henry Draper Extension* (1925–49) classifying around 350,000 stars. (Henry Draper was an American doctor and amateur astronomer, famous for his pioneering work in astrophotography, the use of photographs in astronomy.) In addition, Annie published many short papers and her output provided an immense amount of data that could be used by future researchers. To undertake such a tremendous project required a great deal of patience, precision and dedication to the task.

In 1921, Annie became the first woman to receive an honorary doctorate from a European university when she was awarded one in mathematics and astronomy from Groningen University in the Netherlands. In 1925, she was the first woman recipient of an honorary doctorate from Oxford University. She also received honorary degrees from the University of Delaware, Oglethorpe University and Mount Holyoke College. In 1931, she was the first woman to receive the Henry Draper Medal, awarded every four years by the United States National Academy of Sciences 'for investigations in astronomical physics'. In 1929 she was chosen as one of the 'greatest living American women' by the League of Women Voters, and in 1933 she represented professional women at the World's Fair in Chicago.

Annie Cannon passed away, at age seventy-seven, on 13 April 1941 in Cambridge, Massachusetts, after being ill for over a month. She never married. She is buried in Lakeside Cemetery in Dover. She was showered with prizes and awards for her work, and, in honour of her, the American Astronomical Society presents the Annie Jump Cannon Award annually to female astronomers for distinguished work in astronomy. In 1943, this was won by one of her former colleagues, Antonia Maury. The lunar crater Cannon is named after her, as is the asteroid 1120 Cannonia.

Rachel Louise Carson

Born: 22 May 1907 in Springdale, Pennsylvania, USA
Died: 14 April 1964 (aged 56)
Field: Marine biology
Awards: Presidential Medal of Freedom (post., 1980)

Rachel Louise Carson was born on 22 May 1907 in Springdale, Pennsylvania, USA, and raised on a family farm near Springdale. Her parents were Maria Frazier (McLean) Carson and Robert Warden Carson, an insurance salesman. Being somewhat of a loner, Rachel had a love of the outdoors, especially the birds and plants that she observed around the property, finding a fossilised fish and native fauna and flora. When she was only eight years old, Rachel wrote a book, *The Little Brown House*, about a pair of wrens seeking shelter, along with contributions to a children's literature magazine, *St Nicholas*, which ran from 1873 to 1940. A recurring theme in her stories was nature, especially the ocean.

Rachel attended Springdale's small school, where she completed tenth grade before continuing to graduate top of her class of forty-five students at a nearby high school in 1925. From there she went to Pennsylvania College for Women (now Chatham University), having won a senatorial scholarship. She initially selected English as her major as she had the ambition of becoming a writer, all the while contributing to the school's newspaper and literary supplement, but, in January 1928, she switched to biology and was admitted to the graduate programme at Johns Hopkins University. However, financial difficulties forced Rachel to remain at her college, where she graduated *magna cum laude* (with great distinction) the following year.

From there, Rachel spent a summer and did part-time work at the Marine Biological Laboratory, but by now she was also a graduate student in genetics and zoology at Johns Hopkins, where she commenced in the autumn of 1929. Twelve months later, with her finances dwindling, she switched her studies to part-time and accepted a position in the laboratory of the American biologist Raymond Pearl, where she worked with rats and *Drosophila*, a genus of flies. In June 1932, she graduated with a master's degree in zoology with a thesis on the embryonic development of the *pronephros* (the most basic of excretory organs, like a kidney) in fish.

Although she had intended to enrol for a doctorate at Johns Hopkins in 1934, the poor circumstances of her family caused her to seek a full-time teaching position to support them. But her father passed away suddenly and Rachel was then left to care for her elderly mother. She found a temporary position with the US Bureau of Fisheries. While working there, she wrote radio scripts for a series of weekly educational broadcasts, 'Romance Under the Waters', each of fifty-two minutes, that were designed to spark the public's interest in fish biology and provide favourable publicity for the Bureau. Her work was a resounding success, and after sitting for the civil service exam she obtained the highest score. In 1936, Rachel became only the second woman hired in a full-time position by the Bureau – as a junior aquatic biologist. The following year she published an article, 'The World of Waters', on aquatic creatures for *Atlantic* magazine. This led to her first book, *Under the Sea Wind*, in 1941. Although it received encouraging reviews, it sold only around 2,000 copies. All the while she continued to publish her articles in journals including *Nature, Sun Magazine* and *Collier's*.

By 1945, Rachel had been promoted in the ranks of the US Fish and Wildlife Services, becoming chief editor of publications and supervising a small writing group. She wrote pamphlets on conservation and natural resources and edited scientific articles. In 1948, she began collecting material for a second book on the life history of the ocean. Published in 1951 by Oxford University Press, *The Sea Around Us* won the National Book Award for nonfiction. This book turned out to be one of the most successful volumes about nature ever written. The following year she resigned from government service.

The award and the sales of her books enabled her to move to Southport Island, Maine, in 1953, to concentrate on her writing. There Rachel began field research on the ecology of the Atlantic Shore, and two years later, in 1955, she completed the third volume of her sea trilogy, *The Edge of the Sea*. It was published in the *New Yorker* in two instalments and received very favourable reviews. Rachel was now approaching the peak of her popularity, with numerous speaking invitations and even fan mail. She had also begun a relationship with Dorothy Freeman, a married summer resident of Maine.

Rachel had been disturbed by the use of synthetic chemical pesticides after the Second World War, and in late 1957 Federal proposals to commence a widespread spraying of pesticide to eradicate gypsy moths drew her interest. The spraying involved the use of chlorinated hydrocarbons and organophosphates, such as DDT, mixed with fuel oil. The dangers of the overuse of environmental poisons became the focus for the remainder of Rachel's research activities. She collected evidence, including twenty-seven species of dead fish in the Colorado River and livestock that had been accidentally poisoned. To this end, she commenced a four-year project on examining the damage caused by DDT, also enlisting others to her cause.

When a friend wrote to her about the loss of bird life after pesticide spraying, Rachel was inspired to write *Silent Spring*. Launched in September

1962, it challenged the practices of agricultural scientists and government, focusing on the effect of pesticides on ecosystems as well as the impact on humans. In accusing the chemical industry of spreading misinformation and public officials of uncritically accepting the claims of industry, she heralded the beginning of an environmental movement that was largely directed at the chemical industry. Rachel then became the target of a smear campaign, labelling her as an hysterical woman or a Communist. Rachel expected a backlash and fierce criticism of her work and was concerned that she might be sued for libel. But shortly after its release, she was diagnosed with breast cancer and was weakened by undergoing radiation therapy. She died from the disease on 14 April 1964.

Her work, and the report of President John F. Kennedy's Science Advisory Committee, which validated her research, made pesticides a major public issue. Ultimately, *Silent Spring* was successful in its aims and her work inspired the National Environmental Policy Act of 1970 'to prevent or eliminate damage to the environment and to stimulate the health and welfare of man'. Rachel left a legacy that could be considered the foundation of modern environmentalism. She received medals from the National Audubon Society and the American Geographical Society and was inducted into the American Academy of Arts and Letters.

Rachel Carson is remembered for her outstanding work though the naming of many awards, prizes and places after her, including the Rachel Carson Conservation Park in Maryland, The Rachel Carson National Wildlife Refuge in Maine, the Rachel Carson Prize (founded in Norway in 1991) for women who have made a contribution in the field of environmental protection. In 1980, Rachel was posthumously awarded the Presidential Medal of Freedom. The Rachel Carson Book Prize has been presented by the Society for Social Studies of Science since 1998, and in 2016 the University of California, Santa Cruz, renamed one of its colleges as Rachel Carson College. Her homes are considered national historic landmarks.

Prior to her death in 1964, much of Rachel's correspondence with Dorothy Freeman was destroyed. However, what was left was published by Dorothy's granddaughter in 1995, *Always Rachel: The Letters of Rachel Carson and Dorothy Freeman, 1952–1964: An Intimate Portrait of a Remarkable Friendship.*

Mary Lucy Cartwright

Born: 17 December 1900 in Aynho,
Northamptonshire, England
Died: 3 April 1998 (aged 97)
Field: Mathematics
Awards: FRS (1947); Sylvester Medal
(1964); De Morgan Medal (1968);
Honorary Fellow, Royal Society of
Edinburgh (HonFRSE) (1968); DBE (1969)

Mary Lucy Cartwright was born in Aynho, Northamptonshire, England, on 17 December 1900, where her father, William Degby Cartwright, was curate and later rector. Her mother was Lucy Harriette Maud Cartwright. Mary was the third of five children and, through her grandmother Jane Holbech, was descended from the poet John Donne and William Mompesson, the Vicar of Eyam.

Initially taught by governesses, when aged eleven Mary's secondary education began at Leamington High School (1912–15). She then went to Gravely Manor School in Boscombe (1915–16), before completing her schooling at Godolphin School in Salisbury (1916–19). Her best subject at school was history, but in her final year she was encouraged to study mathematics, and it became her discipline of choice at university. In October 1919, she attended St Hugh's College, Oxford, being one of only five women in the entire university who were studying mathematics.

It was a difficult time to attend university, as the First World War had only just ended the year before and there were many men who had resumed their interrupted university studies or were enrolling for the first time. The lecture halls were often so crowded that Mary could not get into them, relying on copies of the lecture notes instead. After two years, she sat for the Moderations Examinations and was awarded a second-class pass. She was so disappointed at her result that she considered returning to history, but decided to stay with mathematics, at least for the time being. But her love of history remained with her, with many of her subsequent research papers including historical perspectives to add further context to her findings. Mary's breakthrough reportedly came in her third year at a party on a barge during Eights Week (the four-day regatta constituting Oxford's main intercollegiate rowing event of the year), when V. C. Morton, later professor of mathematics at Aberystwyth, suggested that she should attend evening classes of the eminent mathematician and number theorist Godfrey Harold (G. H.) Hardy, then Savilian Professor of Geometry. Inspired, she went on to obtain a first-class degree in Final Honours in 1923, only the second year in which women took final degrees at Oxford.

Mary then spent four years as a schoolteacher, first at Alice Ottley School in Worcester and then at Wycombe Abbey School in Buckinghamshire, the latter carrying the position of assistant mistress. In 1928, Mary returned to Oxford to read for her Doctor of Philosophy degree under Hardy. At one time, when she produced an obviously wrong result, he is reported as saying, 'Let's see. There's always hope when you get a sharp contradiction.'

While Hardy was on leave at Princeton during the academic year 1928–29, Edward Charles 'Ted' Titchmarsh took over the supervision role. Mary's thesis, 'Zeros of Integral Functions of Special Types', was examined by John Edensor Littlewood, who later became a major collaborator with her for many years.

In 1930, Mary was awarded a Yarrow Research Fellowship and attended Girton College in Cambridge, where she continued to research the topic of her doctoral thesis. While attending one of Littlewood's lectures, she solved an open problem which he had posed, her solution now being known as 'Cartwright's Theorem'. To prove the theorem, she used a new approach, applying a technique introduced by the Finnish mathematician Lars Valerian Ahlfors. This was a turning point in her career.

In a long series of papers, Mary continued to explore the theory of complex (especially entire) functions. Among other results, her research described the phenomena which can appear near fractal boundaries, finding new applications in this field. In 1934, Mary was appointed to an assistant lectureship in mathematics at Cambridge and as a part-time lecturer the following year. She became director of studies in mathematics at Girton College in 1936.

In 1938, Mary commenced a new project that had a major impact on the direction of her research. The Radio Research Board of the Department of Scientific and Industrial Research had produced a document outlining particular differential equations that had come out of their modelling of radio and radar work. They contacted the London Mathematical Society for assistance in finding a mathematician who could examine these problems and Mary took an interest, as the dynamics which lay behind the problems were unfamiliar to her. With the assistance of Littlewood, she began studying the equations. He wrote, 'For something to do we went on and on at the thing with no earthly prospect of "results"; suddenly the whole vista of the dramatic fine structure of solutions stared us in the face.' What he described is known today as the 'butterfly effect' which has greatly influenced the direction of modern theory of dynamical systems. Then in 1945, Mary simplified the French mathematician Charles Hermite's elementary proof of the irrationality of π.

At the end of the Second World War, Mary published her discoveries, but as 'nobody paid much attention', she went on to other things, becoming renowned as a pure mathematician. Her work with Littlewood is said to have 'anticipated some of the geometrical ideas that are fundamental to chaotic dynamics', representing 'an important milestone in the evolution of our thinking about dynamic complexity'.

In 1947, largely based on her remarkable contributions in the collaboration with Littlewood, she was elected a Fellow of the Royal Society – the first woman to be awarded that honour. In 1949, she was appointed Mistress of Girton College, which had only been fully incorporated into Cambridge University the year before. She is said to have provided 'quiet, unassuming and clear-headed leadership', always finding time to interview entrance candidates 'in batches of five, lumped together quite irrespective of the subject for which they were applying'.

Despite her administrative roles, Mary continued her research, with an emphasis on cluster sets. In 1956, she was a member of the Royal Society delegation that visited the Soviet Union as guests of the Academy of Sciences. In 1959, she became a Reader in the Theory of Functions at Cambridge, a position she held until 1968. She spent lengthy periods at various institutions, including the academic year 1968–69 at Brown University, then Claremont Graduate College in the USA, along with the University of Wales and Poland in 1969–70. In 1969, she was appointed a Dame of the British Empire (DBE).

Mary broke her hip in a bicycling accident, but continued her work, in 1971 accepting the role of visiting professor at Case Western Reserve University in Cleveland, Ohio. She then returned to England where she spent the next ten years in collaboration with H. P. F. Swinnerton-Dyer.

Dame Mary Cartwright received many honours for her outstanding work, including being the first female winner of the Sylvester Medal of the Royal Society in 1964, 'in recognition of her distinguished contributions to analysis and the theory of functions of a real and complex variable', and four years later the London Mathematical Society awarded her the De Morgan Medal, its premier award. She was well informed in a variety of areas, including painting and music. On 3 April 1998, at age ninety-seven, she passed away in Midfield Lodge Nursing Home in Cambridge. In one of the many tributes to her on her death, a friend and colleague described her as 'a person who combined distinction of achievement with a notable lack of self-importance'. Consistent with her lifelong modesty, she left strict instructions that there were to be no eulogies at her memorial service. She never married.

Yvette Cauchois

Born: 19 December 1908 in Paris, France
Died: 19 November 1999 (aged 90)
Field: Physics
Awards: Ancel Prize from the *Société chimique de France* (1933); Henri Becquerel Prize (French Academy of Sciences, 1935); Gizbal-Baral Prize (French Academy of Sciences, 1936); Henry de Jouvenel Prize for selfless scientific activity (Ministry of National Education, France, 1938); Jerome Ponti Prize from the French Academy of Sciences (1942); Triossi Prize from the French Academy of Sciences (1946); Medal of the Czechoslovak Society of Spectroscopy (1974); Gold medal of the University of Paris (1987)

Yvette Cauchois was born on 19 December 1908 in Paris. She attended school there and was attracted to science from childhood. On completing her secondary education, she enrolled for a degree in physical sciences at the Sorbonne, graduating with a *Licence en Sciences* (Bachelor of Science) in July 1928 and a *Diplôme d'Études Supérieures* (graduate degree) in 1930. With the support of a National Fund for Science studentship, she undertook graduate studies at the Laboratory of Physical Chemistry and was awarded her doctorate in 1933. The thesis topic was 'Extension of the X-ray spectrography. Spectrography focusing by curved crystal; X emission spectra of gases'.

After completing her PhD, Yvette was appointed as a research assistant in the laboratory of Jean Perrin, a French physicist and Nobel Prize winner, at the *Centre national de la recherche scientifique* (National Centre for Scientific Research). She soon established the fundamental principles of a new high-resolution X-ray spectrometer that was named after her and, from 1934, used it to observe gas emissions and multiplets. The new technique was used around the world for the analysis of X-rays and gamma rays, inspiring an increased interest in the research of radiation studies. Yvette also pioneered developments in X-ray imaging and found that, with the aid of curved crystal, X-ray radiation could be focused for use in monochromators and X-ray scattering. Her efforts in soft X-ray distributions were the catalyst in determining the photo-absorption spectra, in which she studied the electronic structure of materials using the radiation reflected from crystals.

In 1937, Yvette was promoted to research associate and, in the same year, participated in the launch of the science museum the *Palais de la Découverte* (Discovery Palace). In January 1938, she was appointed acting head of the physical chemistry laboratory in the Faculty of Sciences of Paris, and later that year she and the Romanian nuclear physicist Horia Hulubei claimed to have discovered element 93, which they named 'sequanium'. However, their assertion was criticised at the time, on the grounds that element 93 did not occur naturally. It was subsequently discovered in 1940 by Edwin McMillan and Philip H. Abelson, who called it 'neptunium' (atomic symbol Np and atomic number 93), after the planet Neptune. In later years, it was found that trace amounts of a number of neptunium isotopes (neptunium-237 to neptunium-240) can occur naturally as the result of either neutron-capture or beta-decay in uranium-containing ores. This means that Yvette and Hulubei may have been correct all along.

After the outbreak of the Second World War in 1939, Yvette continued her role at the laboratory, also acting as head of studies when Jean Perrin fled to the United States after Germany invaded France in 1940. She was promoted to professor at the Sorbonne in 1945, and in 1954 to the Chair of Chemical Physics and head of the laboratory, succeeding the French physicist Edmond Bauer. By 1960, it was apparent that the laboratory needed to expand, so she founded the *Centre de Chimie Physique* (chemical laboratory) at Orsay. Yvette directed this organisation for ten years as well as continuing her research at the Sorbonne and, from 1962, initiated a research programme in collaboration with the *Istituto Superiore di Sanità* (Higher Health Institute) at the *Laboratori Nazionali di Frascati* (National Laboratory of Frascati) in Italy to explore the depths of synchrotron research. She was the first person in Europe to realise the potential of the radiation emitted by electrons rotating in the synchrotron as a source for understanding the properties of matter. In 1970, she produced X-ray images of the sun.

Following the division of the Sorbonne in 1971, Yvette joined the University of Paris VI. In the early 1970s, she embarked upon experiments at the *Laboratoire pour l'utilisation des radiations électromagnétiques* (Laboratory for the use of Electromagnetic Radiation). Between 1975 and 1978 she chaired the French Society of Physical Chemistry, being only the second woman, after Marie Curie (*see entry*), to do so. From 1978 until her retirement in 1983, she had the role of Professor Emerita at the University of Paris VI and was still research-active in her laboratory until 1992, when she was eighty-three. In 1993, she was conferred an honorary doctorate by the University of Bucharest.

Over her lifetime, Yvette Cauchois published more than 200 research papers that are still cited today. She also took a keen interest in assisting both young and underprivileged people. She enjoyed poetry and music and was an accomplished pianist. After meeting a priest from the monastery of Bârsana, she decided to be baptised in the Orthodox religion, doing so in 1999 on a trip to Maramures, Romania. It was on this journey that she contracted bronchitis, passing away on 19 November 1999, a few days after returning to Paris. She never married. She was buried in the Monastery Bârsana in Romania, to whom she left her estate. Her name is given to a street of the new university area of Moulon in Gif-sur-Yvette, along with a street in Tomblaine (Meurthe-et-Moselle).

Edith Clarke

Born: 10 February 1883 in Ellicott City, Howard County, Maryland, USA
Died: 29 October 1959 (aged 76)
Field: Electrical engineering
Awards: Society of Women Engineers Achievement Award (1954); National Inventors Hall of Fame (post., 2015)

One of nine children, Edith Clarke was born on 10 February 1883 in Ellicott City, Howard County, Maryland, USA, into a prosperous family. Her mother was Susan Dorsey (Owings) Clarke, and her father was John Ridgely Clarke. Edith's father died when she was seven and her mother when Edith was twelve years old. Now orphaned, Edith's uncle became her legal guardian. Her upbringing was typical for girls of the day with her status, the emphasis being on proper grooming in preparation for marriage and motherhood. To this end, Edith attended Briarley Hall, a boarding school for girls in Montgomery County. There she studied Latin, history and English literature, while receiving a rigorous training in arithmetic, algebra, and geometry, which laid the groundwork of her future profession.

Inheriting money from her parents' estate at the age of eighteen, Edith enrolled to study mathematics and astronomy at Vassar College in New York, from which she graduated in 1908 and was invited to join *Phi Beta Kappa*, America's oldest academic honour society. Then she taught mathematics and physics at a private school in San Francisco and Marshall College in Huntingdon. Edith also developed a passion for engineering and decided to pursue formal qualifications in the field. In 1911, she enrolled in the civil engineering programme at the University of Wisconsin-Madison, but at the end of her first year took up a summer position as a computer assistant to an AT&T research engineer, Dr George Campbell, whose research included the application of mathematical methods to the problems of long-distance electrical transmissions. One of her first assigned computational tasks was to calculate the first seven terms of an infinite series that represented a probability function. Edith eventually became the manager of a group of women 'computers' who made calculations for the Transmission and Protection Engineering Department during the First

World War. At this time she also studied radio at Hunter College and electrical engineering at Columbia University. She left her role at AT&T in 1918 to study electrical engineering at the Massachusetts Institute of Technology in Boston, earning a master's degree in 1919, the first electrical engineering degree ever awarded to a woman at that institution. Her thesis, supervised by the Irish American electrical engineer Arthur Edwin Kennelly, was titled 'Behavior of a lumpy artificial transmission line as the frequency is indefinitely increased'.

Not being able to find employment as an engineer, Edith accepted employment at General Electric as a supervisor of computers in the Turbine Engineering Department. In 1921, while employed at General Electric, she invented the 'Clarke calculator', a simple graphical device that solved equations involving electric current, voltage and impedance in power transmission lines. It could solve line equations involving hyperbolic functions ten times faster than previous methods. She filed a patent for the device in 1921, which was granted in 1925. Still unable to obtain a position as an engineer, Edith left General Electric to teach physics at the Constantinople Women's College in Turkey for the following two years.

On her return, she was re-hired by General Electric, now as an electrical engineer in the Central Station Engineering Department. Between 1923 and 1945, Edith published eighteen technical papers and, in doing so, became an authority on the manipulation of hyperbolic functions, equivalent circuits, and graphical analysis. In 1926, she became the first woman to present a paper before the American Institute of Electrical Engineers. Titled 'Steady-state stability in transmission systems', it described a mathematical technique to model a power system and its behaviour, allowing engineers to analyse the longer transmission lines then becoming more common.

Another paper, which she co-authored in 1941, was published in the *Transactions of the American Institute of Electrical Engineers*, receiving the National First Paper Prize of the Year award. It was reported as 'the first published mathematical examination of transmission lines over 300 miles long, and the first published use of the network analyser to obtain data'. Her book *Circuit Analysis of A–C Power Systems*, based on the notes for her lectures to General Electric engineers, and published in 1943, soon became the main textbook for new engineers. A second volume appeared in 1950.

In 1945, at age sixty-two and after twenty-six years of service, Edith retired from General Electric. Two years later, however, Edith was appointed a full professor at the University of Texas, becoming the first female professor of electrical engineering in the country. In 1948, she became the first woman elected as a Fellow of the American Institute of Electrical Engineers (now known as the Institute for Electrical and Electronics Engineers). In an interview in the *Daily Texan* on 4 March that year, she remarked, 'There is no demand for women engineers, as such, as there are for women doctors; but there's always a demand for anyone who can do a good piece of work.'

In 1954, Edith received the Society of Women Engineers Achievement Award 'in recognition of her many original contributions to stability theory and circuit analysis'. Three years later, she retired from her academic position in Texas and was the first female engineer to achieve professional standing in *Tau Beta Pi*, the oldest engineering honour society in the USA.

Edith Clarke passed away in hospital in Olney, Maryland in Baltimore, on 29 October 1959 at the age of seventy-six. She never married. In 2015, she was posthumously inducted into the National Inventors Hall of Fame.

In 1985, a leader in the engineering industry, Dr James E. Brittain, wrote,

Edith Clarke's engineering career had as its central theme the development and dissemination of mathematical methods that tended to simplify and reduce the time spent in laborious calculations in solving problems in the design and operation of electrical power systems. She translated what many engineers found to be esoteric mathematical methods into graphs or simpler forms during a time when power systems were becoming more complex and when the initial efforts were being made to develop electromechanical aids to problem solving. As a woman who worked in an environment traditionally dominated by men, she demonstrated effectively that women could perform at least as well as men if given the opportunity. Her outstanding achievements provided an inspiring example for the next generation of women with aspirations to become career engineers.

Anna Botsford Comstock

Born: 1 September 1854 in Otto, New York, USA
Died: 24 August 1930 (aged 75)
Field: Zoology
Awards: National Wildlife Federation Conservation Hall of Fame (post., 1998)

Anna Botsford was born on 1 September 1854 in Otto, New York, USA, the only child of a Quaker mother, Phebe Irish, and father, Marvin Botsford, the latter a prosperous farming family. She was raised on the farm where she spent much time with her mother studying the birds, wildflowers and trees. Anna attended the Chamberlain Institute and Female College, a Methodist school located in Randolph, New York, before returning to Otto where she taught for a year.

In 1874, Anna continued her education at Cornell University in Ithaca, New York, where she studied literature and modern languages. She became a member of *Kappa Alpha Theta*, founded in 1870 as the first Greek-letter sorority for women. As part of her coursework, she enrolled in an invertebrate zoology class lectured by the twenty-five-year-old John Henry Comstock, who founded the entomology programme. Anna's experience there stimulated her interest not only in entomology but in Comstock himself, with the pair spending time together examining the flora and fauna of New York State's Finger Lakes region, a group of eleven long, narrow, roughly north-south lakes in Central New York.

After two years of study, Anna left Cornell but continued her enthusiasm for science by drawing illustrations of insects and preparing diagrams for Comstock's lectures. She had no formal art training, but simply examined an insect under a microscope and drew it. Anna and John's friendship developed into romance and they were married on 8 October 1878 at her parents' home in Otto. The couple leased a property from the university in which they were surrounded by plants, trees and lakes.

The following year, they moved to Washington DC, where John took the role of chief entomologist at the US Department of Agriculture. His position involved significant travel and time away from home as he investigated reports of insect pests, an activity that made him an expert on the nation's insect problems. Meanwhile, Anna became his office assistant and illustrated his field reports, including his 1880 *Report of the Entomologist* on citrus scale insects.

In 1882, they returned to Ithaca, New York, where she re-enrolled at Cornell, completing her bachelor's degree in natural history in 1885. At this time, she registered for wood engraving classes conducted by the master wood engraver, mentor and instructor John P. Davis, from Cooper Union, a distinguished higher education institution for the Advancement of Science and Art, established in 1859 in New York. Anna was singled out for praise as a result of her 'superlative accuracy'.

In 1888, Anna and her husband travelled to Europe, where they spent a period in Germany before returning to start work on the second volume of his textbook, spending time at both Oxford and Stanford Universities where John lectured during the winter months. Anna's engravings were used to illustrate John's textbook, *An Introduction to Entomology*, published in 1888. This was the same year she was initiated into the national honour society *Sigma Xi*, being one of the first four women to do so.

Anna's reputation as a science educator was spreading, and in 1894 she was elected to the New York Society for the Promotion of Agriculture, an association established by a group of philanthropists with the aim of bringing nature studies to rural schools. For John's 1895 book, *A Manual for the Study of Insects*, Anna produced more than 600 plates, with her work being displayed at exhibitions in New Orleans in 1885, Chicago in 1893, Paris in 1900 and Buffalo in 1903. So outstanding were her creations that she became only the third woman elected to the American Society of Wood Engravers.

Anna lectured on nature study at a number of educational institutions across the USA, including Columbia and Stanford Universities as well as the University of Virginia. She held a position of assistant in nature at Cornell, two years later becoming the first woman in Cornell to reach professorial status when she was made an assistant professor. However, this did not sit well with some of the trustees of the college, and in 1900 she was demoted to the rank of lecturer. After an outcry, she was reinstated to her former title the following year.

Anna took on many roles, including as editor of a nature study magazine, *Boys and Girls*, as well as being the contributing editor (1905–07) and editor (1917–23) of *Nature Study Review*. Some of her illustrated books on nature study, written jointly with her husband, included *Insect Life* (1897) and *How to Know the Butterflies* (1904), while her own works encompassed *Ways of the Six Footed* (1903), *How to Keep Bees* (1905), *Handbook of Nature-Study for Teachers and Parents* (1911), *The Pet Book* (1914) and *Trees at Leisure* (1916). Her most celebrated work was the 1912 *Handbook of Nature Study*, a guide for elementary school teachers, which was a compilation of her previous works. It ran to twenty-four editions until 1947. As a result of the large number of books published, Anna and John formed the Comstock Publishing Company with the motto 'Nature Through Books'.

Although she retired from teaching in 1922, Anna continued to lecture and remained committed to her profession in the activities of the American Nature Study Society, as well as serving as an associate director of the American Nature Association. She was named as one of America's twelve greatest living women by the League of Women Voters, in addition to receiving an honorary degree in 1930 from Hobart College, part of the Hobart and William Smith Colleges that together form a liberal arts college situated on the shore of Seneca Lake in Geneva, New York.

On 5 August 1926, John Comstock suffered a stroke that left him an invalid and Anna nursed him, despite her own failing health. She passed away from cancer on 24 August 1930 at their home in Ithaca, followed seven months later by John on 20 March 1931. Anna's autobiography, *The Comstocks of Cornell: John Henry Comstock and Anna Botsford Comstock*, was published in 1953 by Comstock Publishing Associates in Ithaca, and in 1998, over half a century after her death, Anna Comstock was inducted into the National Wildlife Federation Conservation Hall of Fame. Anna and John did not have any children.

Esther Marly Conwell

Born: 23 May 1922 in the Bronx, New York, USA
Died: 16 November 2014 (aged 92)
Field: Chemistry and physics
Awards: IEEE Edison Medal (1997); National Medal of Science (2010)

The youngest of three sisters, Esther Marley Conwell was born on 23 May 1922 in the Bronx, New York, USA, to a housewife from Austria and a portrait photographer from Russia. Her father impressed upon his daughters the importance of being married, to have a job and earn money, particularly as a professional, in order to support a husband. In 1938, Esther attended Brooklyn College in New York, where she studied biology and physics, but only a few lectures of chemistry which she soon dropped. Her initial ambition was to be a high school physics teacher, as these were the only women in science she had come across. She was often the only female in physics lectures, but did not mind the extra attention this seemed to bring. After graduating in January 1942 with a Bachelor of Science in mathematics and physics, she was encouraged by her professor to undertake graduate studies at the University of Rochester. She received a fellowship and graduated with a master's degree in physics in 1945. The previous year, she had married the writer Abraham Rothberg. Esther then enrolled at the University of Chicago, where she was awarded a PhD in physics in 1948, one of only six women to receive a similar doctorate that year in the USA.

In 1952, she began work at Sylvania Labs (later known as GTE labs) where she began research into the conducting properties of germanium and silicon, later moving on to semiconductor research for telecommunications. This was for Esther a most welcoming environment, remarking that 'it was just between me and the research, how well I did and how well I felt about myself'. She stayed there for the next twenty years, before taking up a research role on semiconductors at Western Webster Research Center in Rochester.

Esther soon became an expert on how electrons travelled through semiconductors, her research including the Conwell-Weisskopf theory proposed as part of her master's thesis. This led to a better understanding of the inner workings of transistors that, in turn, ignited the revolution of computers. Her later research, on charges moving through DNA, contributed to a better understanding of the way potentially cancer-causing mutations arise. She discovered that shining sunlight on a piece of DNA can distort it, a particular distortion in which different parts of the same chain 'dimerised', or come together in a different way than if the sun had not shone on them. This was a defect, many of which can lead to cancer and so the interest then turned to how to eliminate them. It was discovered that this could be achieved by having an electron moving up to a place where the dimerisation has occurred and, when the electron arrives, the mutation gets undone and the DNA returns to its normal self, free of defects.

Esther did not embark on a university academic role until 1990, at the age of sixty-eight, because, 'as a woman in science I figured they wouldn't consider me'. She began as an adjunct professor at the University of Rochester and then later as a full-time professor in the Department of Chemistry after her retirement from Xerox in 1998.

She made great efforts to mentor and encourage young women interested in a career in science, and in 2005 received the Dreyfus Foundation's Senior Scientist Mentor Program Award. Three years later, she received the American Chemical Society's Award for Encouraging Women into Careers in Chemical Sciences. Her success as a mentor opened the door for many other women to follow her path and a successful research career. She was elected to the National Academy of Engineering (1980), National Academy of Sciences (1990) and the American Academy of Arts and Sciences (1992).

The vast majority of Esther's working life spanned forty-seven years of research for industry through Bell Laboratories (1951–52), GTE Laboratories (1952–72) and Xerox Corporation (1972–98). She received many honours for her work, including, in 1997, being the first woman to receive the Edison Medal from the Institute of Electrical and Electronics Engineers. The oldest and most prestigious medal in this field of engineering in the USA, it was awarded to Esther for 'fundamental contributions to transport theory in semiconductor and organic conductors, and their application to the semiconductor, electronic copying and printing industries'. Esther was chosen as one of *Discover* magazine's 'Top 50 Women in Science' in 2002, and in 2010 she received the National Medal of Science, nominated by her friend and colleague, the eminent physicist and engineer Mildred Dresselhaus (herself a previous winner) (*see entry*) and presented by US President Barack Obama. She was also a member of the National Academy of Sciences and the National Academy of Engineering.

Esther Conwell, a devotee of ballet in which she took several classes each week, passed away at age ninety-two on 16 November 2014 at Strong Memorial Hospital in Rochester. She had been walking and was struck by a neighbour's car as he was backing out of his driveway. The university flag was lowered for four days on 20 November in memory of an extraordinary woman who contributed so much to the technological revolution and was still active in her research until the end. Esther had one child, a son, Dr Lewis Rothberg, who followed in her footsteps, becoming a professor of chemistry and physics at the University of Rochester.

Gerty Theresa Radnitz Cori

Born: 15 August 1896 in Prague, Austro-Hungarian Empire
Died: 26 October 1957 (aged 61)
Field: Biochemistry
Awards: Nobel Prize in Physiology or Medicine (1947); Garvan-Olin Medal (1948)

Gerty Theresa Radnitz was born on 15 August 1896 in Prague, in what was then the Austro-Hungarian Empire (now the Czech Republic), the daughter of Martha Radnitz and Otto Radnitz, a chemist who became a manager of sugar refineries. Gerty was home-tutored before enrolling in a lyceum (high school) for girls from which she graduated in 1912 at the age of sixteen. She had already decided that she wanted to undertake a degree in medicine. After graduating from the *Tetschen Realgymnasium* (senior high school) in 1914, she attended the Medical School of the German University of Prague, where she graduated in 1920.

It was there that Gerty met Carl Cori. After serving as a lieutenant in the First World War in the Sanitary Corps of the Austrian Army in Italy, Carl returned to the university and studied alongside Gerty, also receiving his medical degree in 1920. Upon their graduation, Gerty converted to Catholicism, which meant they could be married in the Roman Catholic Church. They were married on 5 August 1920. Carl spent a year at the University of Vienna and a further year as an assistant in pharmacology at the University of Graz, after which he emigrated to the United States in 1922, where he obtained a medical research position at the State Institute for the Study of Malignant Diseases (now the Roswell Park Cancer Institute). Gerty followed six months later and joined him in his research. They both became naturalised US citizens in 1928.

Although the Institute discouraged the couple from working together, Gerty even being threatened with dismissal, they continued their investigation into how glucose is metabolised in the human body and the hormones that regulate the process. During their time at Roswell, they published fifty papers, their first joint investigation examining the fate of sugar in the animal body and the effects of insulin and epinephrine. Gerty also published eleven articles as a sole author. In 1929, they proposed the theoretical cycle that would later win them the Nobel Prize for a process known as the 'Cori cycle', a cycle that describes how the human body uses chemical reactions to break down some carbohydrates such as glycogen in muscle tissue into lactic acid, while synthesising others.

In 1931, they both left Roswell and, although Carl had a number of university job offers, almost all refused to hire Gerty, with one declaring that it was 'un-American' for a married couple to work together. He finally found satisfaction at Washington University in St Louis, Missouri, where they were both offered positions, although Gerty's salary as a research associate was only 10 per cent of that of her husband who was an associate professor. It would be thirteen years before she attained that rank.

While at the university, they continued to collaborate in their work with frog muscle, discovering an intermediate compound that enabled the breakdown of glycogen, called glucose 1-phosphate, now known as the 'Cori ester'. They established the compound's structure, identified the enzyme phosphorylase that catalysed its chemical formation, and demonstrated that the Cori ester is the initial step in the conversion of the carbohydrate glycogen into glucose.

Gerty also studied glycogen storage disease, identifying at least four forms, each related to a particular enzymatic defect. She was also the first to show that a defect in an enzyme can be the cause of a human genetic disease. In 1946, she and Carl were awarded the Midwest Award of the St Louis section of the American Chemical Society (ACS) and many accolades followed, including the Nobel Prize in Physiology or Medicine in 1947, which they shared with Bernardo Alberto Houssay. Half of this prize was awarded jointly to Carl and Gerty 'for their discovery of the course of the catalytic conversion of glycogen'.

Gerty received the Squibb Award in endocrinology in 1947, and in the following year the Garvan Medal of the ACS for women in chemistry, when she was also recognised as a 'Woman of Achievement in Science' by the Women's National Press Club. She was appointed by US President Truman to two terms as a member of the board of the National Science Foundation, while also receiving honorary degrees from Yale University, Boston University, Smith College, Columbia University, and the University of Rochester. On 21 September 2004, the ACS designated the research of Gerty and Carl on the metabolism of carbohydrates at the Washington University School of Medicine, a 'National Historic Chemical Landmark'. The crater Cori on the Moon is named after her, as is the Cori Crater on Venus. She and Carl share a star on the St Louis Walk of Fame.

Gerty Cori was the first American woman to win a Nobel Prize in science, but shortly before receiving it, she was diagnosed with myelosclerosis, a fatal disease of the bone marrow. Despite her illness, she continued her scientific endeavours for the next ten years, only letting go in the final months. She passed away at her home on 26 October 1957. She was cremated and her ashes scattered. She was survived by Carl and their only child, Tom Cori. Carl passed away, at age eighty-seven, on 20 October 1984.

Marie Skłodowska Curie

Born: 7 November 1867 in Warsaw, Russian Empire (now Poland)
Died: 4 July 1934 (aged 66)
Field: Physics, chemistry
Awards: Nobel Prize in Physics (1903); Davy Medal (1903); Matteucci Medal (1904); Elliott Cresson Medal (1909); Albert Medal (1910); Nobel Prize in Chemistry (1911); Willard Gibbs Award (1921)

The youngest of five children of Wladislaw and Bronislava Boguska Skłodowska, Marie Skłodowska was born on 7 November 1867 in Warsaw, in what was then the Russian Empire (now Poland). Her father, a teacher of mathematics and physics, made a series of bad investments in which he lost all his savings, as well as his job. From that point, the family struggled financially and was forced to take boarders into their small apartment. Religious as a small child, in 1876, at the age of nine, Marie rejected her faith after her sister died of typhus. Two years later, she lost her mother to tuberculosis, a disease that attacks the lungs and bones.

Demonstrating a remarkable memory from an early age, at sixteen she won a gold medal at the end of her secondary education at the Russian *lycée* (high school) in 1883. As women could not attend the male-only University of Warsaw in Russian-dominated Poland, her father suggested that Marie spend a year in the country with friends. On her return, Marie assisted with the family finances by undertaking work as a private tutor, while secretly taking part in the underground nationalist 'free or floating university', a group of men and women who discussed various topics behind closed doors. Her earnings also enabled her to assist finance her sister Bronisława's medical studies in Paris, with the understanding that Bronisława would, in turn, later help Marie to get an education.

In early 1886, Marie accepted a job as governess with a family living in Szczuki, Poland, but her isolation and loneliness fuelled her determination to attend university. By now, Bronisława was already in Paris, having successfully passed her medical exams, and in September 1891, Marie moved to join her and they shared an apartment. In early November of the same year, Marie enrolled as a student in physics at the University of Sorbonne.

Three years later, after graduating from the Sorbonne, Marie sought a laboratory where she could work on her commissioned research project, the measurement of the magnetic properties of various steel alloys. Following a suggestion from a Polish friend, she visited the shy French physicist Pierre Curie at the School of Physics and Chemistry at the University of Paris. A romance soon developed between them and they became a scientific team who were completely devoted to one another. They were married on 26 July 1895.

By the middle of 1897, Marie had already made great scientific strides with two university degrees, a fellowship and a published paper on the magnetisation of tempered steel. The couple's first daughter, Irène, was born in that year, and it was then that the couple turned their attention to the radiation from uranium, recently discovered by the French physicist and Nobel laureate Antoine Henri Becquerel. Marie's idea was that the radiation was an atomic property and would therefore also be present in some other elements. She established the fact of a similar radiation from the metallic chemical element thorium, and coined the word 'radioactivity' (the spontaneous release of radium). Pierre then set aside his own work to help Marie with her research.

The Curies began searching for other sources of radioactivity, in particular pitchblende, a mineral well known for its uranium content. They discovered that the radioactivity of pitchblende far exceeded the combined radioactivity of the uranium and thorium contained within it. Within six months, two papers from their laboratory were submitted to the Academy of Sciences. The first, read at the meeting of 18 July 1898, announced the discovery of a new radioactive element, 'polonium', which the Curies named after Marie's native country of Poland. The other paper, announcing the discovery of 'radium', was read at the 26 December meeting.

From 1898 to 1902, Marie and Pierre converted several tons of pitchblende, and also published, jointly or separately, a total of thirty-two scientific papers. One of these announced that diseased, tumour-forming cells were destroyed faster than healthy cells when exposed to radium.

In November 1903, the Royal Society of London awarded the Curies the Davy Medal, one of its highest awards. The following month, three French scientists, A. H. Becquerel and the Curies, were announced as the joint recipients of the Nobel Prize in Physics for 1903. Marie was appointed director of research at the University of Paris a few months later, and in December 1904 their second daughter, Ève, was born.

On 19 April 1906, during a wet afternoon on the busy Rue Dauphine at the Quai de Conti, Pierre slipped and fell in front of a horse-drawn wagon and was killed instantly. He was only forty-six years old. Despite her overwhelming grief, two weeks later Marie took over Pierre's teaching post at the Sorbonne, becoming the institution's first female professor and receiving honours from scientific communities all over the world. She was now a widow with two small children and facing the daunting task of leading the research on radioactivity. In 1908, Marie edited the collected works of her late husband, and in 1910 she published her massive volume,

Traité de radioactivité (Treatise on Radioactivity). Shortly after this work, Marie received her second Nobel Prize, this time in chemistry. Despite being the first person to win two Nobel Prizes, she was still rejected for membership of the *Académie des Sciences* (Academy of Sciences) in 1911, both as a woman and for being Polish.

By the end of the First World War, Marie was in her early fifties and her eldest daughter, Irène, a physicist, was appointed as an assistant in her mother's laboratory. Marie passionately continued her research and, in 1921, published her book *La Radiologie et la guerre* (Radiology and War) about her experiences of the use of radiation during the war.

By the late 1920s, Marie's health was beginning to deteriorate. She was almost blind and her fingers had been badly burned by radium. On 4 July 1934, at the age of sixty-six, she passed away in Sancellemoz, a sanatorium in the town of Passy, France. Her cause of death was aplastic anaemia, believed to be a result of prolonged exposure to high-level radiation for many years. Marie was even known to carry test tubes of radium around in the pocket of her lab coat. Established in 1948, Marie Curie is a registered charitable organisation in the United Kingdom which provides care and support to people with terminal illnesses and their families. As of 2019, Marie Curie was still the only woman to win two Nobel Prizes.

Ingrid Daubechies

Born: 17 August 1954 in Houthalen-
Helchteren, Belgium
Field: Mathematics and physics
Awards: Louis Empain Prize
for Physics (1984); American
Mathematical Society Ruth Lyttle
Satter Prize (1997); NAS Award in
Mathematics (2000); ICIAM Pioneer
Prize (2008); Lars Onsager Medal
(2010); Benjamin Franklin Medal
for Electrical Engineering (2011);
Booker Gold Medal of the URSI
(International Union of Radio
Engineers) (2011); Society for
Industrial and Applied Mathematics
John von Neumann Lecture Prize
(2011); Jack S. Kilby Medal of the
IEEE Signal Processing Society
(2011); Frederic Esser Nemmers
Prize in Mathematics (2012); BBVA
Foundation Frontiers of Knowledge
Award (2012); William Benter Prize
in Applied Mathematics (City
University of Hong Kong) (2018)

Ingrid Daubechies was born on 17 August 1954 in Houthalen-Helchteren,
Belgium, the daughter of Marcel Daubechies, a civil mining engineer, and
Simonne (Duran) Daubechies, a homemaker and later a criminologist. As a
child, Ingrid was already familiar with the concept of exponential growth
and complex shapes such as cones and tetrahedrons. Her parents recognised
her talent for mathematics before she reached the age of six and, excelling
in primary school, she was moved up a class after only three months. After
completing secondary school in the Flemish city of Hasselt, she enrolled
at the *Vrije Universiteit Brussel* (Free University of Brussels) at seventeen.
She obtained her bachelor's degree in physics in 1975 and then undertook
research at the same university for a doctorate in physics. She held the
title of research assistant in the Department of Theoretical Physics and,
during the next few years, made several visits to the Centre of Theoretical
Physics in Marseille. It was there that she collaborated with the Croatian-
French physicist Alex Grossmann and developed the basis for her thesis,
'Representation of quantum mechanical operators by kernels on Hilbert
spaces of analytic functions', for which she was awarded a PhD in 1980.

Ingrid continued her research career at the *Vrije Universiteit Brussel* until 1987. In 1984, she was awarded the Louis Empain Prize for Physics, awarded once every five years to a Belgian scientist on the basis of work done before the age of twenty-nine. In 1985, she was promoted to research associate professor, funded by a fellowship from the *Nationaal Fonds voor Wetenschappelijk Onderzoek* (National Fund for Scientific Research). For most of 1986, Ingrid was a guest researcher at the Courant Institute of Mathematical Sciences, New York University, where she made her most recognised discovery, on 'wavelets' in image compression. Using quadrature mirror filter-technology, Ingrid constructed compactly supported continuous wavelets, needing only a finite amount of processing, thereby introducing wavelet theory to digital signal processing.

Ingrid joined the Murray Hill AT&T Bell Laboratories' New Jersey facility in July 1987, and the next year she published her discovery in the scientific journal *Communications on Pure and Applied Mathematics*. Between 1992 and 1997 she was a Fellow of the MacArthur Foundation, and in 1993 she was elected to the American Academy of Arts and Sciences. In 1994, Ingrid was awarded a Leroy P. Steele Prize by the American Mathematical Society, awarded annually for distinguished research work and writing in the field of mathematics, for her book, *Ten Lectures on Wavelets*, and was invited to give a plenary lecture at the International Congress of Mathematicians in Zürich.

From 1994 to 2010, Ingrid was a professor at Princeton University, where she was active within the Program in Applied and Computational Mathematics. She was the first female full professor of mathematics at Princeton University, and in 1997 was awarded the American Mathematical Society Ruth Lyttle Satter Prize, presented biennially in recognition of an outstanding contribution to mathematics research by a woman in the previous six years. The citation for the award referred to Ingrid's 'deep and beautiful analysis of wavelets and their application'. In accepting this award she said, 'I am particularly grateful that the citation mentions both my theoretical work and my interest in concrete applications. They are both important to me, and it is gratifying to see them both recognised.' She also paid tribute to her many collaborators, saying that working with them had enriched both her mathematics and her life.

In 1998, Ingrid was elected to the US National Academy of Sciences and won the Golden Jubilee Award for Technological Innovation from the Institute of Electrical and Electronics Engineers (IEEE) Information Theory Society. In 1999, she became a foreign member of the Royal Netherlands Academy of Arts and Sciences.

In 2000, Ingrid became the first woman to receive the National Academy of Sciences Award in Mathematics, a prize presented every four years for excellence in published mathematical research. The award honoured her 'for fundamental discoveries on wavelets and wavelet expansions and for her role in making wavelets methods a practical basic tool of applied mathematics'. In January 2005, Ingrid became only the third woman since 1924 to give the Josiah Willard Gibbs Lecture, a prize

Ingrid Daubechies

sponsored by the American Mathematical Society. The purpose of the prize is to recognise outstanding achievement in applied mathematics and 'to enable the public and the academic community to become aware of the contribution that mathematics is making to present-day thinking and to modern civilization'. Ingrid's presentation was titled, 'The Interplay Between Analysis and Algorithm'.

In September 2006, the Pioneer Prize from the International Council for Industrial and Applied Mathematics was awarded jointly to Ingrid and the Austrian mathematician Heinz Engl. The prize recognises pioneering work introducing applied mathematical methods and scientific computing techniques to an industrial problem area or a new scientific field of applications. In the same year, she was the Emmy Noether (*see entry*) Lecturer at the San Antonio Joint Mathematics Meetings. In 2010, Ingrid was awarded an honorary doctorate by the Norwegian University of Science and Technology. In January 2011, she moved to Duke University as the James B. Duke Professor of Mathematics, the same year becoming the first woman to be president of the International Mathematical Union, a role she held until 2014. In July 2011, she was awarded the John von Neumann Lecture Prize by the Society for Industrial and Applied Mathematics, its highest honour, established in 1959 for outstanding contributions to the field of applied mathematical sciences and their effective communication to the community. One month later, she was awarded the Jack S. Kilby Signal Processing Medal from the IEEE.

In 2012, Ingrid received one of the largest monetary awards in the USA honouring outstanding achievements in mathematics, in the Frederic Esser Nemmers Prize in Mathematics. This honoured Ingrid's 'numerous and lasting contributions to applied and computational analysis' and 'the remarkable impact her work has had across engineering and the sciences'. In that year, she also received a BBVA (*Banco Bilbao Vizcaya Argentaria*) Foundation Frontiers of Knowledge Award in Basic Sciences for her work on wavelets, 'which has strongly influenced diverse fields of application ranging from data compression to pattern recognition'.

Ingrid is one of the world's most prolific researchers and cited mathematicians, being widely recognised for her study of the mathematical methods that enhance image-compression technology. Her name is associated with the orthogonal 'Daubechies wavelet' and the bi-orthogonal 'Cohen–Daubechies–Feauveau (CDF) wavelet'. A wavelet from this family of wavelets is now used in the JPEG 2000 standard for image compression. Ingrid's research employs automatic methods from mathematics, technology and biology to obtain information from samples, such as bones and teeth. As she explained about wavelets as mathematical building blocks, 'If you painted a picture with a sky, clouds, trees, and flowers, you would use a different size brush depending on the size of the features. Wavelets are like those brushes.' As part of her research, Ingrid also developed cutting-edge image processing techniques to assist in establishing the authenticity and age of some of the world's most famous art works, including those by Vincent van Gogh and Rembrandt.

In 2018, Ingrid Daubechies was the first female recipient of the William Benter Prize in Applied Mathematics from City University of Hong Kong. Carrying a prize of $100,000 USD, the biennial award recognises 'exceptional mathematical contributions that have had a lasting impact on scientific, business, finance and engineering applications'. Ingrid's pioneering work in wavelet theory was acclaimed by the prize officials, who honoured her 'exceptional contributions to a wide spectrum of scientific and mathematical subjects'. She was also part of the 2019 class of Fellows of the Association for Women in Mathematics.

In 2019, Ingrid became a member of the German Academy of Sciences Leopoldina. In 2020 and 2021, along with the fibre artist Dominique Ehrmann, she led a team of mathematicians and artists who collectively built *Mathemalchemy*, a travelling art installation dedicated to a celebration of the intersection of art and mathematics. In 2020 Ingrid received the Princess of Asturias Award for Technical and Scientific Research, and in 2023 was the first woman to receive the Wolf Prize in Mathematics 'for her work in wavelet theory and applied harmonic analysis'.

Olive Wetzel Dennis

Born: 20 November 1885 in Thurlow,
Pennsylvania, USA
Died: 5 November 1957 (aged 71)
Field: Civil engineering

Olive Wetzel Dennis was born in Thurlow, Pennsylvania, USA, in 1885. When
she was six years old, the family moved to Baltimore, Maryland. As a small
child, her parents gave her dolls to play with, but her engineering ability was
evident at an early age. She built houses and designed furniture for the dolls
instead of sewing clothes for them. At the age of ten, her father gave her a tool
set of her own, as he was tired of his daughter damaging his woodworking
equipment, doing things such as building toys for her brother, including
a model streetcar with trolley poles and reversible seats. After finishing
secondary education at Western High School, she enrolled at Goucher College
in Baltimore in 1908, earning a Bachelor of Arts degree, followed the next
year by a master's degree in mathematics from Columbia University.

Olive then taught at a Washington technical high school for ten years
but, as she said, 'the idea of civil engineering just wouldn't leave me'. She
went to two summer sessions of engineering school at the University of
Wisconsin, and subsequently obtained a degree in civil engineering from
Cornell University in 1920, completing it in just one year, rather than two. In
doing so, Olive became only the second woman to obtain a civil engineering
degree from the institution. It is reported that, as she walked up to receive
her testamur at her graduation, a man in the audience yelled out, 'What the
heck can a woman do in engineering?' It was not surprising then, that being
a woman she found it difficult to find employment as an engineer. After she
was engaged by Baltimore and Ohio (B & O) Railroad, she said, 'There
is no reason why a woman can't be an engineer simply because no other
woman has ever been one. A woman can accomplish anything if she tries
hard enough.'

Her appointment as a draftsman in the engineering department for the
B & O was announced under the newspaper headline, 'Woman Civil Engineer

Enjoys Technical Work'. With respect to her role in designing railway bridges in rural areas, she remarked, 'I helped lay out the railway line at Ithaca last December and I am rather anxious to get out on the road again.' Soon after starting her job, she designed her first railroad bridge, in Painesville, Ohio.

The following year, in 1921, she approached Daniel Willard, the president of B & O, pointing out that, as half of the railway's passengers were women, the task of engineering upgrades in service would be handled best by a female engineer. Her gender in this case became an asset rather than a liability. A result of that meeting was that Olive was told 'to get ideas that would make women want to travel on our line'. She was appointed to a new role, which involved developing ideas to smooth the journey, becoming the first 'Engineer of Service'. She was also the first female member of the American Railway Engineering Association.

To improve the experience of passengers, Olive had to have the customer experience herself. So, for the next few years, she spent much of her time on trains. It is said that she would take a B & O train from the beginning to the end of the line, alight and then get on a train in the opposite direction. She also compared the B & O experience with that of rival train companies. She was very 'hands-on', averaging over 50,000 miles (80,500 km) per year on trains, while sometimes sitting up all day testing how effective the seat designs were. She also tested mattresses. In the course of her career, her journeys totalled up to a half a million miles (approximately 850,000 km).

As the supervisor of passenger car design and service, Olive had a wide-ranging influence in the area of creature comforts, and many of her innovations remain in use today. One of the first changes she made was to the timetable, which she considered overly complex. She made it her business to simplify it, making it easier for passengers to understand it. At the time of taking up her role, trains were smelly, dirty and most unattractive for passengers and she set about changing all that. Her innovations included designing the railroad's famous blue-and-white Colonial dining car china, with scenic locations in the centre and historic trains around the edges. She also introduced larger dressing rooms with paper towels, liquid soap and disposable cups. Although her initial focus was on female passengers, she soon realised that all passengers wanted improvements. After long nights travelling coach class, she introduced and helped design reclining seats, dimmable overhead lights and all-night on-board lunch counters, serving sandwiches and coffee. Other improvements were easy-to-clean upholstery, dining car configurations that removed the need for high chairs for children, and shorter seats so that shorter people, including women, could comfortably rest their feet on the floor.

Olive also suggested that there should be stewardesses, nurses, and other helpers on board to provide services when required. She invented, and held the patent for, the 'Dennis ventilator', which enabled the windows of passenger cars to be controlled by passengers. She was later an advocate of air-conditioned compartments, and in 1931 B & O introduced the world's first completely air-conditioned train. The 'crowning glory of her career', she said, was when B & O put her in charge of designing an entire train, the

Cincinnatian, which incorporated all of her innovations and improvements. It was put into service in 1947. In the years that followed, other rail carriers followed suit, as well as bus companies and airlines, which had to upgrade their level of comfort to compete with the railroads.

In 1940, Olive was named by the Women's Centennial Congress as one of America's '100 outstanding career women'. During the Second World War, she served as a consultant for the Federal Office of Defense Transportation while maintaining her position as Engineer of Service for over thirty years. She was one of the most remarkable women in railroad industry history and did not let her gender stand in the way of advancement, stating, 'No matter how successful a business may seem to be, it can gain even greater success if it gives consideration to the woman's viewpoint.'

She retired in 1951 and was quoted in a *New York Times* article saying, 'sometimes, my assignments would require my riding with the engineer of a locomotive during speed and safety checks. But I never took advantage of being a woman.' Nevertheless, as a woman she was not always accepted by the executives of other lines, but her influence as a woman and technical engineer left a lasting impression on the travel industry nationwide.

Olive Dennis passed away on 5 November 1957 in Baltimore at the age of seventy-one. She never married. Apart from her railroad interests, her hobbies included cryptology and solving puzzles and she regularly spoke to women's groups about her life and career, encouraging women to follow their chosen path. As was written of her some forty years after her death, she was the 'Lady Engineer' who 'took the pain out of the train'.

Mildred Spiewak Dresselhaus

Born: 11 November 1930 in Brooklyn, New York, USA
Died: 20 February 2017 (aged 86)
Field: Electrical engineering
Awards: National Medal of Science (1990); IEEE Founders Medal (2004); Harold Pender Award (2006); Oliver E. Buckley Condensed Matter Prize (2008); Oersted Medal (2008); Vannevar Bush Award (2009); Enrico Fermi Award (2012); Kavli Prize in Nanoscience (2012); Presidential Medal of Freedom (2014)

Mildred Spiewak was born on 11 November 1930 in Brooklyn, New York, the daughter of Ethel (Teichtheil) Spiewak and Meyer Spiewak, who were Polish Jewish immigrants. Mildred wrote, 'My early years were spent in a dangerous, multi-racial, low income neighbourhood. My elementary school memories up through ninth grade are of teachers struggling to maintain class discipline with occasional coverage of academics.' Although her family suffered financial hardship, Mildred and her elder brother, Irving, became talented violinists, winning scholarships to music schools.

Raised in the Bronx, from the age of six Mildred took the subway, travelling long distances, laden with books and musical instruments. She applied to the prestigious Hunter High School in Manhattan and easily passed their entrance exams, their Yearbook declaring, 'In math and science, Mildred Spiewak is second to none.' After graduating from high school, she was awarded a Fulbright Fellowship, as a result of which she spent a year at the Cavendish Laboratory, Cambridge University, from 1951 to 1952.

Mildred planned to become a schoolteacher, until she enrolled in a beginning physics class with Rosalyn Yalow (*see entry*), a future Nobel laureate, who suggested she consider a career in science. Following that advice, she graduated with a master's degree from Radcliffe College and a PhD from the University of Chicago in 1958, with a thesis on superconductivity, supervised by the notable physicist Enrico Fermi. At the time, Mildred described herself as 'pretty lonely as we women were only two per cent of the physics community'. Also in 1958, she married Gene Dresselhaus, a renowned physicist and discoverer of the phenomenon in physics labelled the 'Dresselhaus effect'. They met at the university, married, and had four children – Marianne, Carl, Paul and Elliot. As the mother of four children, Mildred began a Women's Forum, focusing on problems faced by working women.

Mildred spent two years at Cornell University as a postdoctoral student, then moved to Lincoln Lab as a staff member. There she began her career by examining magneto-optics in semiconductors, carrying out a series of experiments leading to a fundamental understanding of the electronic structure of semi-metals, especially graphite. Graphite is a nanomaterial composed of sheets of specially structured carbon atoms called 'graphene' – and graphene sheets can be rolled up to form carbon nanotubes. In 1960, Mildred and Gene took up positions at the Massachusetts Institute of Technology (MIT), one of the few places willing to hire husband-and-wife scientists.

With the introduction of laser technology, Mildred was among the first to use lasers for magneto-optics experiments. Forming a research partnership with Ali Javan, the inventor of the continuous wave laser, and their joint student Paul Schroeder, they undertook a high-resolution magneto-optics experiment, using circularly polarised light from a helium-neon laser, and in 1968 created a new model for the electronic structure of graphite. These laser studies were the beginning of her later work in the field of Raman spectroscopy, a light scattering technique, in which a photon of light interacts with a sample to produce scattered radiation of different wavelengths.

In 1968, Mildred became the first woman to be appointed as a full professor at MIT and was a strong campaigner for the role of women in science. In 1971, she and a colleague organised the first Women's Forum at MIT, exploring female roles. In 1973, she won a Carnegie Foundation grant for her efforts. She used resonant magnetic fields and lasers to map out the electronic energy structure of carbon, along with investigating the patterns that evolve when carbon is interwoven with other materials. She was also a pioneer in research on 'fullerenes', also called 'buckyballs', soccer-ball-shaped cages of carbon atoms that can be used as drug delivery devices, lubricants, filters and catalysts.

Mildred formed the idea of rolling a single-layer sheet of carbon atoms into a hollow tube, that was eventually realised as the 'nanotube' – a versatile structure as strong as steel but just one ten-thousandth the width of a human hair. She researched carbon ribbons, semiconductors, nonplanar monolayers of molybdenum sulphide and the scattering and vibrational effects of tiny particles introduced into ultrathin wires.

Mildred also led the resurgence of research in the field of thermo-electrics. In 2003 she co-chaired both a US Department of Energy study, 'Basic Research Needs for the Hydrogen Economy', and a National Academy Decadal Study of Condensed Matter and Materials Physics.

Throughout her illustrious career, Mildred co-authored over 1,400 publications and was the co-inventor on five US patents. She was also an enthusiastic chamber music player, on both violin and viola.

Mildred's achievements were recognised in numerous awards and honours. In 1977 she won the Society of Women Engineers Annual Achievement Award, and in 1985 she was named Institute Professor of MIT, a lifetime appointment held by no more than twelve active MIT professors at a time. In 1990, Mildred was awarded the National Medal of Science, an honour bestowed by the President of the United States to individuals in science and engineering who have made important contributions to the advancement of knowledge.

In 2012, at the age of eighty-one, she was the second female recipient of the Enrico Fermi Award, an award of $50,000 USD honouring scientists of international stature for their lifetime achievement in the development, use or production of energy. She also received a Kavli Prize in Nanoscience (2012). A partnership between the Norwegian Academy of Science and Letters, the Kavli Foundation (US) and the Norwegian Ministry of Education and Research, the biennial Kavli Prizes, each of $1 million USD, recognise scientists for pioneering advances in our understanding of existence at its biggest, smallest and most complex scales, in the fields of astrophysics, nanoscience and neuroscience. In 2013, the Materials Research Society gave her its highest honour, the Von Hippel Award, in acknowledgment of 'those qualities most prized by materials scientists and engineers – brilliance and originality of intellect, combined with vision that transcends the boundaries of conventional scientific disciplines'. In 2014, Mildred was awarded the Presidential Medal of Freedom, one of the highest civilian honours in the USA, and was inducted into the National Inventors Hall of Fame. The following year, she was given the Institute of Electrical and Electronics Engineers (IEEE) Medal of Honor, its highest award. In addition, Mildred received honorary doctorates in the USA, Europe, Israel and Hong Kong (Honorary Degree of Doctor of Science, the Hong Kong Polytechnic University).

In 2017, Mildred was featured in an advertisement for General Electric, in which young girls played with 'Millie Dresselhaus dolls' and dressed up in 'Millie Dresselhaus wigs and sweaters'. It asked the question, 'What if female scientists were celebrities?' and aimed to increase the number of women in STEM (science, technology, engineering and mathematics) roles in its ranks.

On 20 February 2017, at the age of eighty-six, Mildred Dresselhaus passed away at Mount Auburn Hospital in Cambridge, Massachusetts. Known as the 'Queen of carbon science', her career at MIT spanned fifty-seven years and she devoted much of her time to supporting efforts to promote increased participation of women in physics. Reflecting on her achievements, she said, 'All the hardships I encountered provided me with the determination, capacity for hard work, efficiency, and a positive outlook on life that have been so helpful to me in realising my professional career.' She was survived by her husband, Gene.

(Gabrielle) Émilie du Châtelet

Born: 17 December 1706 in Paris, France
Died: 10 September 1749 (aged 42)
Field: Mathematics and physics

Gabrielle Émilie (Émilie) le Tonnelier de Breteuil was born on 17 December 1706 in Paris, France, to a well-connected noble family. She had every privilege afforded to a young girl of her time. Her father, Louis Nicholas le Tonnelier de Breteuil, baron of Preuilly and chief of protocol at the royal court, was born in Paris and both he and his wife, Gabrielle Anne de Froullay, had relatives and friends who could help to advance the family's interests. Louis was also said to be a favourite of the king, Louis XIV.

From the age of six or seven, Émilie was surrounded by the best governesses and tutors. Although her father was aware that she was very talented, he was less impressed with her physical appearance, pledging to 'help his clothes-conscious ugly duckling prepare for life as a spinster'. This may have been somewhat unfair, as one of her male friends wrote, in contrast, of her 'beautiful big soft eyes with black brows, her noble, witty and piquant expression'. Whatever the case, in 1725 when she was eighteen, it was arranged for her to marry into one of the oldest lineages of Lorraine, a semi-independent duchy in north-eastern France. Her husband, the thirty-four-year old Marquis du Châtelet, was made governor of Semur-en-Auxois, a commune of the Côte-d'Or department in eastern France, as a wedding gift, although he had little in the way of wealth.

For the first years of her marriage, the new marquise lived a very traditional life, giving birth to Francoise Gabriel Pauline in June 1726, Louis Marie Florent in November 1727 and Victor Esprit in April 1733, although the latter passed away in late summer the following year. Émilie ran their first household in Semur-en-Auxois, but on occasions she experienced Parisian life, dressing elegantly, going to the theatre and the opera, as well as gambling at the houses of her noble friends.

In 1733, while in Semur awaiting the birth of Victor, Émilie again became interested in mathematics, also reading widely in other academic fields, including philosophy. Her interest was intensified when she became reacquainted with François-Marie Arouet, more commonly known as Voltaire, who had been a guest in her parents' house when she was a child. He had only just published his work *Lettres philosophiques* (Philosophical Letters), a volume considered so outrageous by French authorities that it was banned. Émilie and Voltaire became very close, as they shared common interests, including reportedly a physical relationship. She invited him to live with her on her tolerant husband's estate at Cirey, a commune in the Haute-Saône department in the region of Bourgogne-Franche-Comté, also in eastern France. Voltaire was to become her long-time companion, and by all accounts they enjoyed living together, sharing a great respect and liking for each other. She studied and published scientific articles, with the pair collaborating scientifically as well as setting up a laboratory together in her home.

Émilie returned to Paris to take up serious study of René Descartes' analytical geometry, first with Pierre-Louis Moreau de Maupertuis and then with his colleague, a young mathematics prodigy, Alexis-Claude Clairaut. For the most part, however, she continued to care for her family's interests, especially when the marquis had to re-join his troops for the wars of the 1730s and 1740s. During this period, she found time to continue reading, studying and finally writing and publishing her own scientific works on what was then known as 'natural philosophy'.

Émilie's academic and intellectual strengths were in mathematics and physics, although she shared with Voltaire a keen interest in metaphysics and ethics. She became a devotee of the 'Newtonian philosophy' of Isaac Newton, his theory of the mechanism of the universe, particularly of the motions of the heavenly bodies, their laws and their properties. Eventually she became a follower of Gottfried Leibnitz, who declared that every event has a reason or cause in the prior state of the world. In effect, Newton and Leibnitz had philosophical views that were incompatible. In 1736, the Académie des Sciences (Academy of Sciences) announced an essay contest on the subject, attracting not only Voltaire to enter but Émilie, who did so without his knowledge. Her entry was titled 'on the nature and propagation of fire'. While neither was successful, Voltaire arranged to have both of their works published alongside the winners in 1739. Before Émilie's entry was published, however, she modified her opinion on Newton's philosophies and, although being refused submission of a revised version, was allowed to add a series of *errata* that reflected the ideas of Leibnitz.

In 1744, Émilie published a revised Leibnitz version of her entry and also began to translate Newton's classic work *Philosophiæ Naturalis Principia Mathematica* (Mathematical Principles of Natural Philosophy) into French. Her annotated translation, published in part in 1756, was the first time it was available to the public in that language. As a result, she was accepted as a member of the learned *Respublica literaria* (Republic of Letters). Her standing increased when she won an argument with the executive director of the Academy of Sciences on the issue of the proper formula for kinetic energy. Her writings on science were translated into Italian and German,

(Gabrielle) Émilie du Châtelet

and she was elected to the Bologna Academy of Science. Her first book, *Institutions de physique* (Foundations of Physics), in 1740, revealed that she had read widely in Latin, English and Italian, along with fields as diverse as moral philosophy, chemistry, physics, theology, mathematics, metaphysics, natural and experimental philosophy.

In May 1748, Émilie commenced an affair with the poet Jean François de Saint-Lambert and, at the age of forty-two, became pregnant by him, confiding in a letter to a friend her fear that she would not survive the pregnancy. On the night of 4 September 1749, Émilie gave birth to a daughter, Stanislas-Adélaïde, and her concerns were realised as Émilie did indeed die, just six days later, on 10 September, at Lunéville, from a pulmonary embolism. Her daughter died twenty months later.

In 1759, Émilie Châtelet's translation of Newton's *Principia* into French, still the only one to this day, was published in final complete form as part of the anticipation of the return of Halley's Comet – calculating a comet's orbit had been one of the main proofs of gravitational attraction. Her role as a scientist is significant, not only for her translation and writings but for what she achieved as a woman of her time, as she read, studied, wrote, published and gained recognition in a learned world dominated by men.

Alice Eastwood

Born: 19 January 1859 in Toronto, Canada
Died: 30 October 1953 (aged 94)
Field: Botany
Award: Medal of Achievement (American Fuchsia Society) (1949)

Alice Eastwood was born on 19 January 1859 in Toronto, Canada. Her parents were Eliza Jane (Gowdey) Eastwood and Colin Skinner Eastwood, the superintendent of the Toronto Asylum for the Insane, on the grounds of which the family resided until Alice was six years old. When her mother passed away in 1865, Colin tried to set up as a storekeeper, placing his three children in the care of his brother, the physician Dr William Eastwood. William owned an estate and was a keen gardener and amateur botanist. Alice learned the scientific names of plants from him and could roam his property freely, further sparking her interest in them.

Meanwhile, her father's store failed and Colin moved to Colorado, taking with him his son, while the two daughters, including Alice, were sent to Oshawa Convent near Denver. While at the convent, Alice befriended a priest, Father Pugh, who had planted an experimental orchard for the use of the residents. This led to her interest in botanical work, along with an appreciation of music that she developed thanks to a nun. At fourteen, Alice and her sister were sent to live with their father who was now building a new house. For an income, she worked as a nursemaid for a French family, taking the opportunity to devour books in their large library and study plants on family trips to the mountains.

Upon completion of her father's house, Alice returned to become his housekeeper and attend a public school. With the encouragement and help of teachers, she was able to fill in the gaps in her education sufficiently to be accepted into East Denver High School. However, her studies were disrupted when a family financial crisis forced her to take a break and work in a millinery factory, but she was able to stay on top of her school work and graduated as class valedictorian in 1879. Although she had a burning desire to be a botanist, she decided to support herself, at least in the short term, by taking up a teaching role at the school from which she had just graduated. Although poorly paid, she was able to purchase botanical books and fund expeditions to the Rocky Mountains, where she collected plants that eventually became the centrepiece of the University of Colorado Herbarium in Boulder.

In 1887, Alice met the British biologist Alfred Russell Wallace, accompanying him on a trip up the 14,278 feet (4,352 metre) Gray's Peak, the tenth-highest summit of the Rocky Mountains, during the alpine flowering season. On a visit to the California Academy of Sciences in San Francisco, she became acquainted with Katharine Brandegee, the curator of botany, and her husband, the American botanist Townshend Stith Brandegee. In 1892, Katharine offered Alice a position with a $75 USD monthly salary to become a writer for their magazine, *Zoe*, and to help organise the herbarium at the Academy. To do this, Alice took leave from her teaching role and began the task of organising the botanical collection. In 1893, at her own expense, Alice published a book, *Popular Flora of Denver, Colorado*, and in the same year, when the Brandegees left the Academy, she assumed both the role of curator of botany and editor of *Zoe*, as well as being director of the San Francisco Botanical Club for a number of years.

During this period Alice went on collecting expeditions, including one to Dawson in the Yukon Territory. It was here that she made a 300-mile (approximately 483-km) trip from Whitehorse to Dawson in an open carriage over snow and frozen rivers. As well as editing *Zoe*, she was an assistant editor for the journal *Erythea* and founded, with John Thomas Howell, the magazine *Leaflets of Western Botany*.

In 1906, the devastating San Francisco earthquake destroyed the Academy building. It was due to Alice's extraordinary efforts and determination that a considerable number of the Academy's irreplaceable botanic specimens were preserved. She had separated type specimens from the main collection and she was able to save 1,497 of them from the fire that was overtaking the city. In an effort to restore the collections that had been lost, Alice spent six years at the Smithsonian Institution in Washington, the Arnold Arboretum in Boston, Kew Gardens, the British Museum in London, Cambridge University and the *Jardin des Plantes* (Botanical Garden) in Paris.

A new Academy building was constructed in Golden Gate Park and Alice returned to her role as curator of the herbarium. Being responsible for reconstructing the collections that had been destroyed, she built up the collection, often exchanging duplicates with other institutions. It was said by the American botanist Leroy Abrams that she contributed 'thousands of sheets to the Academy's herbarium, personally accounting for its growth in size and representation of western flora'.

In 1932, Alice was struck by a car at the entrance to Golden Gate Park, permanently damaging her knee. That same year, she launched *Leaflets of Western Botany* with her assistant, the American botanist and taxonomist John Thomas Howell, providing a highly regarded forum for botanical research. To celebrate Alice's eightieth birthday, proceeds from a banquet held in her honour, provided the initial contributions to establish the 'Alice Eastwood Herbarium' at the Academy.

Alice finally retired in 1950 at the age of ninety and was given the title Curator Emeritus. At this time she was also invited to serve as honorary president of the Seventh International Botanical Congress in Stockholm. She flew to Sweden and occupied the chair of Carolus Linnaeus, the 'father of modern taxonomy', for formalising the modern system of naming organisms, binomial nomenclature. This was one of the high points of her career.

Although Alice had little forming training, she became one of the foremost systematic botanists of her era, specialising in the flowering plants of the Rocky Mountains and the Californian coast. She contributed to the popular literature and added to the knowledge of plant taxonomy, publishing over 300 articles during her career. She worked to save a redwood grove in Humboldt County, later named the 'Alice Eastwood Memorial Grove' in her honour. There are seventeen currently recognised species named for her, as well as the genera *Eastwoodia* and *Aliciella*.

Although being of advanced age, Alice Eastwood was in good health and lived independently and alone, never having married, in a small cottage in San Francisco. In May 1953, she fell and broke her hip and, although she was recovering and in good spirits, in September complications arose. She passed away on 30 October 1953 at the age of ninety-four. She is buried in the Toronto Necropolis cemetery, Canada.

Elsie Eaves

Born: 5 May 1898 in Idaho Springs, Colorado,
USA
Died: 27 March 1983 (aged 84)
Field: Civil engineering
Awards: Award of Merit (American Association
of Cost Engineers, 1966); George Norlin Silver
Medal (University of Colorado, 1974); Honorary
Lifetime Member, International Executive
Service Corps (1979)

Elsie Eaves was born in Idaho Springs, Colorado, on 5 May 1898, the
daughter of Edgar Alfred and Katherine (Elliot) Eaves. She attended the
University of Colorado, and in 1920, at the age of twenty-two, she became
its first woman graduate with a degree in civil engineering. For the next
five years she worked for the US Bureau of Public Roads, the Colorado
State Highway Department and the Denver and Rio Grande Railroad. In
1926, Elsie commenced employment with McGraw-Hill in New York City
in the *Engineering News-Record* (a weekly trade magazine) department,
as well as being the publication and sales manager of McGraw-Hill's
Construction Daily.

In 1927, Elsie became the first woman to be a full member of the
American Society of Civil Engineers (ASCE) and decided to make use of
her engineering education by joining the *Engineering and News Reports*
(ENR) as an assistant manager for market surveys. She eventually became
the manager of the Construction Economics department, directing ENR's
measurement of 'Post War Planning' in the construction industry. These
statistics were used by ASCE and the Committee of Economic Development
to determine what work could be commenced immediately after the end
of the Second World War in 1945. Elsie converted the information into
the first continuous database of construction in the planning stages. She
became the manager of *Business News*, holding this position for a further
eighteen years until her retirement in 1963.

In 1956, Elsie was a delegate to the Technology Societies Council,
New York, and the following year she was the first woman to join the
American Association of Cost Engineers (AACE), acting as chairman of the
coordinating committee, 1961–65.

She retired in 1963, but continued practising as an adviser – both to the National Commission on Urban Affairs, on the subject of housing costs, and to the International Executive Service Corps, about construction costs in Iran. In 1964, Elsie was the public relations chairman of the First International Conference of Women Engineers and Scientists in New York. She served as administrative vice president of the North Shore Science Museum, Plandome Manor, Nassau County in New York, 1970–72, then correspondent secretary, 1974–76, followed by a position as trustee from 1976.

In 1974, Elsie received the George Norlin Award, the highest alumni award given by the University of Colorado, given to outstanding alumni 'who have demonstrated a commitment to excellence in their chosen field of endeavour and a devotion to the betterment of society and their community'. In 1979, she was the first woman to receive an Honorary Lifetime Membership to the ASCE. Elsie was also the first woman elected as Chapter Honor Member of *Chi Epsilon* (the US civil engineering honour society) and the first to receive the International Executive Service Corps 'Service to Country' Award.

In 1983, Elsie wrote an article that appeared in the June issue of *US Woman Engineer*, about the changes she had witnessed in her lifetime. However, she passed away on 27 March 1893, prior to its publication, at St Francis Hospital, in Roslyn, New York. She was eighty-four. She never married.

Gertrude Belle Elion

Born: 23 January 1918 in New York City, USA
Died: 21 February 1999 (aged 81)
Field: Pharmacology and biochemistry
Awards: Garvan-Olin Medal (1968); Nobel
Prize in Physiology or Medicine (1988);
National Medal of Science (1991); National
Inventors Hall of Fame (1991); Foreign Member
of the Royal Society (1995); Lemelson-MIT
Prize (1997)

Gertrude Elion was born on 23 January 1918 in New York City, USA. Her father, Robert Elion, had emigrated from Lithuania to the USA when he was twelve, later becoming a dentist. Her mother, Bertha (Cohen) Elion, had arrived from Russian-ruled Poland at fourteen. Gertrude spent the first seven years of her family life living in a large apartment in Manhattan, above her father's dental surgery. In 1924, shortly after the birth of her brother, Herbert, the family moved to the Bronx, where Gertrude attended a public school within walking distance.

In 1933, Gertrude entered Hunter College at New York University where she majored in science, graduating *summa cum laude* (with the highest distinction) in chemistry in 1937. It was difficult for her to find employment as many laboratories refused to hire female chemists, although she obtained a three-month position to teach nurses in the New York Hospital School of Nursing. Upon completing her contract, she worked for no salary as a laboratory assistant to a chemist, with a view to gaining practical experience; she ultimately kept this role for three and a half years, by the end of which she was paid a paltry $20 USD per week.

In the autumn of 1939, she attended New York University and was the only female student in her chemistry class – something that nobody seemed to mind. She commented later that she 'did not feel at all strange'. One year into the programme, Gertrude worked as a substitute high school teacher of chemistry, physics and general science for two years, and in 1941 she graduated with a Master of Science degree in chemistry. By this time, the Second World War was well under way and there was a shortage of chemists in industrial laboratories. Although Gertrude found a position in one at a major food company, it did not involve research. In this role she undertook analytical quality control work for eighteen months.

During this time, Gertrude learned a great deal about instrumentation, next securing a research job in a laboratory at Johnson & Johnson in New Jersey. When the laboratory closed down six months later, she accepted a position as an assistant to the chemist Dr George Herbert Hitchings. Gertrude became very involved in microbiology, expanding her horizons into biochemistry, pharmacology, immunology and, eventually, virology. She found the work fascinating from the outset, as very little was known about nucleic acid biosynthesis or the enzymes involved in it.

She was assigned to work on purines (heterocyclic aromatic organic compounds), which almost completely occupied the remainder of her research. Purines can be found in food and drinks as part of a normal diet and exist in the nucleus of any plant or animal cell. She considered each series of studies as like a 'mystery story', in which she and her team were constantly trying to deduce what the microbiological results meant, with little biochemical information available to assist them.

Her partnership with Hitchings spanned forty years, during which they embarked on an unusual course of creating medicines by studying the chemical composition of diseased cells. Rather than relying on trial-and-error methods, they used the differences in the biochemistry between normal human cells and pathogens (disease-causing agents) to design drugs that would block viral infections. Gertrude and her team developed drugs to combat leukaemia, herpes, malaria and acquired immune deficiency syndrome (AIDS), as well as discovering anti-rejection drugs for kidney transplants between unrelated donors. In particular, she patented the leukaemia-fighting drug '6-mercaptopurine', which she created by replacing one sulphur atom with an oxygen atom.

Gertrude officially retired in 1983, but continued to remain active, holding the title of 'scientist emeritus' and acting as a consultant. She also served as an adviser to the World Health Organisation and the American Association for Cancer Research. In 1988, Gertrude shared the Nobel Prize in Physiology or Medicine with Dr Hitchings and Sir James Black, 'for their discoveries of important principles for drug treatment'. Throughout her career, Gertrude developed forty-five patents in medicine and was awarded twenty-three honorary degrees, despite never receiving a formal PhD. She received numerous awards for her work, including the National Medal of Science in 1991 awarded by US President George H. W. Bush, in the same year becoming the first woman to be inducted into the National Inventors Hall of Fame. In 1995, she was chosen as a Foreign Member of the Royal Society (UK). In 1997, she was granted the Lemelson-MIT Lifetime Achievement Award. Endowed in 1994 by Jerome H. Lemelson, and administered through the School of Engineering at the Massachusetts Institute of Technology, the winner receives $500,000 USD, the largest cash prize for invention in the USA.

Gertrude listed her hobbies as listening to music, opera, ballet and theatre, as well as photography and travel that peaked her curiosity about life. She never married, although was once engaged, but in the 1940s her fiancé died of a bacterial heart infection before they could wed. Only a few years later, he could have been saved by penicillin. In her final years, Gertrude lived at Chapel Hill in North Carolina, where she passed away on 21 February 1999 at the age of eighty-one.

Gertrude Elion's research findings left an extraordinary legacy, the results of which still resonate today. In a 1997 interview, she said that, after receiving her bachelor's degree, she ran into a 'brick wall', as nobody took her seriously. She remarked, 'They wondered why in the world I wanted to be a chemist when no women were doing that.'

Thelma Austern Estrin

Born: 21 February 1924 in New York City, New York, USA
Died: 15 February 2014 (aged 89)
Field: Biomedical engineering
Awards: Achievement Award from the Society of Women Engineers (1981), IEEE Centennial Medal (1984); IEEE Haraden Pratt Award (1991)

Thelma Austern, an only child, was born on 21 February 1924 in New York City, New York, USA, and was raised there. She displayed an outstanding talent for mathematics and attended Abraham Lincoln High School. In 1941, at the age of seventeen, she began studying accounting in business administration at the City College of New York because, as she later said, she was 'always good in math'. It was there that she met Gerald 'Jerry' Estrin, whom she married later that year, before her eighteenth birthday. When Jerry joined the Army in 1942, Thelma undertook a three-month war training course for engineering assistants at the Stevens Institute of Technology and began working at the Radio Receptor Company, which built electronic devices. It was there that Thelma developed an interest in engineering. 'If the war had not happened,' she later said, she and her husband would 'probably be accountants somewhere'. Her work during the war inspired her to become an engineer. Thelma also said that Jerry was committed to her career.

As her husband also had a fascination with engineering, they moved to Madison, Wisconsin, where she enrolled in electrical engineering at the University of Wisconsin at the end of the Second World War. After the completion of her studies, Thelma had earned a Bachelor of Science (1948), a Master of Science (1949) and a PhD (1951), with the thesis 'Determination of the capacitance of annular-plate capacitors by the method of subareas' under the supervision of Thomas J. Higgins.

In the early 1950s, Jerry and Thelma moved to Princeton, New Jersey, where he became part of John von Neumann's group at the Institute for Advanced Study (IAS). As Thelma later said, for Jerry it was 'a great place to be'. 'Einstein was there', and 'they were building a new computer'. Thelma still had another year to finish her doctorate and so, as she said, she 'commuted' from Wisconsin and then, because Princeton 'was not into hiring a woman engineer at the time', she 'got into biomedical engineering'. She joined the Electroencephalography Department of the Neurological Institute of New York at Columbia Presbyterian Hospital. At the time, this was a relatively new field, electronics for medicine.

Through his work at the IAS, Jerry was invited by the Weizmann Institute of Science in Israel to direct the Weizmann Automatic Computer (WEIZAC) Project to build their first computer. Although Thelma and Jerry were Jewish, they 'knew very little about the country' of Israel. They found it 'very exciting'. It was 'a new country just getting started, and very different from the United States'. Thelma was able to work part-time with her husband. She had a young child and her second child was born in Israel. As a result, Jerry and Thelma spent over a year there, working on the machine that was hailed in the mid-1950s as the first electronic computer in the Middle East, representing a landmark in electronic and computer engineering. As Thelma later said, 'everything was much closer' in Israel and 'it was easier to get help'.

In 1956, returning to the IAS after two years, Jerry became an associate professor in the Computer Science Department at the University of California, Los Angeles (UCLA). Speaking of this move fifty years later, Jerry observed that one of his 'biggest mistakes' was the move to Los Angeles. While he had a 'wonderful career' there, it 'wasn't straightforward' for Thelma. Thelma taught engineering part-time at Valley Junior College in Los Angeles, as well as undertaking some consulting work, although she wanted to get back into biomedical engineering. As Jerry later said, 'Thelma was the mainstay … a super mom, it was the story over and over again'.

In 1960, Thelma joined the new Brain Research Institute (BRI) at UCLA, where she organised their data processing laboratory the following year. She served in the role of director of the laboratory between 1970 and 1980, by the end of which she became a Professor in Residence in the Computer Science Department at UCLA. She also served as director of the Engineering and Mathematics Division of UCLA Extension. As she later said, 'the Medical School was sort of one of the first to really use computers to analyse medical data' and she 'kind of fell into that type of position'. Jerry was in computing and her background was in engineering, but she 'picked up the biomedical part' because, when Jerry was working at the IAS, she 'didn't want to work at the same place'.

During her time at UCLA, Thelma designed and implemented one of the first systems for analogue-digital conversion of electrical activity of the nervous system. Her system was a forerunner to the use of computers in medicine. She published a number of notable papers, including her 1961 article, 'Recording the Impulse Firing Pattern', a work that analysed digital techniques to explore spike patterns originating in neurons. She also published research papers involving the use of computers to map the brain, at a time when the internet was still in its infancy and not widely used.

In 1975, Thelma developed a computer network between UCLA and University of California, Davis, as well as being prominent in the Institute of Electrical and Electronics Engineers (IEEE). For her achievements she was named in 1977 as an IEEE Fellow, 'for contributions to the design and application of computer systems for neurophysiological and brain research'. In 1982, she became the first female IEEE vice president, along with assuming the role as president of the IEEE Engineering in Medicine and Biology Society. In 1984, she received the IEEE Centennial Medal and the National Science Foundation gave her a Superior Accomplishment Award. While Thelma did not form the Society of Women Engineers, she was glad that it was established in 1950. She acknowledged the benefit of being in a society made up of a 'bunch of friends'.

In 1989, Thelma received an honorary Doctor of Science degree from the University of Wisconsin-Madison, with a citation that read,

> Refusing to be daunted by prejudice, she demonstrated through the undeniable quality of her work that talent is not tied to gender. She has been a model for other women who have entered and enriched the field of engineering, including two of her daughters.

The following year she retired, becoming a Professor Emerita, two years later receiving the IEEE Haraden Pratt Award, which recognises individuals who have rendered outstanding service to the IEEE.

Thelma was an enthusiastic supporter and actively involved in the promotion of women's careers in engineering and science. This is exemplified in her 1996 paper, 'Women's Studies and Computer Science: Their Intersection', for the IEEE *Annals of the History of Computing*, for which she wrote, 'women's studies implies that we expand the world of science and technology from its patriarchal history, which consider these disciplines as inherently masculine'. She added that women's studies seek to 'understand the elements of gender in the social and political situations' and it was essential in order to 'widen women's access to technology'. In 1999, Thelma was inducted into the Women in Technology Hall of Fame, while another of her many awards included the Pioneer in Computing Award from the Grace Hopper (*see entry*) Conference for Women in Computing.

Thelma was the first woman to join the board of trustees of the Aerospace Corporation, her leadership inspiring many women to undertake careers in aerospace engineering. She spent a great deal of time supporting the application of science and technology to the advancement of women in computer science fields. When asked about the difference between scientists and engineers, she replied, 'In science one deals with things as they are. In engineering, one deals with things as they ought to be.'

Thelma Estrin passed away at her home in Santa Monica on 15 February 2014 at the age of eighty-nine, some two years after her husband Jerry, who died on 29 March 2012. They had been married for seventy years. Their three daughters all became high achievers in their chosen fields: Margo as a medical doctor; Judith, the CEO/President of Packet Design and former CTO and senior vice president of Cisco Systems, while Deborah became a professor of computer science at UCLA.

Thelma's mother had wanted her only child 'to be something in the world'. When asked what she saw as her own greatest contribution in engineering, Thelma said, 'Just saying that women can be as successful as men in the field, and that there really isn't any sex difference between the two, other than the general psychological things that people find. I think it's mostly the environment we live in. Women in the past, I guess because of the lack of technology, and the ease to have a professional life and have a home life, just never really existed before. And it came about, I suppose, in my generation. I think you see a lot of women realising they can do both things at the same time, even though it might be a little more difficult. But I think before, the technology wasn't there. Now the technology is here, and it's possible for women to do both things.'

Margaret Clay Ferguson

Born: 29 August 1863 in Orleans, New York, USA
Died: 28 August 1951 (aged 87)
Field: Botany

Margaret Clay Ferguson was born the fourth of six children on 29 August 1863 in Orleans, New York, USA, to Hannah Mariah (Warner) Ferguson and Robert Bell Ferguson, both farmers. At the age of fourteen, Margaret was a student at Genesee Wesleyan Seminary in Lima, New York, as well as being a teacher in the local public school. After graduating from the Seminary in 1885, Margaret continued her teaching role and two years later was promoted to assistant principal.

The following year, she enrolled in Wellesley College's 'teacher special' programme for working teachers who wished to further their careers as educators. Her studies centred on botany and chemistry and she continued in this role until 1891, when she took up a position as head of the science department at Harcourt Place Seminary in Gambier, Ohio. She stayed there for two years, returning to Wellesley as a botany instructor.

In 1896, Margaret left Wellesley to tour Europe for twelve months. The following year she enrolled at Cornell University, from which she received a Bachelor of Science in 1899 and a PhD in botany in 1901. In 1904 she published her seminal work, *Contributions to the knowledge of the life history of Pinus with special reference to sporogenesis, the development of the gametophytes and fertilization.* At this time, she became an associate professor of botany and head of the department at Wellesley College until 1906, when she was promoted to professor and head, roles she held for the next twenty-four years.

Margaret served as the vice president of the Botanical Society of America, and in 1929 became its first female president. Her interests lay in an analysis of a variety of systems, including fungi, pine and petunia; her study of the latter revealed how plant flower colour and pattern do not follow Mendelian laws of inheritance.

During her tenure at Wellesley, Margaret encouraged many women botanists where laboratory work was a major part of her teaching. She became renowned for advancing scientific education in the field of botany, with a particular contribution on the life histories of North American pines.

Between 1930 and 1932, she was the Director of Botany and a research professor of botany at Wellesley College. At the end of this tenure, at the age of sixty-nine, Margaret retired, although she continued with her research for a further six years. She received many honours throughout her life, including being a member of *Sigma Xi* (the international honour society of science and engineering), Botanical Society America, California Academy of Sciences, American Association of University Professors, American Genetic Association, American Society of Naturalists, American Association of University Women, Science League of America, Massachusetts Horticultural Society, Eugenics Society of the United States, and American Micros Society (vice president, 1914). In 1937, she received an honorary Doctor of Science from Mount Holyoke College.

In her later years, Margaret Ferguson spent time in Florida before moving to San Diego, where she died of a heart attack on 28 August 1951 at the age of eighty-seven. Greenhouses in the Wellesley College Botanic Gardens are named in her honour. She never married.

Lydia Folger Fowler

Born: 5 May 1823 in Nantucket, Massachusetts, USA
Died: 26 January 1879 (aged 56)
Field: Medicine

Lydia Folger was born on 5 May 1823 in Nantucket, Massachusetts, to Eunice (Macy) Folger and Gideon Folger. Lydia was one of seven children. Her father was a businessman and farmer whose own father had settled on Nantucket Island in the 17th century. It was a prominent clan, with Lydia being a member of the Starbuck whaling family through her paternal grandmother Elizabeth Starbuck Folger, while her mother was a member of the Macy family of Nantucket, whose descendants later founded Macy's department stores.

Lydia was educated in local schools. At sixteen she attended Wheaton Seminary in Norton, Massachusetts, where she was later a teacher from 1842 to 1844. During that period, Lorenzo Niles Fowler, a prominent phrenologist and an avid promoter of the field, visited her uncle Walter Folger Jr, a famous astronomer, mathematician and navigator in Nantucket. Lorenzo and Lydia had an immediate attraction to each other and wed on 19 September 1844.

Developed around 1800 by German physician Franz Joseph Gall, phrenology involved a study of the shape and size of the skull, as a determinant of a person's character and mental abilities, based on the assumption that thoughts and emotions are located in specific parts of the brain. It is reported that Lorenzo examined the heads of many distinguished men, including Charles Dickens, Baron Rothschild, Sir Henry Irving and Edgar Allan Poe.

After marriage, Lydia became involved in phrenology herself, giving lectures and writing several books, *Familiar Lessons on Physiology* (1847), *Familiar Lessons on Phrenology* (1847) and *Familiar Lessons on Astronomy* (1848), for the family publishing firm of Fowler & Wells. The couple had two daughters: Amelia in 1846 and Lydia in 1850. Sadly both passed away at an early age.

In 1849, Lydia enrolled at Central Medical College in Syracuse, New York, as one of eight women entering the first co-educational medical school in the USA. When the college moved to Rochester, New York, while Lydia was in her second term, she was appointed to serve as principal of the 'Female Department'. She graduated in June 1850 and, in doing so, became the first American-born woman, and only the second woman, after Elizabeth Blackwell (*see entry*), to receive a medical degree in the USA. In 1851, she was appointed professor of midwifery and diseases of women and children at the college, becoming the first female professor in an American medical school.

When the school closed in 1852, for the next eight years Lydia lived and practised in New York City, frequently lecturing to women on hygiene and physiology, along with championing the further opening of the medical profession to women. She now became involved with women's rights activists, assuming the position of secretary to several of their conventions, as well as being a temperance advocate, including a role as the presiding officer at the Women's Grand Temperance Demonstration held in Metropolitan Hall in New York City during 1853. Lydia also frequently lectured to mainly female audiences on health matters. In 1855, *The New York Tribune* described one of her lectures at a P. T. Barnum-sponsored programme on motherhood: 'She was dressed in a very broadly striped silk, which was anything but a bloomer. Her hair was done up in a French twist with curls in front. Her face is pleasant, she has sunny blue eyes and a sweet mouth. She waved an elegantly embroidered handkerchief as she read her lecture. Quite a number of the little exhibited [babies] were present and contributed their full share to the festivities, at times almost drowning her voice, which is scarcely strong enough for a lecturer.'

In 1860, her third and only surviving child, Jessie Allen Fowler, was born. She later also became a phrenologist. Lydia and Lorenzo travelled and lectured both in the USA and Canada, as well as journeying to England for a lengthy and successful lecture tour. During 1860–61, Lydia studied medicine in Paris and London, the following year becoming an instructor in clinical midwifery at the New York Hygeio-Therapeutic College in New York City.

In 1863, Lydia and Lorenzo moved to London, opening an office on Fleet Street. In the same year, Lydia published a temperance novel, *Nora: The Lost and Redeemed*, and a guide for parents on the physical and mental rearing of children, *The Pet of the Household*. Two years later, Lydia produced *How to Solve It*, a collection of lectures on child care, and *Heart Melodies*, a collection of poems that included the lines: 'That day when Christ in a manger lying long ago was found – Christ, the Light, the Way, Who gilds the world with love's enkindling ray, Who comforts all those on Him relying.'

Lydia was a prolific author and teacher, publishing lectures under a variety of titles directed towards women and women's health. She estimated that she had lectured to, and taught, nearly a quarter of a million American and European women. Lydia became involved in charitable work by means of visiting poor families on behalf of the church, as well as continuing her work practising medicine and teaching women about health, education, and parenting.

She contracted blood poisoning, and during her nine-week illness her vital organs gradually deteriorated. In late 1878, she became ill with pneumonia and passed away, at age fifty-six, on 26 January 1879. She is buried in Highgate Cemetery in London. Lydia Fowler made an outstanding contribution to medicine through her support for the opening up of the medical profession to women and her contribution to women's health and education.

Rosalind Elsie Franklin

Born: 25 July 1920 in Notting Hill, London, England
Died: 16 April 1958 (aged 37)
Field: Chemistry
Awards: Honorary Louisa Gross Horwitz Prize (Columbia University, 2008) and many others

Rosalind Franklin was born on 25 July 1920 in Notting Hill, London, the second child and eldest daughter of five children in a prominent Jewish family. Her father, Ellis Arthur Franklin, was a politically liberal London merchant banker who taught at the city's Working Men's College, and her mother was Muriel Frances (Waley) Franklin. Rosalind's great-uncle Herbert Samuel (later Viscount Samuel), was the Home Secretary in 1916 and the first practising Jewish person to serve in the British Cabinet, while her aunt, Helen Caroline Franklin, was married to Norman de Mattos Bentwich, the Attorney-General in the British Mandate of Palestine. During the Second World War, Rosalind's parents were instrumental in assisting Jewish refugees from Europe to settle in Great Britain, particularly those from the *Kindertransport*, which rescued Jewish refugee children.

In 1931, at the age of eleven, Rosalind attended St Paul's Girls' School, this being one of the few institutions to offer physics and chemistry for girls. At fifteen, Rosalind made the decision to be a scientist, but was actively discouraged by her father as he considered that, as a woman, it would be very difficult for her to carve out a career. She completed high school in 1938 and won a university scholarship, which she donated to a refugee student. Now aged eighteen, Rosalind was accepted into Cambridge University, studying natural sciences with a major in chemistry. With the Second World War now beginning in Europe, her father refused to pay for her second year and implored her to postpone her education and concentrate on the war effort. He relented after her mother's intervention and Rosalind completed her degree, graduating with second class honours and obtaining the top marks in physical chemistry for which she was awarded a research fellowship.

In 1942, Rosalind began her investigation of coal utilisation for London Coal, an activity that was essential for driving the British war effort.

She classified coals according to their porosity and related this to their performance as fuels with great accuracy. One finding was that coal can act as a molecular sieve, with its fine structure capable of separating mixtures of molecules. This research was the basis for her PhD thesis and she was awarded a doctorate by Cambridge in 1945.

In 1947, Rosalind, now aged twenty-seven, moved to Paris where she joined the team of Jacques Mering, a Russian-born naturalised French engineer, which was undertaking pioneering research in X-ray diffraction studies of amorphous solids such as coal and she used this time to study its atomic structure. It is reported she enjoyed French culture and French conversation, being credited by her French colleagues with speaking 'the best French any of them had ever heard in a foreign mouth'.

In May 1950 she discovered that there were two forms of deoxyribonucleic acid (DNA), and the following year was offered a three-year research scholarship at King's College in London. It was there that she was engaged to progress the X-ray crystallography unit at King's College where the physicist and molecular biologist Maurice Wilkins was already using X-ray crystallography to try to solve the DNA problem. Unfortunately, Franklin arrived while Wilkins was absent and, on his return, he assumed that she was hired to be his assistant and this poor beginning to their relationship never improved.

Despite this, Rosalind worked instead with a student, Raymond Gosling, and was able to get two sets of high-resolution photos of crystallised DNA fibres by using two different fibres of DNA, one more highly hydrated than the other. This enabled her to deduce the basic dimensions of DNA strands, and the likely helical structure. She also found that when DNA was exposed to high levels of moisture, its structure changed. She named the high moisture form 'B DNA', while the drier form became 'A DNA'.

Rosalind presented her findings at a lecture in King's College, which was attended by the DNA enthusiast James Watson who, in his book *The Double Helix*, confessed that he did not pay sufficient attention, so that he was unable to fully describe the lecture nor the results to his colleague Francis Crick. Watson and Crick were at the Cavendish Laboratory and had been working on solving the DNA structure, although Rosalind did not know Watson and Crick as well as Wilkins did. She did not collaborate with them and it was Wilkins who showed Watson and Crick the X-ray data Franklin obtained, confirming the 3-D structure that Watson and Crick had speculated about for DNA. In 1953, both Wilkins and Franklin published papers on their X-ray data in the same issue of *Nature*, with Watson and Crick's paper on the structure of DNA. In 1953, Franklin left Cambridge and joined the University of London, Birkbeck College, laboratories where she headed the research group that investigated the 3D structure of tobacco mosaic virus and other viruses.

In mid-1956, while on a trip to the USA, it was discovered that she had two tumours in her abdomen that required periods of hospitalisation and treatment. She continued her research, publishing seven papers in 1956, and six others in 1957, but by the end of the year she became quite ill and, although briefly returning to work in January 1958, she relapsed in March and died in Chelsea, London, on 16 April 1958 of bronchopneumonia, secondary carcinomatosis and ovarian cancer. She was buried in the family plot at Willesden United Synagogue Cemetery at Beaconsfield Road in the London Borough of Brent.

In 1962, the Nobel Prize in Physiology or Medicine was awarded to James Watson, Francis Crick and Maurice Wilkins for solving the structure of DNA. Watson suggested that Rosalind, along with Wilkins, should be awarded a Nobel Prize for Chemistry, but the Nobel Committee does not make posthumous nominations. In *The Double Helix*, Watson outlined how the two had become friends while working together. He also remarked that he would never have won a Nobel Prize or published a famous paper if it were not for Rosalind. Described as 'a deft experimentalist, keenly observant and with an immense capacity for taking pains', her work was a crucial part in the discovery of the DNA structure and Rosalind Franklin was undoubtedly one of most important pioneers in the field.

(Marie) Sophie Germain

Born: 1 April 1776 in Rue Saint-Denis, Paris, France
Died: 27 June 1831 (aged 55)
Field: Mathematics and physics

(Marie) Sophie Germain was born on 1 April 1776 in a house on Rue Saint-Denis, in Paris, France. She was reportedly a shy child, the second of three daughters of a wealthy Parisian silk merchant Ambroise-Francois Germain (who some believed was a goldsmith), an elected representative of the bourgeoisie to the *États-Généraux* (Estates General) in 1789. Her mother was Marie-Madeleine (Gruguelu) Germain. During the French Revolution, which began in 1789 when she was thirteen, Sophie's interest in mathematics reportedly began when she was confined to home because of the danger of revolts on the streets. She pursued this interest with determination and passion and using her father's extensive library.

Her first biographer was a family friend, Count Guglielmo Libri Carrucci dalla Sommaja, who wrote how Sophie's father confiscated her candles and clothes and removed any heating to discourage his daughter from studying such an 'unfeminine subject' as mathematics. But she did so anyway, covered in layers of blankets and by the light of candles that she managed to smuggle in. She wanted to attend the *École Polytechnique*, a French public institution of higher education and research, established in 1795 in Palaiseau, a suburb southwest of Paris. However, women could not be admitted. Instead, Sophie befriended male students and acquired their lecture notes, submitting a report to a faculty member, the renowned mathematician Joseph-Louis Lagrange, under the assumed identity of a former male student, Monsieur Antoine-August Le Blanc. Lagrange was impressed, but became astounded when he later learned it had been written by a woman. Using her hypotheses, he was able to deduce the correct partial differential equation for the vibration of electric plates. She continued to correspond with the foremost mathematicians of the day, including the prominent mathematician Adrien-Marie Legendre, and her interest in number theory peaked in 1798.

In 1808, a German physicist, Ernst Chladni, arrived in Paris and experimented with vibrating elastic plates. His experiments supported a metal or glass rectangular plate horizontally by a stand affixed to its centre. Then he sprinkled a fine powder, such as sand, on it and made it vibrate by drawing a violin bow rapidly along the edge. The powder would then be thrown from the moving points to the nodes, these being the points that remained stationary. The resulting lines or curves were known as 'Chladni figures' and formed different patterns as the notes changed. This phenomenon caused so much excitement that Napoleon authorised an extraordinary prize for the best mathematical explanation and a contest was announced in 1811 by the Paris Academy of Sciences. Sophie's was the only valid entry. Although she had used the right approach, however, it contained some mathematical errors that resulted in her effort being rejected.

In 1813, the contest was re-opened and she now entered a revised paper that included experimental verification of her theory, but it received only an honourable mention. Her interest in number theory was revived in 1815 when she noticed a prize for a proof of the so-called 'Fermat's Last Theorem' (a problem that was only solved by Andrew Wiles in 1995). Sophie wrote to Carl Gauss in which she stated that her preferred field was number theory and provided a general outline towards a substantial proof. The theorem was Fermat's conjecture that the equation $X^n + Y^n = Z^n$ has no positive integer solutions for n greater than 2. Her contribution was to demonstrate the impossibility of positive integer solutions if X, Y and Z are primes to one another (share no common factors) and to n, where n is any prime number less than 100. This did not completely solve the theorem but was headed in the right direction. Unfortunately, Gauss never replied to her letter.

The following year there was a third and final contest announced by the Academy of Sciences for the Chladni problem, and this time she enlisted the assistance of a number of mathematicians to produce a new application. In 1816 she submitted a paper under her own name and won the grand prize, it being one of the high points of her career. However, the Academy had a policy of excluding women other than member's wives and so she was unable to attend their sessions. This was rectified seven years later when the secretary of the Academy, mathematician Joseph Fourier, obtained tickets for her.

She continued her work on elastic surfaces and produced three publications, in 1821, 1826 and 1828, expanding the field of the physics of vibrating curved elastic surfaces and examined the effect of variable thickness.

Sophie also had a keen interest in philosophy and psychology. Two of her works were published posthumously. The first, *Oeuvres philosophiques – Pensées diverses* (most likely written when she was younger), contained personal opinions, comments on physicists throughout and summaries of scientific topics while the second, *Considérations générales sur l'état des sciences et des lettres aux différentes époques de leur culture*, was a scholarly discourse on the theme of unity of thought between the sciences and the humanities.

In 1829, Sophie discovered that she had breast cancer. She nevertheless continued her work and, two years later, published her paper on the curvature of elastic surfaces, along with another that involved the principles of examination that later led to the discovery of laws of equilibrium and the movement of elastic solids.

Having never married, Sophie passed away on 27 June 1831 at the age of fifty-five, her death certificate describing her simply as *'rentière – annuitant'* or property holder. Sophie was buried in the Père Lachaise Cemetery in Paris.

In 1837, six years after her death, Carl Gauss remarked that Sophie 'proved to the world that even a woman can accomplish something worthwhile in the most rigorous and abstract of the sciences and for that reason would well have deserved an honorary degree'. Although she did not receive an honorary degree, in 2003 the Academy of Sciences established the *Prix Sophie Germain* (Sophie Germain Prize), to honour a French mathematician for research in the foundations of mathematics. One hundred years after her death, a street and a girls' school were named after her, and a plaque was placed at the house where she died. The school also houses a bust commissioned by the Paris City Council.

Catherine 'Kate' Anselm Gleason

Born: 25 November 1865 in Rochester, New York, USA
Died: 9 January 1933 (aged 67)
Field: Engineer

The eldest of four children, Catherine 'Kate' Anselm Gleason was born on 25 November 1865 in Rochester, New York, to Ellen and William Gleason who had emigrated from Ireland. William owned a machine tool company, later named Gleason Works, which became one of the world's most important makers of gear-cutting machine tools and was still in operation in 2019. When she was eleven, Kate's stepbrother Tom died of typhoid fever. Tom had been an invaluable employee, which caused hardship at her father's company, and so at the age of twelve Kate began helping her father and did so for the next seven years.

In 1884, now aged nineteen, Kate enrolled in the Cornell Mechanical Arts programme in Ithaca, New York, becoming the first female student to study engineering there. She had to leave before the end of her first academic year, however, as she needed to help her father. The firm was in financial difficulties and William could no longer afford to pay the man he had hired to replace her. While Kate never completed the requirements for a degree, through training and self-learning she earned the title of engineer and is recognised for her accomplishments in the field. Later, she was able to undertake some further education at Sibley College of Engraving and the Mechanics Institute, now known as Rochester Institute of Technology (RIT).

Between 1890 and 1901, Kate was actively involved as the treasurer and saleswoman for Gleason Works. In 1893, she toured Europe to expand the company's business – one of the first times an American manufacturer had tried to globalise its operation. As of 2019, international sales make up almost three-quarters of the company's business. Kate played a pivotal role in the firm's early operations, with brothers James and Andrew. Through her example, she showed that a bias against women was not justified, and that women were more than capable of holding senior positions in mechanical engineering and sales. In one instance, after engineers at Packard developed spiral bevel gears, Kate pioneered the machine tools to mass-produce them cheaply and quickly, with automotive differentials being the primary market. Consequently, Gleason Works became a leading US producer of cutting-edge machinery. The car giant Henry Ford gave Kate full credit for the invention, calling it 'the most remarkable machine work ever done by a woman'. The eminent American machinist, technical journalist, author and editor Fred Herbert Colvin described Kate in his memoirs as 'a kind of Madame Curie of machine tools' and that 'she knew as much as any man in the business'.

In 1913, due to family disputes, Kate left Gleason Works at the age of forty-eight and found employment at the Ingle Machining Company, commencing work there on 1 January 1914. In that year, she also became the first woman elected to full membership in the American Society of Mechanical Engineers (ASME), representing the society at the World Power Conference in Germany. When Ingle was struggling financially, Kate was appointed as the Receiver in Bankruptcy for the company, reportedly the first woman ever to be appointed by a court to this position. Kate managed to guide the company back into solvency, repaying outstanding debts, and returning it to the shareholders before the end of 1915 as a profit-making enterprise.

Kate's next role, between 1917 and 1919, was as the first female president of the First National Bank of Rochester, replacing the incumbent who had resigned to fight in the First World War. In this position, she supervised a problem loan and used it to finish the housing complexes left by the previous loan holder. To further her humanitarian efforts in Rochester, she helped to finance and build eight factories for various companies, including a construction company that built houses for the middle class. Following this success, Kate experimented with concrete to build cheap fireproof houses in East Rochester at an affordable cost, using a pouring method she developed. In 1921, Kate described her methods in an article she wrote for a trade magazine, *Concrete*, titled 'How a Woman Builds Houses to Sell at a Profit for $4,000'. In the 1920s, Kate left Rochester to study adobe buildings in California. In 1924, she was asked by the city of Berkeley, California, to help them rebuild after a fire.

Kate was an avid supporter of women's suffrage, encouraged by her mother's friend Susan B. Anthony, a leading US suffragette. In 1906, Kate and her father had hosted Susan's final birthday party at their home, which included an orchestra of women. A report of the 1912 National American Woman Suffrage Association Convention mentions her as having promised $1,200 USD to the suffrage movement, one of its largest pledges. Many of her personal writings speak of the contributions that she and her father made to women's suffrage.

Kate was deeply moved by the plight of those affected by the First World War, in particular the French village of Septmont, which had been destroyed by the conflict. Kate gave a considerable amount of her own funds and her tireless energy into rebuilding the village hall, community centre and market place, an act that earned her a medal from the French government for her unselfish efforts on behalf of the French people. She also rebuilt a castle in Septmont for herself.

Although she reportedly received over 200 letters proposing marriage, Kate considered that marriage would hinder her professional life and so she remained unmarried. On 9 January 1933, Kate Gleason died of pneumonia at the age of sixty-seven and is interred in Riverside Cemetery in Rochester. Much of her considerable $1.4 million USD estate was bequeathed to cancer research and institutions in the Rochester area, including libraries, parks, and the Rochester Institute of Technology. The Kate Gleason Fund for charitable and educational causes was also established. In 1920, she had purchased land on the Sea Islands off the coast of South Carolina where she planned to build a resort for artists and writers, but at the time of her death only ten units had been completed. The project was later completed by her sister.

The Kate Gleason College of Engineering at RIT is named in her honour, as is a hall of residence, and a bust of her is located in the hallway. In August 1949, Kate's sister, Eleanor, witnessed the opening of the Gleason Memorial Pool in East Rochester, while in 2010, RIT Press published a collection of Kate's letters. In 2019, Gleason Works remains an international organisation, retaining a strong connection with RIT. In 2011, the Gleason Family Foundation honoured her legacy by establishing the Kate Gleason Award of the ASME to recognise a woman who is either a highly successful engineering entrepreneur or represents a lifetime of achievement in an engineering discipline.

Gertrude 'Trude' Scharff Goldhaber

Born: 14 July 1911 in Mannheim, Germany
Died: 2 February 1998 (aged 86)
Field: Physics
Awards: Fellow of the American Physical Society (1947); National Academy of Sciences (1972); Long Island Achiever's Award in Science (1982); Outstanding Woman Scientist Award, New York Chapter of the Association for Women Scientists (1990)

Gertrude ('Trude') Scharff was born on 14 July 1911 in Mannheim, Germany, into an upper-middle-class Jewish family. Her father, Otto, was the third generation to lead the family business of imported foodstuffs. Her mother, Nelly (Steinharter) Scharf, was the daughter of a furrier who died when she was young. As a young girl, Trude attended a public school where she developed an interest in science. Shortly after the First World War, the family moved to the bigger city of Munich, to establish new headquarters for Otto's growing business. However, after Germany's defeat in the war there was a period of hyperinflation, with prices doubling daily. Trude remembered her father bringing home suitcases filled with German marks just to pay for the next day's groceries. Suppliers resorted to adulterating foodstuffs, including flour. Trude remembered eating bread with sawdust in it.

By about 1928, Trude matriculated and attended the University of Munich. She also learned to drive. Although this was initially against her father's wishes, he gave her a car, a convertible that she drove to university, while continuing to live at home – a rare luxury at the time. She soon developed an attraction to physics, although her father suggested that she study law instead – a better course for someone destined to manage a business. She recalled telling him that law did not interest her: 'I wanted to understand what the world is made of. That actually turned out to be a good thing because when the Nazis came – I'm Jewish – I had to leave the country, and physics was something that could be done anywhere.'

Like many other students, Trude spent time at several other universities, including the University of Freiburg, the University of Zürich and the University of Berlin, before returning to the University of Munich. It was at the University of Berlin that Trude met Maurice Goldhaber, an Austrian-born physicist, and began a correspondence that continued over the next few years.

With the rise of the Nazi party in Germany in 1933, Trude's parents left for Switzerland, taking their liquid assets with them. But Trude and her sister were reluctant to leave at this time, which meant their parents had to return to support them. Otto was arrested and was only released after he agreed to repatriate his money from Switzerland. Perversely, the economy stabilised after the Nazis took power, which also meant that Otto's business grew. While this meant that Trude's physics studies could continue, the anti-Semitism of the Nazis was also intensifying. She felt increasingly ostracised, but remained to finish her thesis with the support of her supervisor, the German physicist Walter Gerlach, examining the magnetic behaviour of stretched nickel wires. She obtained her PhD in 1935, publishing her results the following year.

Maurice was now at Cambridge University, and encouraged Trude also to leave for England, which she did as soon as she completed her doctorate. She later said that she should have left Germany earlier, but felt that she had to finish her thesis. The anti-Semitic laws meant that she could not take money out of Germany, so she took what she could in personal possessions. She lived for six months on the money she made from selling her Leica camera, as well as earnings made from translating German to English, eating mostly potatoes to save money, while she looked for a physics job in London. She eventually found a one-year fellowship in the laboratory of the English physicist George Paget Thomson (winner of the Nobel Prize for Physics in 1937) at Imperial College, where she researched electron diffraction.

In April 1938, Maurice was offered a job at the University of Illinois, Urbana. He wrote to Trude proposing marriage. She accepted his mailed offer of marriage, and he returned to England in May 1939 so they could be married and leave for the USA. As he had a faculty position there, state rules forbade Trude to be employed there as well. Maurice was very supportive of her work, and so she worked in his laboratory as an unpaid assistant with no space allocated to her, undertaking her own research, now in nuclear physics as the lab was devoted to this research. It was during this period that the couple had two sons: Alfred and Michael.

In 1941 she studied neutron-proton and neutron-nucleus reaction cross sections, and in the following year she investigated gamma radiation emission and absorption by nuclei. She found that neutrons were released spontaneously with nuclear fission. Although this was always a theoretical result, she was the first to demonstrate it. As her work was classified during the war, it was only published in 1946, after the war had ended. In 1947, Trude was elected a Fellow of the American Physical Society, and in the following year she and Maurice determined that beta rays were identical with electrons, a finding of great interest to their peers. Trude used three-dimensional plots to explain her findings to others, which later evolved into a conventional method with computers so that 'its origins may have been forgotten', as a biographical memoir about her recalls: 'Nowadays this kind of visualization is done with computer graphics and so has become much easier, if not as tangible.'

In 1950, the family moved from Illinois to Long Island in New York State, where Trude and Maurice both joined the staff of Brookhaven National

Laboratory in Upton. She was the first woman with a PhD on the scientific staff there, and this was her first regular, long-term, paid job since the completion of her doctorate. At some point in the 1950s, Trude was at a physics conference in Florida and was interviewed by a local newspaper. At the end of the interview she was asked how she managed to combine her career with raising two boys. She was reported as saying, 'That's simple; I neglect them.'

In 1960, Trude commenced a series of monthly lectures known as the 'Brookhaven Lecture Series', which were still being held in 2019. She also created a training institute at Brookhaven for pre-college science teachers, and in 1979 co-founded Brookhaven Women in Science, describing itself as 'a diverse and inclusive community that promotes equal opportunity and advancement for all women in support of world-class science'.

In 1972, Trude was elected to the National Academy of Sciences, only the third woman physicist to be honoured in this way. She served on their Committee on the Status of Women in Physics of the American Physical Society where she strived to end obstacles for women in science. In 1977, Trude had to retire at the Brookhaven National Laboratory at age sixty-six, returning in 1985 as a research collaborator until 1990. In 1982, she was awarded the Long Island Achiever's Award in Science, and in 1984 was a visiting scholar of *Phi Beta Kappa*, the oldest honour society in the US, sending her to eight campuses for the year. In 1990, the New York Chapter of the Association for Women Scientists gave Trude their Outstanding Woman Scientist Award. In 1992, Brookhaven Women in Science established the 'Gertrude Scharff-Goldhaber Prize', which was still awarded in 2019.

On 2 February 1998, Trude passed away, at age eighty-six, in Brookhaven Memorial Hospital in Patchogue, New York. She was survived by her husband Maurice, who lived to the age of 100, as well as her two sons, both theoretical physicists, Dr Alfred Goldhaber of Setauket, New York and Dr Michael Goldhaber of Oakland, California.

Gertrude 'Trude' Scharff Goldhaber was an outstanding physicist who made significant inroads into the understandings of nuclear fission and the structure of atomic nuclei, working in areas that included long-lived isomers (different versions of elements) and heavy ions. A biographical memoir says that her work 'played an integral part in unfolding the story of nuclear structure, alerting experimentalists to regions of the periodic table of importance and confronting theorists with the realities of nature'.

Apart from her deep and abiding interest in science, Trude was interested in how mythology reflects a civilization. She was also a keen tennis player and mountain climber. Her legacy is an outstanding contribution to science and education, along with greater recognition and equality for women. She declared,

> The vicious cycle which was originally created by the overt exclusion of women from mathematics and science must be broken ... [I]t is of the utmost importance to give a girl at a very early age the conviction that girls are capable of becoming scientists.

Anna Jane Harrison

Born: 23 December 1912 in Benton City,
Missouri, USA
Died: 8 August 1998 (aged 85)
Field: Chemistry

Anna Jane Harrison was born on 23 December 1912 in Benton City, Missouri, USA, to Mary Katherine (Jones) Harrison and Albert Harrison, who were both farmers. One of Anna's earliest memories of science education was during her early primary school years, when she was asked to go home and 'learn about caterpillars'. It is reported that, when she asked her father, he told her all about Caterpillar tractors, which were 'revolutionising farm life at the time'. Anna was seven when her father passed away and her mother was left to care for her and her elder brother while tending to the farm alone, which she undertook for the next forty years.

Anna's early school years took place in a one-room schoolhouse in Benton City, Audrain County, Missouri. Her interest in science was heightened when she attended high school in Mexico, Missouri, and, it is said, that she, 'like many future scientists, had especially good science teachers'. From there, she enrolled at the University of Missouri in Columbia, Missouri, graduating with a Bachelor of Arts in chemistry in 1933 and in education in 1935. Anna then returned to Benton City to teach at the small school she had attended as a girl, but two years later returned to her studies at the University of Missouri where she undertook graduate work in physical chemistry.

After obtaining a master's degree in 1937, Anna published a paper on photovoltaic effects (voltages generated by interactions between dissimilar chemicals when struck by light) in solutions of Grignard reagents (organic derivatives of magnesium). Soon after, Anna enrolled in Missouri's doctoral programme with a dissertation considering the association of sodium ketyls (a type of radical). She graduated with her PhD in 1940 and accepted a position as an instructor of chemistry at Sophie Newcomb Memorial College, the women's college of Tulane University in New Orleans.

In 1942, Anna took leave from her teaching during the Second World War to conduct secret wartime research at the University of Missouri. In particular, during 1944, she conducted research on toxic smoke for the National Defense Research Committee, the A. J. Griner Co., in Kansas City, Missouri, and Corning Glass Works in Corning, New York. This essential work involved the creation of smoke-detecting field kits for the US Army. For her outstanding research she received the Frank Forrest Award from the American Ceramic Society.

In 1945, Anna joined the faculty at Mount Holyoke College, a women's school in South Hadley, Massachusetts, as an assistant professor, to work with the renowned professor and researcher Emma Perry Carr, whom she already knew. Anna was later promoted to full professor status in 1950 and subsequently held the chair of the department of chemistry between 1960 and 1966. At Mount Holyoke, Anna continued her research on ultraviolet light and photolysis, the breakdown of molecules via light absorption. In 1969, she received the Manufacturing Chemists Association Award in College Chemistry Teaching. Her research investigated the structure of organic compounds and their interaction with light, particularly in the ultraviolet and far ultraviolet bands. The Petroleum Research Fund Advisory Board of the American Chemical Society (ACS) awarded Anna a grant for 'an experimental study of the far ultraviolet absorption spectra and photodecomposition products of selected organic compounds'.

During the 1970s, Anna advocated for improvement in the communication of science to the public, especially to public officials. In 1971, she became chair of the ACS Division of Chemical Education. She was a member of the National Science Board between 1972 and 1978, and in 1977 received the James Flack Norris Award for Outstanding Achievement in the Teaching of Chemistry, ACS Northeastern Section. The following year, she became the first woman president of the ACS and continued her advocacy of the advancement of scientific knowledge by the chemical profession. In 1982, Anna received the ACS George C. Pimentel Award in Chemical Education, recognising outstanding achievements in chemical education.

In 1979, Anna retired from Mount Holyoke, being awarded the title of Professor Emeritus, taking up a teaching role at the US Naval Academy in Annapolis, Maryland. She remained active in the promotion of science, along with serving on a number of scientific councils, including in 1983–84 as president of the American Association for the Advancement of Science. From 1988 to 1991, Anna was a member of the Board of Directors of *Sigma Xi*, the scientific research honour society, while also being on the editorial boards of the *Journal of College Science Teaching* and *Chemical & Engineering News*. As a representative of her scientific organisations, she travelled widely – to India, Antarctica, Japan, Spain and Thailand.

In 1989, with co-author Edwin S. Weaver, a former Mount Holyoke colleague, Anna published *Chemistry: A Search to Understand,* described as suitable for students who were 'intellectually curious but not professionally driven' in the field of chemistry.

Anna Jane Harrison

On 8 August 1998, Anna Harrison died in Holyoke Hospital, at age eighty-five, from a stroke. She never married. Anna was the recipient of twenty honorary degrees and was voted at a Mount Holyoke College reunion of the class of 1968, as 'one of the people who has had the greatest impact on my life'. She was known as a skilled classroom teacher, able to make complicated things clear and much of her life was devoted to obtaining increased funding from state and federal agencies for science education and promoting the cause of women in science. As a result, Anna Harrison was an inspiration for many young women to pursue a scientific career.

Caroline Lucretia Herschel

Born: 16 March 1750 in Hanover, Germany
Died: 9 January 1848 (aged 97)
Field: Astronomy
Awards: Gold Medal of the Royal Astronomical Society (1828); Prussian Gold Medal for Science (1846)

Caroline Lucretia Herschel was born on 16 March 1750 in Hanover, Germany. Her father, Isaac Herschel, was a talented musician, an oboist with the Hanoverian Foot Guards. Lacking in formal education himself, he nevertheless did his best to provide an education for his six children, although his wife, Anna Ilse (Moritzen) Herschel, felt it necessary to educate only the four boys, as the two daughters would be destined to become household help for the family. Caroline's sister Sophie, some twenty-three years older than Caroline, accepted her fate, but Caroline developed a keen interest in her father's activities and he involved her in scientific discussions, including astronomy. She wrote in her notes how they went 'on a clear frosty night into the street, to make me acquainted with several of the beautiful constellations, after we had been gazing at a comet which was then visible'.

While her father's army duties against the French took him away between 1757 and 1760, Anna took charge and enrolled Caroline in an inferior garrison-run school, as well as a knitting school. She knitted stockings for her brothers and wrote letters for her mother, who was illiterate. At the age of ten, Caroline contracted typhus, which permanently affected her growth. She was never to grow taller than 4 feet 3 inches (130 centimetres) and her parents concluded that she would never marry – an assumption that turned out to be correct. Caroline's father died when she was seventeen and she remained with her mother, carrying out domestic duties for a further five years. By now, one of her brothers, William, was an organist and orchestra leader in Bath and, much to the chagrin of their mother who was about to lose a servant, he spirited Caroline away to live with him and become his housekeeper. It was there that she took lessons in singing, arithmetic, bookkeeping, English, and spherical geometry, which she was keen to put to practical use.

Meanwhile, William was fascinated with astronomy and by the uncharted skies, deciding that he would undertake a survey of the heavenly bodies. Caroline was forced to abandon her music studies to help him build a telescope, something she deeply resented. As her brother's granddaughter Constance Lubbock wrote in her 1933 book, *Herschel Chronicle*, Caroline's summer of 1775 was 'taken up with copying Music and practising, besides attendance on my Brother when polishing [telescopic mirrors]'. Constance wrote that 'it required all Caroline's devotion to overcome the dismay with which she found herself swept along in such an unexpected direction'. She became indispensable. Lubbock quoted her as declaring that 'every leisure moment was eagerly snatched at for resuming some work which was in progress, without taking time or changing dress, and many a lace ruffle ... was torn or besmattered with molten pitch. ... I was even obliged to feed him by putting the victuals by bits into his mouth – this was once the case when at the finishing of a 7 feet mirror he had not left his hands from it for 16 hours together'.

In 1781, William found fame when he discovered a new comet he named *Georgium sidus*, which was later recognised as the planet Uranus. As his musical activities wound down, his fame now spread, to the extent that the Royal Society arranged for him to demonstrate his telescopes to the Royal family. With an annual stipend of £200, William left Bath and made his way to Windsor Castle where he and Caroline set up their telescopes, an activity in which she now became more deeply involved. Within two years, she had discovered three new nebulae (clouds of dust and gas), resulting in William presenting her with a telescope she called a 'Newtonian small sweeper'. By December 1783, she found herself 'entirely attached to the writing desk and had seldom an opportunity of using my newly acquired instruments', but still managed to discover eight comets in the period 1786–97. William IV was so impressed that he granted her a salary of £50 per year – the first money she had ever earned in her own right.

Caroline now set about a new task, rewriting the star catalogues of the first Astronomer Royal, John Flamsteed, finding numerous discrepancies and errors. He had published his original observations in a volume that was separate from the catalogue and his cross-indexing left a lot to be desired. As a result of considerable and painstaking efforts, her *Catalogue of Stars* was published by the Royal Society in 1798, containing a list of 'upwards of 560 stars that are not inserted in the British Catalogue'.

When William passed way in 1822, Caroline was heartbroken and made the spur-of-the moment decision to return to Hanover, a choice that she constantly regretted. However, William's son John, now aged thirty, had commenced his own career as an eminent astronomer, physicist and chemist and she now took a keen interest in his work. To assist him, she assembled a new catalogue of nebulae that she arranged in zones, using the information in William's *Book of Sweeps* and *Catalogue of 2,500 Nebulae*. Although this work was not published, in 1828 the Royal Astronomical Society awarded her a gold medal. No woman would be awarded it again until Vera Rubin in 1996.

In 1835, Caroline and Mary Somerville (*see entry*) became the first women to be awarded honorary memberships of the Royal Society, and three years later Caroline was elected to the membership of the Royal Irish Academy in Dublin. In 1846, on her ninety-sixth birthday, she was awarded the Gold Medal of Science from the King of Prussia for her lifelong achievements.

Caroline Herschel is remembered for her important contributions to scientific discovery, particularly in astronomy, where she is generally credited with finding eight comets and locating a number of new nebulae and star clusters. One of these, the periodic comet 35P/Herschel-Rigollet, bears her name, while the asteroid 281 Lucretia, discovered in 1888, bears her middle name, and the crater C. Herschel on the Moon is named after her.

She spent the final years writing her memoirs and lamenting her body's limitations, which restricted her ability for further discoveries. Caroline passed away on 9 January 1848 at the age of ninety-seven. She was buried at 35 Marienstrasse in Hanover in the cemetery of the Gartengemeinde, next to her parents and with a lock of William's hair. Her gravestone inscription reads: 'The eyes of her who is glorified here below turned to the starry heavens'.

Beatrice Alice Hicks

Born: 2 January 1919 in Orange, New Jersey, USA
Died: 21 October 1979 (aged 60)
Field: Engineering
Awards: Woman of the Year in Business (*Mademoiselle* magazine, 1952); Society of Women Engineers Achievement Award (1963); National Women's Hall of Fame (post., 2002)

Beatrice Alice Hicks was born on 2 January 1919 in Orange, New Jersey, USA, to Florence (Benedict) Hicks and William Lux Hicks who was an American chemical engineer. At thirteen, inspired by the Empire State Building and the George Washington Bridge, Beatrice said to her father that she, too, would become an engineer. Although her parents took a neutral stand on her career choice, some of her teachers and classmates tried to discourage her as they considered engineering a 'socially unacceptable role for a woman'.

After she completed secondary education at Orange High School in 1935, she went to Newark College of Engineering (now New Jersey Institute of Technology), receiving a Bachelor of Science degree in chemical engineering in 1939. Of the 900 students in her class, only two were women. During her time in college, Beatrice worked as a telephone operator in the treasury office of an Abercrombie & Fitch store, and also found work in the university library. Beatrice served in the role of research assistant at Newark for the next three years, studying the history of Edward Weston's inventions and taking additional classes at night. In 1942, she obtained a position with Western Electric in Kearny, New Jersey, becoming the first woman to be hired as an engineer by the company. During this time, she worked on technology for both telephone and aircraft communication, while also designing and testing quartz crystal oscillators. In 1948, Beatrice married Rodney Duane Chipp, a fellow engineer.

When her father passed away, Beatrice left Western Electric and in 1955 purchased the control of the company Newark Controls Company from her uncle. A metalworking firm which her father had started in Bloomfield, New Jersey, it designed and manufactured environmental sensing equipment, of great utility in the space programme. She served as vice president and chief engineer and eventually the president of the company.

During her time at Newark, Beatrice designed and patented a gas density switch on 24 July 1962, which was later used in the US space programme, including the Apollo moon landing missions. She was also a leader in the field of sensors to detect when devices were reaching structural limits. She wrote several technical papers on the gas density switch and enrolled in a master's degree in physics at the Stevens Institute in Hoboken, New Jersey, which she obtained in 1949.

In 1950, Beatrice and about sixty other women formed an organisation with the goal of advancing female engineers and increasing female participation in engineering. Two years later, it was incorporated as the Society of Women Engineers, with Beatrice serving as the inaugural president for two consecutive terms, from 1950 to 1952.

In 1960, the National Society of Professional Engineers selected Beatrice and her husband Rodney for a research and speaking tour of South America, to foster international cooperation between engineers across the Americas. Between 1960 and 1963, Beatrice also served on the Defense Advisory Committee for Women in Services, one of the oldest federal advisory committees of the Department of Defense. Composed of civilian men and women appointed by the Secretary of Defense, it provided advice and recommendations on matters and policies relating to servicewomen in the Armed Forces. Beatrice was also the director of the First International Conference of Women Engineers and Scientists, representing the USA at four International Management Congresses. In 1963, the Society of Women Engineers presented Beatrice with their highest honour, the Society of Women Engineers Achievement Award. She toured the USA as a champion of female engineers.

When Rodney died in 1966, the now widowed Beatrice, aged just forty-one, sold Newark Controls Company and took over his consulting business. She earned many honours over the course of her career, including honorary doctorates from Hobart and William Smith College in 1958 and the first such award given by the Rensselaer Polytechnic Institute in 1965. She was named *Mademoiselle* magazine's 'Woman of the Year in Business' in 1952 and the Newark College of Engineering's Alumna of the Year in 1962. Worcester Polytechnic Institute also awarded her an honorary doctorate.

Beatrice was a member of the National Society of Professional Engineers (NSPE), American Society of Mechanical Engineers (ASME), Institute of Electrical and Electronics Engineers (IEEE) and *Eta Kappa Nu*, the international honour society of the Institute of Electrical and Electronic Engineers, and in 1978 was elected to the National Academy of Engineering. The Society of Women Engineers, that she helped to found, has grown from its initial membership of 60 to over 35,000 in 2019.

On 21 October 1979, Beatrice Hicks passed away at age sixty, in Princeton, New Jersey and was buried in Arlington National Cemetery, Virginia, in Plot Section 13 Site 16405. In 2002, she was posthumously inducted into the National Women's Hall of Fame. A memorial tribute to her said,

Dr Hicks knew how to be effective in advising small business and also was an able adviser to larger organizations as well as international operations. She could design complex systems and keep them operating. In spite of the positions of strength and value that she attained, Dr Hicks retained her charming personality and was always helpful and understanding of others; she had a heart of gold.

All her friends deeply miss her.

Dorothy Crowfoot Hodgkin

Born: 12 May 1910 in Cairo, Egypt
Died: 29 July 1994 (aged 84)
Field: Biochemistry
Awards: Fellow of the Royal Society
(1947); Royal Medal (1956); Nobel Prize in
Chemistry (1964); Order of Merit (1965);
Copley Medal (1976); Lomonosov Gold
Medal (1982)

The eldest of four daughters, Dorothy Mary Crowfoot was born on 12 May 1910 in Cairo, Egypt, to Grace Mary (Hood) Crowfoot (known as Molly), an expert in ancient textiles, and John Winter Crowfoot, who worked for the country's Ministry of Education. During the hot seasons in Egypt, the family returned to England. They were in England as the First World War began, and it was there that Molly left her three daughters – Dorothy, aged four; Joan aged two; and Elisabeth, two months – before returning to Cairo with John, where she gave birth to her fourth daughter, Diana.

On occasions, Dorothy visited her parents on archaeology sites in the Sudan and developed an interest in science, particularly chemistry, from a young age. Her passion was fuelled by her mother, who was an eminent botanist.

In 1920, the family settled in Beccles, Suffolk. Dorothy's biographer, Georgina Ferry, wrote that she began her scientific career 'in a small private class for the children of parents of modest means and independent views' run by the Parents' National Educational Union (PNEU), whose motto was 'I am, I can, I ought, I will', as an alternative to the other local schools. It trained young women to be governesses for primary school aged children, with a set syllabus, including courses in physics and chemistry, at a time when 'nature study' was the limit of most primary school teachers. At ten years old, Dorothy briefly attended a small class taught by a Miss Fletcher, who had been trained by the PNEU and introduced the subject of chemistry to her young students. Dorothy's biographer, historian Georgina Ferry, wrote that

> The progressive educators who designed the course had grasped the importance of practical demonstrations in catching the imagination of the young. Dorothy and her fellow students made solutions of alum and copper sulphate from which to grow crystals. Over the days that

followed they watched as the solutions slowly evaporated. Gradually the crystals appeared, faceted like jewels, twinkling in the light. Dorothy was enchanted. 'I was captured for life,' she later wrote, 'by chemistry and by crystals.'

In 1921, when she was eleven, Dorothy attended a state secondary school, the Sir John Leman School in Beccles, Suffolk. It is reported that she fought to be allowed to participate in the chemistry classes, normally reserved for boys despite the fact that the chemistry teacher, Criss Deeley, was a woman. Dorothy's mother fostered her daughter's interest in chemistry, giving her the published Royal Institution 'Christmas lectures' of Sir William Henry Bragg in 1923 and 1925, especially prepared for children. It was here that Dorothy learned of the X-ray techniques used to observe the arrangement of atoms in crystals.

Dorothy was also encouraged by geologists and the chemist A. F. Joseph, a friend of the family, who also worked on the sites. At thirteen, Dorothy found a mysterious mineral on the ground and, with the aid of her chemistry set, correctly identified it as ilmenite crystal. This triggered a love of crystallography, with a desire to analyse atomic and molecular structures.

While Dorothy achieved outstanding results in her leaving certificate exams in 1927, she needed Latin for entry into the University of Oxford or Cambridge. The headmaster of Sir John Leman School gave her private tuition, which enabled her to pass the entrance examination to the University of Oxford where, in 1928 and aged eighteen, she began studying chemistry at Somerville College, named after Mary Somerville (*see entry*). Dorothy was soon recognised as an exceptional student. Her enthusiasm for laboratory work included analysing samples of ancient coloured glass that her parents sent her from their archaeological digs. Dorothy was also a keen illustrator and developed her own interest in archaeology by completing a detailed illustration of a Byzantine mosaic for one of her father's publications. For her honours year in her bachelor's degree, Dorothy used X-rays to study the structures of certain salts. At the time, X-ray crystallography was a groundbreaking method to examine the structure of molecules – a process that could take years of observation using complex mathematics before these structures could be fully understood.

In 1932, Dorothy was awarded a first-class honours degree at the University, only the third woman to achieve this distinction in chemistry at Oxford. She then enrolled in the Department of Mineralogy at Cambridge University as a PhD student in the laboratory of John Desmond Bernal. Bernal had trained with Sir William Bragg and now headed the X-ray crystallography laboratory, which was in great demand, pioneering the use of the technique to study biological molecules. Bernal was often away, so Dorothy was very busy. Not long after she had gone to Cambridge, Somerville College offered her a research fellowship. Dorothy was hesitant to leave the exciting environment of Bernal's laboratory, but Somerville was happy for her to remain there for the first year of her fellowship. So she returned to Oxford in 1934, to continue her doctoral work under the supervision of Herbert 'Tiny' Powell.

It was in 1934 that Dorothy experienced pain in her hands, which was diagnosed as rheumatoid arthritis. It became progressively worse and more debilitating over time, ultimately leading to deformities in her hands and feet. In her final years, she spent much of the time in a wheelchair, although she nevertheless remained active in research.

In 1936, she completed her Cambridge PhD, an X-ray diffraction analysis of crystals of steroids. That same year she was awarded a permanent fellowship at Somerville College. Her research during this time included mapping the architecture of cholesterol and examining the structure of penicillin, which had been discovered in 1928. This was essential to create synthetic versions of it.

In 1937, Sir William Bragg invited Dorothy to use the exceptional X-ray equipment at the Royal Institution to obtain higher quality photographs of her insulin crystals. While in London, she stayed with Margery Fry, the former principal of Somerville. Staying in the house at the same time was Margery's cousin Thomas Lionel Hodgkin, a graduate in history who had been Dorothy's exact contemporary at Oxford. He had recently lost his job as personal secretary to the British High Commissioner in Palestine through his vociferous support of the Arabs, had become a communist, and was now reluctantly being trained as a schoolteacher. It is said that at this time, Dorothy was in love with John Bernal, who was not only married but involved in at least one other serious involvement. However, after only one or two further meetings, she and Thomas agreed to marry. By the time of their wedding on 16 December 1937, Thomas had found a new career in adult education and was teaching history to unemployed miners in Cumberland. Dorothy, with the support of both families, retained her fellowship at Somerville, which had by this time been made permanent, and continued her research. Thomas returned to Oxford on weekends. They had three children: Luke (b. 1938), Elizabeth (b. 1941) and Toby (b. 1946). However, she continued to publish as Dorothy Crowfoot until 1949 when, it was said, she 'bowed to social pressure' and added her married name to write as Dorothy Crowfoot Hodgkin.

During the Second World War, Dorothy received a large grant from the Rockefeller Foundation to continue her work on the structure of insulin, taking over equipment evacuated from Bernal's lab and two of his research assistants, Harry Carlisle and Käthe Schiff. With Carlisle, Dorothy solved the complete three-dimensional structure of cholesterol iodide – the first crystallographic study she was able to completely undertake, and the first anywhere of such a complex organic molecule. Dorothy's research into the structure of penicillin also intensified, in collaboration with Oxford scientists, leading to the first major publication on the structure of penicillin. Howard Florey and Ernst Chain demonstrated the efficacy of penicillin against bacterial infections in animals and humans during 1940 and 1941, but its chemical formula was unknown.

In 1944, Dorothy was shortlisted by Oxford for the readership in chemical crystallography, but the post went to her former supervisor, Powell. Two years later, she was appointed to the lesser post of university demonstrator, which nevertheless doubled her income from her college fellowship. In 1947, Dorothy was elected as a Fellow of the Royal Society, at the relatively early age of thirty-six.

Her work on penicillin led to many industrial contacts. In 1948, Lester Smith of the British pharmaceutical company Glaxo gave Dorothy some dark red crystals of the anti-pernicious anaemia factor, vitamin B12. She then continued her research on the structure of vitamin B12. While she had been actively encouraging Oxford to establish computing facilities, an offer from the University of California, Los Angeles, provided the assistance she needed to solve the full structure of vitamin B12 by 1957. It is said that the fact she had succeeded moved Lawrence Bragg (the son of Sir William Bragg, whose lectures had inspired young Dorothy) to describe her achievement as 'breaking the sound barrier'.

From the late 1950s onwards, Dorothy's research focused on the structure of insulin, which she had photographed in 1935. She encouraged and fostered a group of researchers working on insulin in her laboratory. Between the 1950s and 1970s, she established lasting relationships with fellow international scientists, including at the Institute of Crystallography in Moscow, as well as in India and in China, working in Beijing and Shanghai on the structure of insulin. In fact, after her initial visit to China in 1959, she travelled there seven more times over the following twenty-five years. In 1958, the American Academy of Arts and Sciences elected her a 'Foreign Honorary Member'. Dorothy spent the early 1960s in Africa at the University of Ghana, where her husband was the director of the Institute of African Studies.

In 1969, working with Guy Dodson and M. Vijayan, Dorothy finally built the first model of the insulin molecule, working almost non-stop over a single weekend. Dodson recalled the moment vividly: 'It was a triumphant occasion in which Dorothy, though suffering from swelling ankles and forced into wearing slippers, worked with concentration and wonderful spirits'. Dorothy's biographer, Ferry, wrote that, in the case of each of the three projects for which she is best known – penicillin, vitamin B12, and insulin – 'Dorothy pushed the boundaries of what was possible with the techniques available. Her distinction lay not in developing new approaches, but in a remarkable ability to envisage possibilities in three-dimensional structures, grounded in a profound understanding of the underlying chemistry.' She was, in the words of former colleague Lord Phillips, not just a scientist's scientist but 'a crystallographer's crystallographer'.

In 1964, having been proposed twice before, Dorothy won the Nobel Prize for chemistry for 'her determinations by X-ray techniques of the structures of important biochemical substances'. She was only the third woman to have won a Nobel Prize for chemistry after Marie Curie (*see entry*) and her daughter Irène Joliot-Curie, and the fifth woman to win any science Nobel Prize. As of 2019, she was the only British woman scientist to have been awarded a Nobel Prize in any of the three sciences it recognises.

In 1965, Dorothy was the second woman in sixty years, after Florence Nightingale, to be appointed to the Order of Merit (OM) by a British monarch. Although she is said to have 'disliked titles' and had reportedly told her husband that she would refuse the title of Dame Commander of the British Empire (DBE) if offered one, she regarded the OM as 'rather different really'. Established in 1902 by Edward VII to recognise distinguished service, admission into the order remains the personal gift of the sovereign, and is restricted to a maximum of twenty-four living recipients from the Commonwealth. Dorothy also accepted the first 'freedom of Beccles', an honour 'hastily invented for her by the town in which she spent her schooldays'.

Dorothy was the president of the International Union of Crystallography between 1972 and 1975, and from 1970 to 1988 served as the chancellor of the University of Bristol. In 1976, she won the prestigious Copley Medal which, since 1731, has been presented by the Royal Society for exceptional achievements in research in any branch of science. Her award read, 'in recognition of her outstanding work on the structures of complex molecules, particularly Penicillin, vitamin B12 and insulin'. As at 2019, she was still the only woman to receive this award. Between 1977 and 1983, Dorothy was a Fellow of Wolfson College at Oxford. In 1977, she was conferred the title of Professor Emeritus by Oxford University. In addition to her scientific work, Dorothy remained actively dedicated to the cause of world peace throughout her life, her political sensibilities encouraged from an early age by her mother. In 1957, she was a founder of the Pugwash Conference on Science and World Affairs, an international organisation bringing together scholars and public figures to work towards reducing the danger of armed conflict and to seek solutions to global security threats.

On 29 July 1994, Dorothy Crowfoot Hodgkin died of a stroke, aged eighty-four, at her home at Shipston-on-Stour, in Warwickshire, central England. Her husband Thomas had passed away on 25 March 1982. She was honoured posthumously, in August 1996, as one of five 'Women of Achievement' selected to appear on a set of British postage stamps. All but Dorothy were DBEs. Of a set of stamps issued in 2010 to celebrate the 350th anniversary of the Royal Society, she was the only woman included.

There is an asteroid, Hodgkin, named as a tribute to her and the Royal Society awards the Dorothy Hodgkin Fellowship in her honour for outstanding scientists at an early stage of their research. There are also buildings at the University of York, Bristol University and Keele University named after her, along with the science block at her former school, John Leman High School. Since 1999, the Oxford International Women's Festival has presented an annual memorial lecture, usually in March, to celebrate her extraordinary achievements and, marking the fiftieth anniversary of her winning the Nobel Prize, 2014 was designated the 'International Year of Crystallography'.

In his obituary of Dorothy, molecular biologist, Nobel laureate and Copley Medal winner Max Perutz wrote,

> She pursued her crystallographic studies, not for the sake of honours, but because that was what she liked to do. There was a magic about her person. She had no enemies, not even among those whose scientific theories she demolished or whose political view she opposed. Just as her X-ray cameras bared the intrinsic beauty beneath the rough surface of things, so the warmth and gentleness of her approach to people uncovered in everyone, even the most hardened scientific crook, some hidden kernel of goodness.

Erna Schneider Hoover

Born: 19 June 1926 in Irvington, New
Jersey, USA
Field: Mathematics
Awards: National Inventors Hall of Fame (2008)

Erna Schneider was born on 19 June 1926, in Irvington, New Jersey. Her father was a dentist and mother a former teacher. She had a younger brother who died of polio when he was five. Erna and her younger siblings were raised in South Orange, where she attended public school. In her youth, she enjoyed swimming, sailing and canoeing in the Adirondacks. Developing a keen interest in science at an early age, her inspiration was the biography of Marie Curie (*see entry,*) which argued that, despite the prevailing views of the time about gender roles, women could succeed.

Erna attended the prestigious Wellesley College in Massachusetts from which she graduated with honours and a Bachelor of Arts in medieval history in 1948. She was a member of *Phi Beta Kappa*, the oldest academic honour society in the USA, as well as a Durant Scholar, a recognition of Wellesley's highest achieving students. In this period, most female university graduates studied education, English or foreign languages, but Erna attended Yale University where she enrolled for a PhD in philosophy and the foundations of mathematics. Her dissertation, a study of logic, was titled 'An Analysis of Contrary-to-Fact Conditional Sentences' and in 1951 she became one of the five per cent of PhDs in mathematics in the USA that were awarded to women at that time.

Erna's first teaching role after graduation was as a professor in philosophy and logic at the private liberal arts institution, Swarthmore College, in Pennsylvania, in 1951. In 1953, Erna married Charles Wilson Hoover Jr, who encouraged her in her academic career, but she was unable to obtain tenure as a married woman. In 1954, she resigned to relocate to New Jersey when Charles found employment there. It was then that Erna immediately took a role as a senior technical associate for Bell Laboratories, which had just begun investigating the development of electronic switching systems.

These were for use not only by the public telephone systems but also for private business exchange systems and their aim was to improve the ability to take a greater number of calls. Call centres had a particular problem if they received a large number of calls in a short amount of time, overloading the unreliable electronic relays, causing the entire system to come to a halt.

Promoted in 1956, the following year Erna enrolled in Bell Laboratories internal training programme, considered to be the equivalent of a master's degree in computer science. The symbolic logic included in her dissertation had applications in switching circuit design and her knowledge of feedback theory and analysing traffic and pattern statistics enabled her to design the stored programme control. Using computer programming, she was able to apply her theory to eliminate the danger of overload in processing calls and to facilitate more effective service during peak calling times. As she said in an interview in 2008, 'To my mind it was kind of common sense ... I designed the executive programme for handling situations when there are too many calls, to keep it operating efficiently without hanging up on itself. Basically it was designed to keep the machine from throwing up its hands and going berserk.'

After applying in 1967, in 1971 she became one of the first women in the USA to receive a software patent when she was issued Patent No. 3,623,007 for a computerised telephone switching system, 'Feedback Control Monitor for Stored Program Data Processing System'. Prior to her discovery, most businesses relied on hardwired or mechanical switching systems that could not cope with many simultaneous incoming calls.

Erna's revolutionary switching system was the first reliable device to use a computer, including transistor circuits and memory-stored control. She reportedly developed the initial design while she was in hospital for the birth of her second child and later the company's lawyers who were managing the patent application had to go to her home so that she could sign the papers while she was on maternity leave. During her time at Bell Laboratories, Erna also developed computer programmes designed to ensure that outdoor telephone lines remained in good working order. In addition, her research into radar control programmes of the Safeguard Anti-Ballistic Missile System, designed to intercept intercontinental ballistic missile warheads, was a factor in the ending of the Cold War. In 1978, she was the first woman promoted to the role of Technical Department Head at Bell Laboratories. Her group applied artificial intelligence techniques, large databases and transaction systems to support telephone operations through the use of computer software. The first stored programme control system was installed in a private business in 1963, and by 1983 some 1,800 of the switching systems were serving 53 million subscriber lines. She was heavily involved in a number of high-level applications, including artificial intelligence methods, large databases, and transactional software to support large telephone networks.

In 1987, at the age of sixty-one, Erna retired from Bell Laboratories and began promoting the importance of K–12 (kindergarten to 12th grade) education and highlighting the lack of women in teaching science disciplines.

She developed the first 'Conference for the Expanding Your Horizons' programme, in association with the American Association of University Women and Girl Scouts of America. In 1990, Erna received the Wellesley College Alumni Achievement Award, recognising her as a 'champion of affordable, quality public education'. From 1980, she served as chairperson of the Trenton State College Board of Trustees. The president of the college described Erna as 'a tenacious, energetic leader,' and credited her with much of the college's progress, particularly in the area of attracting and supporting female professors and staff. In 2008, she was inducted into the National Inventors Hall of Fame, which recognises 'the enduring legacies of exceptional US patent holders', whose influence 'can inspire the future and motivate the next great innovator to transform our world'.

In 2020, after her time on their board, the College of New Jersey awarded Erna an honorary degree for services to higher education in New Jersey. She was also the recipient of the National Center for Women & Information Technology's Pioneer Award in 2023. By any measure, Erna is an inspiration for women in the field of computer technology. On 19 June 2023, she turned ninety-seven.

Grace Brewster Murray Hopper

Born: 9 December 1906 in New York City, New York, USA
Died: 1 January 1992 (aged 85)
Field: Technology and Computing
Awards: American Campaign Medal (1944); World War II Victory Medal (1945); US Naval Reserve Medal (1953); National Defense Service Medal (1953 and 1966); Armed Services Reserve Medal (1963, 1973 and 1983); Society of Women Engineers Achievement Award (1964); Legion of Merit (1967); Meritorious Service Medal (1980); Defense Distinguished Service Medal (1986); Presidential Medal of Freedom (post., 2016) and many others

Grace Brewster Murray was born on 9 December 1906 in New York City, New York, USA. Her father, Walter Fletcher Murray, and mother, Mary Campbell (Van Horne) Murray, were of Scottish and Dutch descent. Grace's great-grandfather Alexander Wilson Russell was an admiral in the US Navy and fought in the Battle of Mobile Bay during the American Civil War. Grace was the eldest of three children

Grace's preparatory school education was at Plainfield, New Jersey. Her mother, Mary, had developed an interest in mathematics as a young woman, but was unable to study anything beyond geometry as it was not considered appropriate for a woman of her day. As a result, she encouraged Grace not to limit her horizons based simply on her gender. In this she was supported by Walter, who gave an equal education to their two daughters, Grace and Mary, and their son Roger. Grace was initially rejected for early admission to Vassar College at sixteen as her test scores in Latin were not high enough. However, she was admitted the following year, graduating in 1928 with a bachelor's degree in mathematics and physics. She was admitted to *Phi Beta Kappa*, the oldest US academic honour society. In 1930, she earned a master's degree in mathematics at Yale University and married New York University professor Vincent Foster Hopper.

Grace began teaching mathematics at Vassar in 1931, and in 1934 she obtained her PhD under the supervision of the Norwegian mathematician Øystein Ore. Her dissertation, 'New Types of Irreducibility Criteria', was published that year.

Grace tried to enlist in the US Navy early in the Second World War, but was rejected as, at thirty-four, she was considered too old. Moreover, her weight to height ratio was too low. Her job as a mathematician and mathematics professor at Vassar College was also valuable to the war effort. In 1941, she was promoted to the rank of associate professor. In 1943, she obtained leave from Vassar and joined the Navy Reserve to serve in the all-female division, Women Accepted for Volunteer Emergency Service (WAVES), which had been formed in 1942. After obtaining an exemption to enlist, as she was 15 lbs (6.8 kg) below the Navy minimum weight of 120 lbs (54 kg), she trained at the Naval Reserve Midshipmen's School at Smith College in Northampton, Massachusetts.

In 1944, Grace topped her class and was assigned as a lieutenant, junior grade, to the Bureau of Ships Computation Project at Harvard University, where she served on the programming staff of the Mark I computer. Also known as the 'Automatic Sequence Controlled Calculator', it weighed over 5 tons (4,500 kg) and was the first large-scale automatic calculator and a precursor of electronic computers. The team was led by American physicist and computing pioneer Howard H. Aiken, with whom Grace co-wrote three papers. After a moth got into the circuits of Mark I, Grace coined the term 'bug' to refer to unexplained computer failures. (The remains of the moth have been retained in the group's log book at the Smithsonian Institution's National Museum of American History in Washington, D.C.)

At the end of the Second World War in 1945, Grace and Vincent divorced. She then requested a transfer to the regular Navy, but was denied due to her age, as she was now thirty-eight. She continued to serve in the Navy Reserve and remained at the Harvard Computation Lab until 1949, turning down a full professorship at Vassar in favour of working as a research fellow at Harvard under a Navy contract.

Grace became one of the first programmers in computing history, believing that programming languages should be as easily understood as English. Consequently, she was highly influential in the development of one of the first programming languages called COBOL (COmmon Business-Oriented Language), which became one of the most widespread languages in the business of its time.

In 1949, Grace joined the Eckert-Mauchly Computer Corporation, where she designed an improved compiler that translated a programmer's instructions into computer codes. She remained working there when it was taken over by Remington Rand in 1951 and by Sperry Rand Corporation in 1955. During 1957, her division developed 'Flow-Matic', the first English-language data-processing compiler.

In 1964, Grace was awarded the Society of Women Engineers Achievement Award, the Society's highest honour, 'In recognition of her significant contributions to the burgeoning computer industry as an engineering manager and originator of automatic programming systems'.

On 31 December 1966, aged sixty, Grace retired from the Navy with the rank of Commander but was recalled to active duty in August the following year to help standardise the Navy's computer languages. She retired again in 1971, but was again recalled to active duty in 1972, attaining the rank of Captain in August 1973 and Commodore in December 1983. On 8 November 1985, at the age of seventy-eight, she was made a Rear Admiral (lower half). By age seventy-nine, she was the oldest officer on active US naval duty when she retired for the final time on 31 August 1986.

Throughout her illustrious career Grace won many awards, including the inaugural Data Processing Management Association Man of the Year award (now called the Distinguished Information Sciences Award). In 1969, the annual Grace Murray Hopper Award for Outstanding Young Computer Professionals was established by the Association for Computing Machinery in her honour.

In 1973, she became the first American, and the first woman of any nationality, to be made a Distinguished Fellow of the British Computer Society, and in 1986, upon her retirement, she was the first woman to be awarded the Defense Distinguished Service Medal.

On 1 January 1992, Grace Hopper passed away in Arlington, Virginia, at the age of eighty-five. She never remarried after her divorce in 1945, nor did she have any children. In 1994, the Grace Hopper Celebration of Women in Computing was founded, and has since been organised annually by the Anita Borg Institute for Women and Technology and the Association for Computing Machinery.

In 1996, the USS *Hopper* (DDG-70) was launched, nicknamed 'Amazing Grace' – one of the very few US military vessels named after women. The Department of Energy's National Energy Research Scientific Computing Center named its flagship system 'Hopper' in 2009, the same year that the Office of Naval Intelligence created the Grace Hopper Information Services Center. On 22 November 2016, she was posthumously awarded a Presidential Medal of Freedom for her accomplishments in the field of computer science. A year later, Hopper College at Yale University was named in her honour. During her lifetime she was awarded over thirty honorary degrees from universities worldwide.

Grace Hopper had some memorable quotes, including:

The most important thing I've accomplished, other than building the compiler, is training young people. They come to me, you know, and say, 'Do you think we can do this?' I say, 'Try it'. And I back 'em up. They need that. I keep track of them as they get older and I stir 'em up at intervals so they don't forget to take chances.

She also declared, 'A ship in port is safe; but that is not what ships are built for. Sail out to sea and do new things.' Grace Hopper was one of the outstanding pioneers of computing and left a legacy that stands to this day and her nickname of 'Amazing Grace' is a testament to the way she improved technology and brought the world of computers to such great heights.

Margaret Lindsay Murray Huggins

Born: 14 August 1848 in Dublin, Republic of Ireland
Died: 24 March 1915 (aged 66)
Field: Astronomy

Margaret Lindsay Murray was born on 14 August 1848 in Dublin, Ireland, one of two children to Helen (Lindsay) Murray and solicitor John Murray. John had trained at King's Inns in Dublin and then established his own legal practice there. The family lived in Dun Laoghaire, County Dublin. In 1857, when Margaret was only nine years old, her mother passed away. Her father later remarried, and his second wife, Elizabeth (Pott), gave birth to two sons and a daughter.

Much of Margaret's early education took place privately at her home, and included art, classics, literature, languages and music. She also spent some time at a boarding school in Brighton, England, and was greatly influenced by her Scottish grandfather, Robert Murray, who was a wealthy bank officer at the Bank of Ireland, but also had expertise in astronomy. As a result, Margaret also developed a keen interest in the subject. From a young age, he took her outdoors in the afternoons and evenings to teach her about the constellations and how to identify them, and it was on such occasions that she was inspired to be an astronomer. She studied sunspots from the age of ten.

In 1873, in a sustained effort to educate herself, Margaret read the religion magazine *Good Words*, which often published items on general subjects as well as science. It was there that she discovered an article by the president of the British Association for the Advancement of Science (now the British Science Association) about the recent work done by the renowned astronomical spectroscopist William Huggins. Irish astronomer and historian of science Mary Brück wrote that even before Margaret met William in person 'she was an ardent admirer' of the man who would become her husband. Margaret eventually met William, 'one of the towering figures of his age', through a mutual acquaintance, Howard Grubb, a Dublin telescope maker. The couple were married at the Monkstown Parish Church on 8 September 1875. They immediately devoted themselves to their research, resulting in a number of important astronomical findings.

Margaret worked alongside her husband at an observatory at Tulse Hill, initially being documented as his assistant, but later investigation of their observatory notebooks revealed that she also conducted many of her own research projects. Historian of science Barbara J. Becker wrote that: 'She was more than an able assistant, amanuensis and illustrator, whose work conformed to her husband's research interests. Her very presence and expertise both strengthened and shaped the Tulse Hill research agenda.' Theirs was a truly collaborative partnership. They were the first to observe and identify the series of hydrogen lines in the spectrum of the star Vega (*Alpha Lyrae*). Their detailed notes led to their first paper, in 1889, discussing the studies of the 'spectra of planets'. This was followed by a fundamental study of the Orion Nebula. They were also among those who observed the nova of 1892, *Nova Aurigae*. Margaret took charge of the visual observations, while they both collected photographic spectra over several nights. Their research work placed them at the forefront of astronomical spectroscopy.

Margaret is especially known for her investigation of the spectrum of the Orion nebula. She and William were able to show that some nebulae are full of stars while others, such as the Orion nebula, consist completely of dust and gases. In 1889, Margaret, by now also a pioneer in the field of spectroscopy, and William co-authored *An Atlas of Representative Stellar Spectra*. Their work had a major influence on their daily lives and their home resembled a workplace rather than a family residence. In 1903 they published their final piece of joint scientific research, on the spectra of certain radioactive substances.

Astonishingly, Margaret received no formal training in astronomy, rather relying on her own reading and research of popular astronomy books, including Sir John Herschel's *Outlines of Astronomy*, published in 1878. In her spare time she had developed a hobby in photography, and was later influential in prompting William's successful programme in photographic research, soon rising to the forefront of spectroscopic astrophotography.

In 1890, the British Astronomical Association was founded to provide an opportunity for amateur astronomers who were dissatisfied with the restrictions of the Royal Astronomical Society, as women were ineligible for fellowship. Margaret supported the new association and was one of four women elected to the council of forty-eight.

In 1897, William was created a Knight Commander of the Order of the Bath. The citation was 'for the great contributions which, with the collaboration of his gifted wife, he had made to the new science of astrophysics'. As Mary Brück wrote, 'Those honoured were all men; this reference to Margaret, now Lady Huggins, made her the only woman even remotely mentioned in the honours list.' She was granted honorary membership of the Royal Astronomical Society in 1903. Only three women had been so honoured before in its history, since 1828, including Caroline

Herschel (*see entry*) and Mary Somerville (*see entry*). William was president from 1900 to 1905. Reflecting on this period, Margaret wrote, 'I thank God fervently for the success and usefulness of the last five years, I had the responsibility of them; and I have been Sir William's sole Privy Counsellor, all through, I am aware that the [Royal Society] has been good enough to do me honour. It recognises me as Sir William's faithful and sole, I trust, pretty capable assistant.'

Barbara Becker said that Margaret 'deliberately cloaked herself in the invisible garb of the proper Victorian lady, taking care that her collaborative assistance did not contradict or interfere with the image she had helped to create of her husband as innovator and principal observer.'

They worked together closely for three decades. In their retirement, William and Margaret collected and edited their scientific papers, which were published in 1909, with Margaret also providing pen drawings to illustrate them. William passed away the following year and, after thirty years of dedication to science, Margaret now felt that her best work was behind her. She planned to write William's biography, but it did not eventuate as she fell ill and underwent various surgeries and spent some time in hospital.

Margaret was a great admirer of the achievements of American women in the academic world and was an avid supporter of women's education. She was also an expert on the history of early musical instruments, even writing a book on the violinmaker Magnini, a predecessor of makers such as Stradivari. Margaret also developed expertise on the history of astrolabes and armillaries, writing articles on these for the *Encyclopaedia Britannica*'s 11th edition.

She died on 24 March 1915 at the age of sixty-six and was cremated, her ashes being laid next to William's at Gold Green Cemetery. In her will, she directed that a memorial was to be erected in St Paul's Cathedral, London, in William's honour. This memorial consists of a pair of medallions which are inscribed, 'William Huggins, astronomer 1824–1910', and the other, 'Margaret Lindsay Huggins, 1848–1915, his wife and fellow worker'. They had no children. In 1997, a plaque was installed to mark the house she grew up in, in Monkstown, Dublin.

Margaret Huggins' obituary in the London *Times* mentioned that the noted English astronomer Richard Proctor referred to her as the 'Herschel of the Spectroscope', in deference to the renowned astronomers Sir William and Sir John Herschel. Margaret bequeathed to Wellesley College in Massachusetts, USA, and their Whitin Observatory, some of her astronomy collection including cherished astronomical artefacts. Her relationship with William is said to be one of the most successful husband and wife partnerships in the whole of astronomy.

Frances Betty Sarnat Hugle

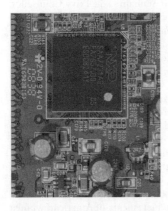

Born: 13 August 1927 in Chicago, Illinois, USA
Died: 24 May 1968 (aged 40)
Field: Engineer, scientist and inventor

Frances Sarnat was born on 13 August 1927 in New York to Nahum 'Nathan' Sarnat (Sarnatzky) and Lilyan (Steinfeld) Sarnatzky, both immigrants. She attended Hyde Park High School, being the first girl to win the Wilson Jr College Math Tournament, in which she represented her school at age sixteen in 1944. After graduating from secondary school, Frances enrolled in the Bachelor of Philosophy programme at the University of Chicago, earning her degree in philosophy and chemistry in 1946, after just two years and at the age of eighteen. She then entered the University of California Medical School, but left there in 1947.

In 1947, during her studies, she married fellow student, William ('Bill') Bell Hugle, and the couple lived in Chicago where together they founded several research and development companies. These included Hyco-Ames Labs, where she assumed the role of director of research and developed materials, processes and specialised equipment that would pave the way for her future investigations. The company was set up in her parents' apartment and focused on developing gem-quality star sapphires and rubies. Although never producing the gems, they concentrated on funding and developing the necessary equipment. Frances designed and built a completely automatic Verneuil furnace, 6 cubic feet large, that could reach the required temperature of 3,600 degrees Fahrenheit (approximately 1,982 degrees Celsius). In 1948, they secured financial support from New York lawyer John G. Broady and Hyco-Ames became Stuart Laboratories, where she worked from October 1949 until February 1951.

In March 1951, Frances accepted a position with the Standard Electronics Research Corporation, where she worked on classified material. In August 1952 she moved to the Baldwin Piano Company, which was interested in the use of transistors in their electronic organs. Frances was the advanced research engineer at Baldwin, with her husband Bill as her supervisor. It was there that they became prolific inventors, filing numerous patent applications, some of which Baldwin formally filed and others that did not come to fruition in Baldwin's engineering division. Among the patents filed in 1956 and 1957 were methods for producing semi-conductive films and printed circuits. In 1957, the University of Chicago awarded her a Bachelor of Science degree in chemistry, based upon the coursework she had completed between 1944 and 1947.

In 1959, Frances and Bill both commenced work at Westinghouse Company in Pittsburgh, Pennsylvania, and the following year the organisation transferred them to southern California to set up an astro-electronics laboratory. In 1960, Frances received a Master of Science degree from the University of Cincinnati.

In late 1961, the couple moved to the Laurelwood Subdivision in Santa Clara, in the San Francisco Bay Area, where they co-founded Siliconix in 1962, with Frances developing its first products and becoming the first director of research and chief engineer. In 1964, they left Siliconix and developed products for two more semiconductor companies, including Hugle Industries and Stewart Warner Microcircuits, where Frances again served as director of research and as chief engineer.

While Bill was both an inventor and an entrepreneur, Frances preferred the engineering side of the business, although she was not always popular with many male engineers who resented having a woman as a supervisor or even a colleague. She is reported as saying, 'I am a woman and an engineer; I am not a woman engineer'; and resented the notion that her gender defined the type of engineer she was.

The Hugles were awarded at least seventeen patents. Frances was a significant pioneer in the development of Silicon Valley, with one of her most notable solo patents being her 1966 process for Automated Packaging of Semiconductors (granted in 1969 after her death) and the development of TAB (tape-automated bonding) for the first time, allowing the miniaturisation that is used today in thousands of products from hearing aids to personal computers. She has also been identified as a pioneer in early flip chip technology (a method for interconnecting semiconductor devices).

Frances was politically and socially active, being a founder of the first Headstart programme in the Santa Clara Valley and protested against the Vietnam War. In January 1966, while an adjunct professor of chemistry at Santa Clara University, she was the moderator for a Conference on 'Abortion and social science'. Frances also played an active role in the Unitarian Church. Her hobbies included hiking and camping in Yosemite, and surfing in Santa Cruz.

On 24 May 1968, Frances Hugle died at her home from stomach cancer, at the age of forty. She is buried at Beverly Cemetery, Blue Island, Cook County, Illinois, USA. She was survived by husband Bill (who passed away in 2003) and their four children, Margaret, Cheryl, David and Linda. The Women in Engineering Committee of the Institute of Electrical and Electronics Engineers offers a scholarship in her name, the Frances B. Hugle Memorial Scholarship, honouring 'her many significant engineering accomplishments, and to help provide the resources for female engineers to follow in her footsteps', for a female student who has completed two years of undergraduate study in an engineering curriculum.

Shirley Ann Jackson

Born: 5 August 1946 in Washington, D.C., USA
Field: Technology
Awards: Candace Award for Technology (1982); National Women's Hall of Fame (1998); Richtmyer Memorial Award (American Association of Physics Teachers 2001); Vannevar Bush Award (2007); National Medal of Science (2014)

Shirley Jackson was born 5 August 1946 in Washington, D.C., USA, the second daughter of Beatrice (Cosby) Jackson, a social worker, and George Hiter Jackson, a postal worker. Her parents recognised the value of education and her father encouraged her talent for science by becoming involved in her projects. At Roosevelt Senior High School, after performing well on an IQ test in sixth grade, she was placed on her school's honours track. She then attended accelerated programmes in mathematics and science, completing high school in 1964 as the top student.

The same year, she began at Massachusetts Institute of Technology (MIT) as one of fewer than twenty African American students in her 900-member freshman class, and the only one studying theoretical physics. She later said that she was unprepared for the loneliness, telling *Science* magazine, 'The irony is that the white girls weren't particularly working with me, either.' She said, 'I had to work alone. I went through a down period, but at some level you have to decide you will persist in what you're doing and that you won't let people beat you down.'

Despite this, Shirley developed a love of materials science. While still a student, she volunteered at Boston City Hospital and tutored students in Boston's African American neighbourhood of Roxbury at the YMCA. In 1968, she graduated with a Bachelor of Science degree after writing a thesis on solid-state physics. In the same year, she was offered fellowship support to stay on for her PhD, with a thesis on theoretical elementary particle physics. Her doctoral supervisor was James Young, the first full-time tenured black professor in the physics department.

Shirley obtained her PhD in 1973, the first African American woman at MIT to do so in any academic field. She was also the second African American woman in the USA to earn a doctorate in physics. Aware of her own position, she lobbied MIT to admit more minorities. During the 1970s, she undertook postdoctoral research into subatomic particles at several physics laboratories in the USA and Europe. The first of these positions was as a research associate at the Fermi National Accelerator Laboratory in Batavia, Illinois (Fermilab) where she studied hadrons (composite particles made of quarks).

In 1974, Shirley became a visiting scientist at the European Organisation for Nuclear Research (CERN) in Switzerland, where she examined theories of strongly interacting elementary particles. In 1976 and 1977, she lectured at the Stanford Linear Accelerator Center in physics and was a visiting scientist at the Aspen Center for Physics. Part of her research surrounded the Landau-Ginsburg theories of charge density waves in layered compounds, and she also studied two-dimensional Yang-Mills gauge theories and neutrino reactions. Describing her interests, Shirley said, 'I am interested in the electronic, optical, magnetic, and transport properties of novel semiconductor systems. Of special interest are the behaviour of magnetic polarons in semi-magnetic and dilute magnetic semiconductors, and the optical response properties of semiconductor quantum-wells and superlattices. My interests also include quantum dots, mesoscopic systems, and the role of antiferromagnetic fluctuations in correlated 2D electron systems.'

In 1976, Shirley joined the Theoretical Physics Research Department at AT&T Bell Laboratories, New Jersey, where she examined the fundamental properties of various materials and how they might be used in the semiconductor industry. Two years later, she joined the Scattering and Low Energy Physics Research Department. In 1988, moved to the Solid State and Quantum Physics Research Department. At Bell Laboratories, she examined the optical and electronic properties of two-dimensional and quasi-two dimensional systems.

Shirley served on the faculty at Rutgers University in Piscataway and New Brunswick, New Jersey, from 1991 to 1995, in addition to her consulting work with Bell Laboratories on semiconductor theory. She also concentrated her research on the electronic and optical properties of two-dimensional systems. In 1995, US President Bill Clinton appointed her to serve as chairman of the US Nuclear Regulatory Commission (NRC), becoming the first woman and first African American to hold that position. In this role she had 'ultimate authority for all NRC functions pertaining to an emergency involving an NRC licensee'. She also assisted in the establishment of the International Nuclear Regulators Association.

In 1998, Shirley was inducted into the National Women's Hall of Fame for 'her significant contributions as a distinguished scientist and advocate for education, science, and public policy'. On 1 July 1999, she became the eighteenth president of Rensselaer Polytechnic Institute (RPI), being the first woman and first African American to hold this position. Since her appointment, she has helped raise over $1 billion USD in donations for philanthropic causes. In 2001, she received the Richtmyer Memorial Award, given annually by the American Association of Physics Teachers, in recognition of those who have not only produced important physics research, but also acted as physics educators. She has also received many honorary doctorate degrees.

Shirley has received numerous fellowships, including the Martin Marietta Aircraft Company Scholarship and Fellowship, the Prince Hall Masons Scholarship, the National Science Foundation Traineeship and a Ford Foundation Advanced Study Fellowship. She has been elected to numerous special societies, including the American Physical Society and American Philosophical Society. In spring 2007, she received the Vannevar Bush Award for 'a lifetime of achievements in scientific research, education and senior statesman-like contributions to public policy'. Established in 1980 in memory of Vannevar Bush, who served as a science adviser to President Franklin Roosevelt during Second World War, the award honours truly exceptional lifelong leaders in science and technology who have made substantial contributions to the welfare of the USA through public service activities in science, technology, and public policy.

Shirley Jackson was appointed an International Fellow of the Royal Academy of Engineering (FREng) in 2012, and in 2014 was named recipient of the National Medal of Science, as well as winning the CIBA-GEIGY Exceptional Black Scientist Award. The National Medal of Science was established in 1959 as a Presidential Award to be given to individuals 'deserving of special recognition by reason of their outstanding contributions to knowledge in the physical, biological, mathematical, or engineering sciences'.

Shirley is married to Morris A. Washington, a physics professor at Rensselaer Polytechnic Institute, and has one son, Alan, a Dartmouth College alumnus. She is a member of *Delta Sigma Theta*, a sorority of college-educated women dedicated to public service with an emphasis on programmes that target the African American community.

Shirley and her husband Morris were named to the inaugural class of the Capital Region Philanthropy Hall of Fame in 2019, and in the same year the American Physical Society Forum on Physics and Society awarded her the Joseph A. Burton Forum Award. In 2021, she was presented with the Hans Christian Oersted Medal from the American Association of Physics Teachers, and also in 2021 received, from the UC Berkeley Academic Senate, the Clark Kerr Award for distinguished leadership in higher education.

Sophia Louisa Jex-Blake

Born: 21 January 1840 in Hastings, England, UK
Died: 7 January 1912 (aged 71)
Field: Medicine

Sophia Louisa Jex-Blake was born on 21 January 1840 at 3 Croft Place, Hastings, England, the youngest daughter of Mary (Cubitt) Jex-Blake and Thomas Jex-Blake, a retired lawyer and proctor of Doctors' Commons. She was home-schooled until the age of eight, after which she attended private schools in southern England. Against her parents' wishes, she enrolled at Queen's College, London, qualifying as a teacher. From 1859 to 1861, she tutored mathematics to younger girls, although her father would not allow her to accept a salary as he was appalled at the idea of his daughter working for a living. In 1862, Sophia travelled to Europe and filled a temporary teaching position at the Grand Ducal Institute in Mannheim.

In 1865, Sophia went to the USA to learn more about the education of women. There she met the activist for health and social reform Dr Lucy Sewall, resident doctor at the New England Hospital for Women. This meeting, as well as the time that Sophia spent as an assistant at the hospital, influenced her decision to become a doctor. Harvard would not admit her, as a woman, and before she was able to enrol at a new medical school being set up by Elizabeth Blackwell (*see entry*) in New York, Sophia had to return to the UK on the death of her father. In 1869, she published an essay, 'Medicine as a profession for women', concerning her observations in the USA, in a book edited by Josephine Butler, *Women's Work and Women's Culture*.

No English medical school would accept women, but Sophia pressed her case in Scotland in 1869. Although the Faculty and Academic Senate supported her admission, it was overturned by the University Court, on the basis that the University could not make the necessary arrangements 'in the interest of one lady'. Undaunted, she advertised in *The Scotsman* newspaper, which resulted in six other women joining her cause. Collectively known as the 'Edinburgh Seven', the University Court approved their admission, which made the University of Edinburgh the first university in Britain to admit women. It is reported that she wrote to Lucy Sewall: 'It is a grand thing to enter the very first British University ever opened to women, isn't it?'

The experience of the women was not straightforward. The angry response they evoked when they went to take an examination and their entry to the hall was blocked by a mob of hostile students became known as 'the riot at Surgeons' Hall'. The resulting publicity turned public opinion in the women's favour and, in 1871, Sophia brought a libel action against a member of the university staff whom she accused of starting the riot. As a result, she was awarded a farthing in damages but was left with a legal bill of nearly £1,000. But opposition in the medical profession also became more entrenched and, for two years, the women were excluded from the Edinburgh Royal Infirmary. Then, in January 1872, the University Court decided that even if the women completed the course and successfully passed all the examinations, they could not be granted medical degrees. Sophia took the case to the Scottish court, which initially declared the university's decision invalid, but in early 1873 the appeal court not only supported the university but held that it had never had the power to accept women students. Sophia's group stayed in Edinburgh for another year, gaining what experience they could in the wards of the infirmary.

In March 1874, Sophia took her career plans to London. Her first step was to found the London School of Medicine for Women. With the support of a number of eminent medical men as lecturers, Sophia acted as Secretary of the school, but was also one of its students. When the school opened on 12 October 1874 it had fourteen students, including the Edinburgh women. But it took the support of Members of Parliament to champion a Private Member's Bill, which passed on 11 August 1876, to enable women to qualify in medicine to overcome the resistance they continued to encounter.

Before Sophia took her medical examinations in Ireland, she tested herself by sitting medical examinations in Switzerland. In January 1877, she was awarded an MD at Bern. Four months later, she passed the examination in Dublin, qualifying as Licentiate of the King's and Queen's College of Physicians of Ireland. She became only the third woman in Britain registered with the General Medical Council and the first practising doctor in Scotland. The same year, she was able to secure a hospital affiliation for the London School of Medicine for Women, which meant that British women has access to comprehensive medical training and registration.

In the 1880s, Sophia practised medicine privately in Edinburgh. She founded the Edinburgh Hospital and Dispensary for Women and Children and, in 1886, the Edinburgh School of Medicine for Women. In 1894, the University of Edinburgh admitted women to graduate in medicine.

In 1899, Sophia retired from active work and her hospital moved into her former residence, being renamed Bruntsfield Hospital. She moved to a small dairy farm, 'Windydene', in Rotherfield in rural Sussex, with her close friend, the schoolteacher and author Dr Margaret Todd, nineteen years her junior. Although her heart was failing, she enjoyed entertaining her many friends and continued to campaign for women's suffrage until her death on 7 January 1912. She was buried at St Denys' churchyard, Rotherfield. Margaret Todd later wrote of her friend:

> She was impulsive, she made mistakes and would do so to the end of her life: her naturally hasty temper and imperious disposition had been chastened indeed, but the chastening fire had been far too fierce to produce perfection ... But there was another side to the picture after all. Many of those who regretted and criticised details were yet forced to bow before the big transparent honesty, the fine unflinching consistency of her life.

Sophia Jex-Blake was an outstanding pioneer who fought hard for the rights of women to practise medicine. To honour her commitment, the University of Edinburgh displays a plaque near the entrance to the medical school describing her as a

> Physician, pioneer of medical education for women in Britain, alumnus of the University.

Nalini Joshi

Born: Rangoon, Myanmar (formerly Burma) in 1958
Field: Mathematics
Awards: Fellow of the Australian Academy of Science (2008); Georgina Sweet Australian Laureate Fellow (since 2012); AFR and Westpac top 100 Australian Women of Influence (2015); Officer of the Order of Australia (2016); Eureka Award for Outstanding Mentoring of Young Researchers (2018)

Nalini Joshi was born in Rangoon, Myanmar (formerly Burma) in 1958 and, for the most part, had a happy childhood in the 1960s, declaring that 'what I remember is the wonderful games, the beautiful people, but always feeling a little bit of an outsider because I wasn't part of the dominant group'. Her father, Vinoy, a medical practitioner, was conscripted into the army and posted to the remote Shan States in the west of the country. Nalini recalled that she grew up near jungles with wild animals, but was free to explore as long as she went to school. And so she lived as a child in the 'Golden Triangle', the area where the borders of Thailand, Laos, and Myanmar meet, which was, at the time, one of the biggest opium-producing areas in the world. The winters were cold and they pasted newspapers on the walls to stop the wind. School lessons meant sitting on the floor and writing on small slate boards with chalk. Although her mother, Usha, feared and hated it, Nalini loved that experience.

Vinoy and Usha were both determined to live in a place that would give their children a suitable education and chance of a good life. Nalini said of her father, 'He told me that I used to top most of my classes in primary school, but I never got the top prizes at the end of the year. So he knew that there was something going wrong and he wanted to make sure that his family was in a place that would be better for their future.'

Also, partly to escape the political upheaval in Burma, in 1971 Nalini and her family migrated to Australia, where she attended Fort Street Girls' High School in Sydney. She proved to be an excellent student with an insatiable appetite for reading, devouring numerous books that eventually led to an interest in science fiction and mathematics. At one stage, she wanted to be an astronaut, and discovered that she could study astronomy through mathematics. This was the catalyst for her love of mathematics, later stating that,

> Maths is the underlying framework that allows us to describe everything that we observe in the universe, whether it's outside us or inside us. It's the only logical language that allows us to compare, analyse and check evidence.

Meanwhile, her father took a position at Marrickville District Hospital, after which he went into general medical practice.

After completing high school, Nalini attended the University of Sydney, where she obtained a Bachelor of Science with first class honours and the university medal in applied mathematics in 1982, after which she enrolled at Princeton University in the USA. There she graduated with a master's degree and PhD in computational and applied mathematics in 1987, with her thesis titled, 'The connection problem for the first and the second Painlevé transcendents'. She was only the second woman to obtain a PhD in applied mathematics at Princeton.

On 21 July 1984, while at Princeton, Nalini married the oceanographer Robert Gardiner-Garden and they later had a son and daughter. In 1987, she commenced an appointment as a postdoctoral research fellow and lecturer at the Australian National University in Canberra in the Australian Capital Territory, after which she obtained a lectureship at the University of New South Wales. It was there that she became a founding member of the Women in Research Committee (WIRC), which began new schemes to enable staff with caring responsibilities to attend conferences. She was promoted to senior lecturer, but resigned to move to Adelaide in 1996, due to her husband's employment there.

Nalini was awarded a prestigious Australian Research Council (ARC) Senior Research Fellowship at the University of Adelaide, which she held until 2002 when she returned to the University of Sydney to become the first female professor of mathematics. There, during 2007–2010, Nalini was the first woman to be head of the School of Mathematics and Statistics. She was also director of the Centre for Mathematical Biology between 2006 and 2013.

Nalini has established an international reputation for her outstanding research on the mathematical structure of nonlinear systems. Her results have provided new complex analytical and geometric techniques for understanding their solutions. Her interests lie in non-linear differential and difference equations, with emphasis on integrable systems such as the Painlevé equations.

Nalini has also become a passionate advocate for gender equality in science, remarking that, 'we're losing fifty per cent of the talent group that we have naturally available to us if we say that women shouldn't be answering these questions. That's just a no-brainer.' In 2015, Nalini was recognised as one of the Australian Financial Review and Westpac's 100 'Women of Influence' in Australia. These awards aim 'to increase the visibility of women's leadership in Australia, highlighting the important contribution women make in creating Australia's future' and the awardees are chosen 'to promote Australian women who have dedicated their time and energy to help and encourage other women in their industry, and who are fighting for change every day'. In being selected, Nalini said, 'I hope that this recognition will enable me to spread my message about mathematics more widely in Australia and elsewhere. I would like to tell everyone how human mathematics is. It is not an esoteric and elitist pursuit, but a beautiful creation of the human mind, which has turned out to be useful in all walks of life.'

In 2015, she travelled around the UK from June to July, delivering nine lectures as the special 2015 Hardy Fellow of the London Mathematical Society, its 150th annual Fellow. The following year, in the 2016 Queen's Birthday honours awards, Nalini was appointed an Officer of the Order of Australia, one of Australia's highest civilian honours, 'for distinguished service to mathematical science and tertiary education as an academic, author and researcher, to professional societies, and as a role model and mentor of young mathematicians'.

Nalini Joshi has held a professorship in Applied Mathematics since 2002, and in December 2017 was awarded an inaugural Ruby Payne-Scott (*see entry*) Professorial Distinction. She has supervised numerous PhD and honours students, with several prominent international mathematicians undertaking collaborative research with her. She was recognised in 2018 as an 'Outstanding Mentor of Young Researchers' in the prestigious Australian Museum Eureka Prizes, which reward excellence in the fields including scientific research and innovation, science leadership and science communication.

Nalini was elected as vice president of the International Mathematical Union (2019–2022) and has held major leadership roles in Australian mathematics. Nalini was a member of the Commonwealth Science Council (2014–2018), chair of the Australian Academy of Science National Committee for Mathematical Sciences (NCMS) (2011–2016) and president of the Australian Mathematical Society (AustMS) (2008–2010).

In October 2019, Nalani was recognised in the NSW Premier's Prizes for 'Excellence in Mathematics, Earth Sciences, Chemistry or Physics'. In the following year she was awarded the George Szekeres Medal, given for outstanding research contributions over a fifteen-year period, from the Australian Mathematical Society. Nalani was also presented with the 2021 ANZIAM Medal by Australia and New Zealand Industrial and Applied Mathematics for 'unparalleled contributions to applied mathematics in leadership, gender equity, and promotion of mathematics'.

The encouragement, support and retention of talented women in science is an issue close to Nalini's heart. She hopes that her journey will inspire others and she has certainly been very successful in that endeavour.

Helen Dean King

Born: 27 September 1869 in Owego, New York, USA
Died: 7 March 1955 (aged 86)
Field: Biology
Award: Ellen Richards Prize (shared with Annie Jump Cannon, 1932)

Helen Dean King was born on 27 September 1869 in Owego, New York, to Lenora (Dean) King and William A. King, a wealthy businessman and president of the King Harness Company. Her grandfather was the Reverend William H. King, who preached for temperance and abolition. She attended Owego Free Academy, as her father had. At the age of eighteen, she entered Vassar College and received her Bachelor of Arts degree in 1892. In 1894, she returned to Vassar as a graduate student in biology and assistant demonstrator in the biology laboratory. In 1895, she enrolled in graduate studies at Bryn Mawr College in Pennsylvania, studying under the supervision of the evolutionary biologist and geneticist Thomas Hunt Morgan (who later won a Nobel Prize in 1933). It was he who suggested the topic of her thesis, 'The Maturation and Fertilization of the Egg of *Bufo lentiginosus*' (the common toad). She received a scholarship to do postgraduate work in biology and palaeontology and pursue doctoral studies with Morgan and the geologist Florence Bascom (*see entry*). Helen graduated with a PhD in 1899 and published her dissertation in 1901. Where Morgan said that her 'long and laborious piece of investigation ... promises to give results', Florence reported that 'her work has been done faithfully'. From 1899 to 1907, Helen taught science at Miss Florence Baldwin's School for Girls.

Between 1906 and 1908, Helen had the role of University Fellow for research in zoology at the University of Pennsylvania. Then, in 1908, Helen obtained a position without pay at the Wistar Institute of Anatomy and Biology in Philadelphia as a volunteer assistant, to assist in the institute's technical work, but also allowing her to pursue her own research. Within a year, she was appointed to the staff as an 'Assistant in Anatomy'. Morgan continued as her mentor, encouraging independent research and suggesting new problems.

Her interest in sexual fertilisation, the maturation of the egg and the determination of sex demonstrated the influence of Morgan and the

Entwinklungsmechanik (development mechanics) school. Helen held the view that embryological questions could be solved only through rigorous experimentation and a mechanistic interpretation.

In her work on toads' eggs, Helen was able to increase the proportion of females by slightly drying the eggs or by withdrawing water from them by placing them in solutions of salts, acids, or sugars. In other research, she found that hybridising altered the sex ratio, as did the age of the female, suggesting that the genetic material of cells undergoes changes due to age.

In her work on sex determination, Helen stressed the influence of environmental factors on altering the sex ratio. She believed that there are two kinds of spermatozoa, those carrying the accessory element (the Y chromosome) and those which do not. She opposed the prevailing theory that any egg is capable of fertilisation by any spermatozoon that happens to come in contact with it. She persisted in her theory that ova exercise a kind of selection, accepting one kind of sperm and rejecting the other kind, on the basis of different environmental conditions.

Helen's work on amphibians anticipated her life-long interest in sex ratios and sex determination, as well as her awareness of the impact of the environment in determining such characteristics. In moving to the Wistar Institute, she encountered people who were involved in mammalian genetics and interested in eugenics. This institute was known for its success in breeding a genetically homogeneous laboratory animal, known as the 'Wistar Rat', which the institute claimed was the first animal to be 'standardised' for laboratory use.

In 1909, Helen changed the subject of her research from amphibians to rats. Studying the inbreeding of albino rats, Helen found that she could select for the tendency to produce an excess of males or an excess of females. She concluded that, in rats, the sex ratio is somewhat amenable to selection and that the female has more influence in determining sex than does the male. She suggested that factors such as heredity, environment and nutrition act on ova in such a way as to render them more easily fertilised by one kind of sperm than by the other.

She began a project of animal breeding when she mated two males and two females from a single litter of albino rats. Helen continued this dynasty for more than 139 generations, to develop what would later be called the 'King colony', a part of the famous Wistar Institute stock of white rats that figured in innumerable research projects throughout the world. In 1919, she wrote, 'In the female element, as in the female organism, resides the power to select that which is for the best interests of the species.'

By the mid-1930s, the Wistar Institute rats had become the world's most widely used laboratory animal, leading to the WISTARAT trademark in 1942. By the 1960s, as the focus of the institute continued to evolve, its rat breeding stocks were sold off along with the rights to the trademark, with more than half of the current albino strains used today. With an estimated 117 types, they are believed to have derived from a single Wistar Rat.

The first female scientist to work at the Wistar Institute, Helen remained there until her retirement in 1950, becoming a full professor of embryology in 1927 and a member of Wistar's advisory board in 1928. She served as vice president of the American Society of Zoologists in 1937 and was associate editor of the *Journal of Morphology and Physiology* from 1924 to 1927 and editor of the Wistar Institute's bibliography service from 1922 to 1935.

Helen King passed away on 7 March 1955, aged eighty-six, in Philadelphia, Pennsylvania, having never married and outliving most of her colleagues. She was laid to rest at Evergreen cemetery, Owego, Tioga County, New York. Her Memorial ID is 111073780. Science historian Marilyn Bailey Ogilvie wrote that the 'brief obituary' in the Minutes of the Board of Managers was 'a slim tribute to the woman who had dedicated her life to her science and to the Wistar Institute'.

In her honour, the Wistar Institute now offers the annual Helen Dean King Award in her honour, that 'Recognizes Outstanding Women in Biomedical Research', including a lecture by a prestigious scholar 'to create an opportunity to talk about the persistent challenges for increasing participation of women in the field'.

Maria Margaretha Winckelman Kirch

Born: 25 February 1670 in Panitzsch near Leipzig, Electorate of Saxony (Germany)
Died: 29 December 1720 (aged 50)
Field: Astronomy
Award: Gold medal of Royal Academy of Sciences (Berlin, 1709)

Maria Margaretha Winckelmann was born on 25 February 1670 in Panitzsch, a village about 6 miles (10 km) to the east of Leipzig, in Saxony (now Germany). Her father was a Lutheran minister who home-schooled Maria, believing that she deserved the same education as boys. By the age of thirteen, both of her parents had passed away so Maria's schooling was then taken over by one of her uncles. She had also been receiving a general education from her brother-in-law Justinus Toellner, along with the eminent self-taught astronomer Christoph Arnold (known as the 'astronomical peasant'), who worked as a farmer in Sommerfeld, about 4½ miles (7 km) from her home.

Before long, Maria developed her own interest in astronomy. From an early age, she studied with Arnold, becoming his unofficial apprentice and later his assistant. She lived with him and his family, but the astronomy she learned was somewhat disorganised. It was through Arnold that Maria met the renowned German mathematician and astronomer Gottfried Kirch, some thirty years her senior. They fell in love and married in 1692, having their first child Christfried in 1694 and their second, Christine, in 1696. Gottfried, with his three sisters, had produced calendars that contained critical astronomical information, including the phases of the Moon, the times of sunrise and sunset, and the positions of the planets. Maria joined him in this work and, from 1697, they also began recording weather information, the data they collected proving vital to navigation. Their calendars were sold by the Royal Berlin Academy of Sciences. (Part of the astronomical calendar that they developed is pictured above.)

In 1700, Gottfried was offered the post of Astronomer Royal at the court of the elector of Brandenburg, Frederick III, in Berlin. A new observatory was constructed there but it was not completed until 1711 so in the meantime, the couple worked at the private observatory of the amateur astronomer Bernhard Friedrich, Baron von Krosigk. Maria and Gottfried worked as a team and took turns observing the heavens. At the time, women were not allowed to attend university in Germany, so Maria's astronomical work took place mainly outside the universities. Maria became one of the few women active in astronomy in the eighteenth century and she was largely known by the nickname 'Kirchin', the feminine version of her husband's name.

During her time at the academy, Maria observed the heavens every evening, starting at 9.00 p.m. It was on one of these nights, on 21 April 1702, that she discovered the so-called 'Comet of 1702' (C/1702 H1). Unfortunately, her husband took the credit for it, recording in his notes for that night,

> Early in the morning (about 2:00 a.m.) the sky was clear and starry. Some nights before, I had observed a variable star and my wife (as I slept) wanted to find and see it for herself. In so doing, she found a comet in the sky. At which time she woke me, and I found that it was indeed a comet ... I was surprised that I had not seen it the night before.

It was not until 1710, the year that he died, that he finally admitted the truth, and she was then deemed to be the first woman to have discovered a comet. In 1707, Maria published her observations of the *aurora borealis*, and in 1709 she published a paper, '*Von der Conjunction der Sonne des Saturni und der Venus*' ('On the Conjunction of the Sun, Saturn and Venus').

When Gottfried passed away, Maria asked the Royal Berlin Academy of Sciences that she and her son Christfried should be allowed to continue producing calendars. Maria made special note that, while her husband was ill, it was she who did all the necessary work. The president of the academy, mathematician Gottfried Leibniz, was the only one who supported her petition. It was rejected because other academy members considered that a woman producing its calendar would be an embarrassment. She spent the following eighteen months petitioning the royal court for the position, receiving a final rejection in 1712. An inexperienced astronomer, Johann Heinrich Hoffmann, was appointed Astronomer Royal with the responsibility of producing calendars instead. Expressing her disappointment, Maria declared, 'Now I go through a severe desert, and because ... water is scarce ... the taste is bitter.'

In 1712, Maria moved to Von Krosigk's observatory and in the same year wrote a paper about an upcoming conjunction of Jupiter and Saturn in 1714. Following Von Krosigk's death in 1714, she became an assistant to a mathematician in the town of Danzig and, with Christfried, took over Hevelius Observatory there (named after the famous seventeenth century astronomer Johannes Hevelius) at the request of his family.

Hoffmann died in 1712, and Christfried assumed the position of Astronomer Royal, employing his mother and two of his sisters, Christine and Margaretha, as assistants. The following year, the academy reprimanded

Maria for having too high a profile, especially being 'too visible at the observatory when strangers visit'. She was ordered to 'retire to the background and leave the talking to ... her son'. As she refused to comply, the academy forced her to relinquish her house on the observatory grounds.

Maria Kirch then continued working in private. She died of a fever on 29 December 1720 in Berlin at the age of fifty. After her death, Christfried, who never married, and his sisters continued the tradition of the Kirch family's astronomical contributions. He lived with his three sisters for twenty years until his death from a heart attack in 1740.

Margaret Galland Kivelson

Born: 21 October 1928 in New York City, New
York, USA
Field: Space Physics
Awards: Guggenheim Fellowship
(1973–74); Radcliffe Graduate Society
Medal (1983); Harvard University's 350th
Anniversary Alumni Medal (1986); NASA
Group Achievement Award (1995, 1996);
Fellow of the National Academy of Sciences
(1999); Fleming Medal of the American
Geophysical Union (2005); Alfven Medal of
the European Geophysical Union (2005);
Gerard P. Kuiper Prize of the American
Astronomical Society (2017)

Margaret Galland was born on 21 October 1928 in New York City to a
father who was a physician with a keen interest in physics and a mother who
had an undergraduate degree in physics from an institute where Max Planck
and Albert Einstein, each Nobel Prize winners, were both on the faculty.
Margaret was raised in New York City, in Manhattan, recalling that she
learned about the names of the planets while in grade school, before there
were any spacecraft.

While attending the Horace Mann High School for Girls, Galland aimed
to follow a career in science but was unsure about the job prospects. Her
uncle advised her to become a dietitian, sensing that a physical science career
as a woman would be difficult. However, after her secondary education she
disregarded his suggestion and, in 1946, enrolled in the physics programme
at Radcliffe College, the women's college of Harvard University. She obtained
her bachelor's degree in 1950, with a major in quantum electrodynamics, and
a master's degree in 1952. In 1957, she was awarded her PhD in physics under
the supervision of Julian Schwinger, with a thesis titled, 'Bremsstrahlung of
High Energy Electrons'.

At the time it was a rarity for a woman to study physics, with only one
in forty as part of her undergraduate class and one in sixty in postgraduate
studies being female students. So much so that when she began her first
degree her family teased her that she was really pursuing a 'Mrs' degree.
Before the Second World War, courses at Radcliffe were segregated from
the courses at Harvard. In her first Radcliffe/Harvard subjects after the war,
classes did not return to being separated by gender and she was often the
only woman in her lectures.

While still at college, aged twenty, Margaret married Daniel Kivelson,
later a 1958 Guggenheim Fellow in Physics. Their first child, Steven Allan

Kivelson, was born in May 1954 and their second, Valerie Kivelson, in 1957. Between 1955 and 1971, Margaret worked as a physics consultant at the RAND Corporation in Santa Monica, California, where she researched the interactions of plasmas and electron gases. With her colleague Don F. DuBois, she derived a correction to Lev Landau's relation for the damping excitations of unmagnetised plasma. Landau was a Soviet physicist, who won the 1962 Nobel Prize in Physics.

During 1965 and 1966, she took leave from RAND to join Daniel on his sabbatical leave in Boston and, through a Fellowship from the Radcliffe Institute for Advanced Study, she was able to conduct scientific research in a university setting at Harvard and the Massachusetts Institute of Technology (MIT). In 1967, Margaret joined the University of California, Los Angeles (UCLA), as an assistant research geophysicist, soon progressing through the ranks within the geophysics and space physics departments, becoming a full professor at UCLA's Department of Earth and Space Sciences in 1980.

In 1973, Margaret won a Guggenheim Fellowship to work at the Imperial College in London, stating,

> That fellowship gave me for the first time the sense that I was being taken seriously as a scientist. More than money, it gave me status and increased my self-confidence considerably.

From 1977 to 1983, she was a member of the Board of Overseers at Harvard College and served as chair of the Department of Earth and Space Sciences (1984–87 and 1999–2000), as well as with NASA's Advisory Council (1987–93), the National Research Council's Committee on Solar-Terrestrial Research (1989–92) and co-chaired the UCLA Academic Faculty Senate's Committee on Gender Equality issues (1998–2000). In 2009, Margaret became a Distinguished Professor of Space Physics, Emerita.

From 2010, and concurrently with her appointment at UCLA, Margaret has been a research scientist and scholar at the University of Michigan, her primary research interests including the magnetospheres of Earth, Jupiter, and Saturn, especially Jupiter's Galilean moons. She has published over 350 research papers and is co-editor, with Christopher T. Russell, of a widely used textbook, *Introduction to Space Physics*, which first appeared in 1995, published by Cambridge University Press. Her husband Daniel passed away in 1978 at the age of seventy-three. He was a distinguished UCLA chemistry professor and outstanding teacher who joined the faculty in 1955, chairing the department from 1975–78. The couple had been avid collectors of Turkoman rugs and lent pieces from their collection to shows at the Textile Museum in Washington, D.C. and the Los Angeles County Museum of Art.

During an interview on her life and its biggest surprises, Margaret stated,

> The biggest surprises of course were finding a planetary magnetic field at Jupiter's moon Ganymede and then finding evidence of a global scale ocean on Europa. That was at a time when nobody was aware that there might be liquid water (on these moons). People had talked about

it but there was no evidence of water in a liquid state on any of the moons of Jupiter or Saturn at that time. That was really a very exciting discovery.

She had headed the proposal to put the magnetometer on the American unmanned *Galileo* spacecraft and became principal investigator of the magnetometer. Launched in 1995, *Galileo* captured data in Jupiter's magnetosphere for eight years. It was destroyed when it entered Jupiter's atmosphere on 21 September 2003. Margaret was also a co-investigator on both the Fluxgate Magnetometer of the earth-orbiting NASA-ESA Cluster mission and NASA's Themis mission, as a member of the Cassini magnetometer team, and as a participant in the magnetometer team for the European JUICE (JUpiter ICy moons Explorer) mission to Jupiter.

In 2017, Margaret Kivelson was awarded the annual Gerard P. Kuiper Prize of the American Astronomical Society for outstanding lifetime achievement in the field of planetary science.

Margaret's son Steven is an American theoretical physicist, renowned for several major contributions to condensed matter physics and is currently the Prabhu Goel Family Professor at Stanford University. Her daughter Valerie is the Arthur F. Thurnau, Professor of History at the University of Michigan.

In a talk she gave at Radcliffe College, Harvard University in 1997, Margaret said,

> Creative science has much in common with other creative arts: painting, sculpture, poetry, music. A creative scientist applies tools familiar to the artist, demanding solutions with underlying symmetry, often deliberately broken to produce special effects, demanding elegance, valuing minimalism, and then typically supplementing these constraints with the rigor of mathematical analysis. ... Many scientists share a belief-structure that could well be accepted by contemporary painters and choreographers.

In 2019, Margaret received the Gold Medal of the Royal Astronomical Society, their highest possible honour. In 2020 she was elected as a Foreign Member of the Royal Society, and in the following year was awarded the James Clerk Maxwell Prize for Plasma Physics of the American Physical Society for 'groundbreaking discoveries in space plasma physics and for seminal theoretical contributions to understanding space plasma processes and magnetohydrodynamics'.

Sofia Vasilyevna Korvin-Krukovskaya Kovalevskaya

Born: 15 January 1850 in Moscow, Russian Empire
Died: 10 February 1891 (aged 41)
Field: Mathematics and physics

The second of three children, Sofia Vasilyevna Korvin-Krukovskaya was born in Moscow on 15 January 1850. Her mother, Yelizaveta (Schubert) Fedorovna, was a descendant of a family of German immigrants and her father, Lieutenant General Vasily Vasilyevich Korvin-Krukovsky, a member of the minor nobility, was of Russian-Polish descent and served in the Imperial Russian Army as head of the Moscow artillery. In 1858, when Sofia was eight, he retired to his Krukovsky family estate in Palibino, Vitebsk province.

In her early years, Sofia has a sound education through a Polish private tutor, Y. T. Malevich, and, after displaying an original flair for mathematics, she transferred to a renowned advocate for women's rights, Alexander N. Strannoliubskii, who taught her calculus. He was a mathematics professor at the naval academy and immediately recognised her talent and potential as a mathematician. Although her flair for mathematics was apparent, as a woman she could not attend university in Russia. In order to study in another country, she needed permission from her father, or husband if she had one. As a result, in 1868 she entered a sham marriage to Vladimir Kovalevsky, a young radical palaeontology student who was also a book publisher and the first to translate the works of Charles Darwin in Russia.

They left Russia, spending a brief period in Vienna where Sofia attended lectures in physics at the university, after which, in 1869, the couple moved from Russia to Germany. In April that year, they moved to Heidelberg, Germany, where after some effort she attended courses in mathematics and physics. During this time, Vladimir enrolled in the University of Jena to obtain a doctorate in palaeontology.

After completing a doctoral course in mechanics at Heidelberg, in April 1869 Sofia visited London, where she was invited to attend George Eliot's 'Sunday Salons'. At nineteen, she was introduced to Herbert Spencer and was led, at Eliot's urging, into a debate on 'a woman's capacity for abstract thought', as well as participating in social movements and shared ideas of utopian socialism.

Moving to Berlin in 1870, as a woman she was not even allowed to audit classes, but managed to take private lessons from Karl Weierstrass (known as the 'father of modern analysis'). He was so impressed with her skills that over the ensuing three years he tutored her in the same material he covered in his university lectures. After a brief trip with Vladimir to Paris in 1871, to assist her sister Anyuta, she returned to Berlin and, in 1874, presented three papers to the University of Göttingen as her doctoral dissertation, on partial differential equations, the dynamics of the rings of Saturn and elliptic integrals. With the support of Weierstrass and being able to bypass the normal lectures and examinations, she was awarded a PhD in mathematics, *summa cum laude* (with highest distinction). In doing so she became the first woman to hold that degree, and in her paper on differential equations she included what is now known as the 'Cauchy-Kovalevskaya theorem', which gives conditions to a certain class of those equations.

By the early 1880s, Sofia and Vladimir were in financial difficulty and, although she offered to lecture at the university for free, she was not allowed to do so as a woman. The couple started a house-building enterprise, but as mortgage rates rose in 1879, they went bankrupt. Soon after, they obtained a position helping neighbours electrify street lights, thereby easing their financial woes. They later returned to Russia but could not obtain professorships due to their radical beliefs so they headed back to Germany, where by now Vladimir suffered from severe mood swings. Nevertheless, they decided to get formally married and had a daughter, also named Sofia but known as 'Fufa'. Just one year later, Fufa was placed in the care of Sofia's older sister so Sofia could continue her mathematical work. She also left her husband. In 1883, Vladimir's mental condition worsened and, when it seemed he could be facing prosecution over his role in a commercial swindle, he committed suicide.

During the same year, Sofia secured a position as a *privat-docent* or private teacher at the University of Stockholm, Sweden, where she met the eminent mathematician Magnus Gösta Mittag-Leffler through his sister, the actress, novelist and playwright Anne Charlotte Mittag-Leffler. The two women formed a lifelong friendship. In 1884, Sofia was appointed as a *Professor Extraordinarius* (Professor without Chair) for five years at Stockholm University and became the editor of the journal *Acta Mathematica*. Four years later, Sofia was awarded the *Prix Bordin* (Bordin Prize) of the French Academy of Science for her memoirs, *On the Rotation of a Solid Body About a Fixed Point*, which included the discovery of what is now known as 'Kovalevskaya top'. The judges considered her work so outstanding that they raised the prize money from 3,000 to 5,000 francs.

In 1889, Sofia was appointed a *Professor Ordinarius* (Professor with Chair) at Stockholm University, in doing so becoming the first woman to hold such a position at a university in northern Europe. After a change in the Academy's rules, and much lobbying, she was granted a chair in the Russian Academy of Sciences, although at no stage was she offered a role as a professor in Russia.

In addition to her mathematical works, Sofia wrote literature, including a memoir, *A Russian Childhood*, a partly autobiographical novel, *Nihilist Girl*, in 1890 and plays co-authored with Anne Charlotte Mittag-Leffler. Sofia was intimately involved in the feminist movement of late 19th century Russian nihilism, this posing a risk in her life that was already under strain by the difficulties of a stressed and underpaid career, as well as exile from a number of European countries.

After returning from a vacation in Genoa, as a result of influenza complicated by pneumonia, Sofia Vasilyevna Kovalevskaya died at the age of forty-one on 10 February 1891. She was buried in Solna, Sweden, at *Norra begravningsplatsen* (the Northern Cemetery). The lunar crater Kovalevsky is named in her honour, as are streets in Saint Petersburg and Moscow. The Alexander von Humboldt Foundation of Germany grants a biennial Sofia Vasilyevna Kovalevskaya Award to promising young researchers.

In a letter she wrote in 1890, she said,

> It seems to me that the poet must see what others do not see, must see more deeply than other people. And the mathematician must do the same.

Doris Kuhlmann-Wilsdorf

Born: 22 February 1922 in Bremen, Germany
Died: 25 March 2010 (aged 88)
Field: Metallurgy, tribology
Awards: Medal for Excellence in Research of
the American Society of Engineering Education
(1965 and 1966); Heyn Medal of the German
Society of Materials Science (1988); Society of
Women Engineers Achievement Award (1989);
Ragnar Helm Scientific Achievement Award
of the Institute of Electrical and Electronics
Engineers (1991); Christopher J. Henderson
Inventor of the Year (2001); Fellow of TMS-AIME
(2006)

Doris Kuhlmann was born on 22 February 1922 in Bremen, Germany, to
Adolph Friedrich and Elsa (Dreyer) Kuhlmann. Doris worked as an apprentice
metallographer and materials tester for two years before she entered the
University of Göttingen in 1942. It was then that she developed her love of
science. Doris completed her undergraduate work in metallurgy in 1944, earning
her master's degree in physics two years later. In 1947, she completed her PhD in
materials science. She then continued her research under the physicist Sir Nevill
Francis Mott (who later won a Nobel Prize in 1977) at the University of Bristol.

On 4 January 1950, Doris married Heinz G. F. Wilsdorf, a talented
experimentalist whom she had met at the University of Göttingen. She then
adopted the surname of Kuhlmann-Wilsdorf. Soon after their wedding, they
moved to Johannesburg, South Africa. Heinz was employed as a scientist at the
Council for Scientific and Industrial Research in Pretoria and Doris became a
lecturer in physics at the University of the Witwatersrand. In 1955, she received
an honorary Doctor of Science degree from the University of the Witwatersrand.

In 1956, after the birth of their two children, Gabriele and Michael, the
family moved to the USA. Heinz accepted a position at the Franklin Institute
in Philadelphia, Pennsylvania, while Doris joined the faculty of the University
of Pennsylvania. As part of a dual-career couple, it is reported that Doris
encountered some resentment and sometimes overt discrimination from the
all-male faculty, but in time she came to love the university and, in 1961,
became a full professor.

In 1963, Doris became the first female professor of engineering physics at
the University of Virginia. Heinz was invited to chair the newly established
Department of Materials Science. In 1966, Doris was promoted to the
position of University Professor of Applied Science, the highest academic
rank at the university. She held that position for forty years. However,

during that period, she was devastated by the deaths of her two much-loved children, Gabriele in 1969 and Michael in 1979.

Doris was internationally recognised for her ground-breaking work in plastic deformation, surface physics, and crystal defects, developing a unified theory of plasticity for dislocation behaviour. She discovered why aluminum sheets only crinkle under pressure, whereas those made from other metals break. A major achievement was her development of metal fibre brushes for use as sliding electrical contacts in electric motors, for which she was granted six patents. She also published more than 300 scientific papers and started two companies.

Doris received many honours and awards. In 1988, she was awarded the Heyn Medal from the German Society for Materials Science, its highest accolade, for her work on the theory of metal deformation. In 1991, the Institute of Electrical and Electronics Engineers awarded her the Ragnar Holm Scientific Achievement Award, which recognises contributions in the field of electrical contacts, and in 2001 she won the Christopher J. Henderson Inventor of the Year Award of the University of Virginia, in recognition of her research relating to electrical brushes. In the same year, Gregory H. Olsen, a former student of the Department of Materials Science and Engineering at the University of Virginia, provided most of the funding for a new materials science building to be named in honour of Doris and Heinz.

In 1994, Doris and Heinz funded a professorship in their name at the University of Virginia. In 2004, Doris received an honorary doctorate from the University of Pretoria, and she was also an active member of the Society of Women Engineers. She led university-wide discussions on evolution, apartheid, and honesty.

Doris retired in 2005, following which she taught a university seminar course in science and religion, continued her research, and participated in the 'Semester at Sea' programme. Doris had a thirst for knowledge and a need to understand the world around her and the way it worked, combined with a dynamic personality and ability to persuade others of her views. These led to her many successes and widespread recognition. In 2006, she was made a Fellow of the Minerals, Metals and Materials Society and the American Institute of Mining, Metallurgical and Petroleum Engineers (TMS-AIME).

After Heinz suffered a stroke, Doris devoted herself to his care, designing a rope-and-pulley system to ease his transition into the water-therapy spa. He died in 2000. On 25 March 2010, Doris passed away in Charlottesville, Virginia, at age eighty-eight, after a short illness.

In a memorial tribute written about her by the National Academy of Engineering, she was said to have 'a keen mind and could leapfrog in her thinking so much that it was hard to keep up with her', and the South African physicist Frank R. N. Nabarro spoke of her 'uncanny ability to be right for no apparent reason'. It was also said that 'everyone was her extended family' and that 'her friendships were enduring and interactions frequent', accompanied by 'a wide friendly smile'. Despite the death of her children, and her husband, Doris remained 'warm, friendly, and intellectually engaged throughout her life and touched many people with her charm and wit'.

Doris Kuhlmann-Wilsdorf's papers are held at the Albert and Shirley Small Special Collections Library at the University of Virginia.

Stephanie Louise Kwolek

Born: 31 July 1923 in New Kensington,
Pittsburgh, Pennsylvania, USA
Died: 18 June 2014 (aged 90)
Field: Chemistry
Awards: Chemical Pioneer Award (1980);
Lavoisier Medal (DuPont, 1995); National
Inventors Hall of Fame (1995); National
Medal of Technology (1996); the IRI
Achievement Award (1996); Perkin Medal
(American Chemical Society (1997);
National Women's Hall of Fame (2003)

Stephanie Louise Kwolek was born on 31 July 1923 in the suburb of New
Kensington, Pittsburgh, Pennsylvania, USA, to Polish immigrant parents.
Her mother was Nellie (Zajdel) Kwolek and her father, John Kwolek (Jan
Chwałek). He was a naturalist by profession and as a young child Stephanie
spent hours with him exploring the natural world. He died when Stephanie
was ten years old. She credited her pursuit of science to him and her interest
in fashion to her mother.

In 1946, Stephanie graduated with a Bachelor of Science degree with a
major in chemistry from Margaret Morrison Carnegie College, Carnegie
Mellon University, a private research university based in Pittsburgh. She had
planned to become a doctor and worked in a job in a chemistry-related field
to save enough money to attend medical school.

The chemist (William) Hale Charch offered Stephanie a job at DuPont's
Buffalo, New York, facility in 1946. Stephanie only intended to work
there temporarily, but found it so interesting that, in 1950, she decided to
remain, moving with DuPont to Wilmington, Delaware. In 1959, she won
a publication award from the American Chemical Society (ACS) for her
paper, 'The Nylon Rope Trick', demonstrating a way of producing nylon in a
beaker at room temperature. It is still a common classroom experiment and
the process was later extended to high molecular weight polyamides.

In 1964, with a petrol shortage looming, Stephanie's group began
searching for a lightweight, yet strong, fibre to be used in tyres. The polymers
she had been working with at the time formed liquid crystal while in solution
that then had to be melt-spun at over 200 degrees Celsius, producing weaker
and less stiff fibres. A unique technique in her new projects and the melt-
condensation polymerisation process was to reduce those temperatures to
between 0 and 40 degrees Celsius. In 1965, she invented poly-paraphenylene

terephthalamide, which became known as 'Kevlar', the strong and lightweight fibre that is used in bulletproof vests, among other safety applications.

In a speech in 1993, Stephanie explained that 'the solution was unusually (low viscosity), turbid, stir-opalescent and buttermilk in appearance. Conventional polymer solutions are usually clear or translucent and have the viscosity of molasses, more or less. The solution that I prepared looked like a dispersion but was totally filterable through a fine pore filter. This was a liquid crystalline solution, but I did not know it at the time.' Although solutions such as this were usually discarded, she persuaded a technician, Charles Smullen, who ran the spinneret, a device designed to determine the durability of fibres, to test her solution.

She was shocked to discover that the new fibre did not break when nylon typically would, making it apparent that it was a much stronger material. Her supervisor and the laboratory director understood the significance of her discovery, which developed the field of polymer chemistry. By 1971, modern Kevlar was introduced, and she discovered that the fibres could be made even stronger by heat-treating them. Kevlar is five times stronger than steel on an equal weight basis. The polymer molecules are shaped like rods or matchsticks and are highly oriented, giving Kevlar its amazing strength. In 1980, she received the Chemical Pioneer Award from the American Institute of Chemists, established to recognize chemists or chemical engineers who have made outstanding contributions to advances in chemistry or the chemical profession.

First used commercially as a replacement for steel in racing tyres, Kevlar is used as a material in more than 200 applications, including tennis rackets, skis, parachute lines, boats, airplanes, ropes, cables, and bulletproof vests. It has also been used for car tyres, fire fighter boots, hockey sticks, cut-resistant gloves and armoured cars, and for protective building wear like bomb-proof materials, hurricane safe rooms, and bridge reinforcements.

In 1985, Stephanie and her co-workers patented a method for preparing polybenzobisoxazole (PBO) and polybenzothiazole (PBT) polymers. In 1995, she became the fourth woman to be inducted into the National Inventors Hall of Fame. In the same year, Stephanie was awarded the DuPont Company's Lavoisier Medal for technical achievement, named in honour of Antoine Lavoisier, the 'father of modern chemistry' and a mentor to the founder of the DuPont Company. It is given to DuPont scientists and engineers who have made outstanding contributions to DuPont and their scientific fields throughout their careers. As of 2019, Stephanie is the only female employee to have received that honour. The citation paid tribute to her as a 'Persistent experimentalist and role model whose discovery of liquid crystalline polyamides led to Kevlar aramid fibers'.

In 1986, she retired as a research associate for DuPont, but in her later years acted as a consultant to them, serving on both the National Research Council and the National Academy of Sciences. During her forty years as a research scientist, she filed and received many patents.

In 1996, Stephanie received the National Medal of Technology and Innovation, America's highest honour for technological achievement, which is presented by the US President. The same year, she also received the Industrial Research Institute Achievement Award 'to honor outstanding accomplishment in individual creativity and innovation that contributes broadly to the development of industry and to the benefit of society'. In 1997, she was awarded the Perkin Medal of the American Chemical Society. Considered the highest honour in the US chemical industry, it is given to a scientist residing in America for an 'innovation in applied chemistry resulting in outstanding commercial development'. In 2003, Stephanie was inducted into the National Women's Hall of Fame.

On 18 June 2014, Stephanie Kwolek passed away at the age of ninety in Wilmington, Delaware, and was interred at St Mary's Cemetery, New Kensington, Pittsburgh. During the week of her death, the one millionth bullet-resistant vest made with Kevlar was sold. She was awarded honorary degrees by Carnegie Mellon University (2001), Worcester Polytechnic Institute (1981) and Clarkson University (1997). The Royal Society of Chemistry instituted the biennial Stephanie L. Kwolek Award in 2014 'to recognise exceptional contributions to the area of materials chemistry from a scientist working outside the UK'. She is also featured as one of the Royal Society of Chemistry's 175 Faces of Chemistry. She never married.

Hedwig Eva Maria Kiesler
'Hedy Lamarr'

Born: 9 November 1914 in Vienna, Austria-Hungary
Died: 19 January 2000 (aged 85)
Field: Technology
Awards: The Electronic Frontier Foundation Pioneer Award and the Bulbie Gnass Spirit of Achievement Bronze Award (1997) (post.); a star on the Hollywood Walk of Fame

Hedwig Eva Maria Kiesler (later known as 'Hedy Lamarr') was born on 9 November 1914 in Vienna, Austria-Hungary, the only child of Gertrud 'Trude' (Lichtwitz) Kiesler, a pianist from Budapest, and Emil Kiesler, a prominent bank director. Hedwig was raised a Christian, Trude having come from an upper-class Jewish family that had converted from Judaism to Catholicism.

In the late 1920s, Hedwig was discovered as an actress and taken to Berlin by producer Max Reinhardt. There she trained in the theatre before returning to Vienna to work in the film industry, first as a 'script girl' (a secretary to a motion-picture director who records information about the photographing of each scene, prompts actors, and writes a synopsis for advertising the movie), and then as an actress. In 1933, as a teenager, she married Friedrich ('Fritz') Mandl who was thirteen years her senior and chairman of Hirtenberger Patronen-Fabrik, a leading Austrian armaments firm founded by his father, Alexander Mandl.

According to her 1967 autobiography, *Ecstasy and Me*, Hedwig attended a convent school for girls. Her first role was as an extra in the 1930 German film *Geld auf der Straße* ('Money on the Street'). She performed well enough to be in three more German productions in 1931. However, it would be her fifth film, at age eighteen, which gained her worldwide fame. In 1933, she appeared in the German production called *Ecstasy*, the story of a young woman married to a much older man but who fell in love with a young soldier. In the movie, she is seen swimming and running nude. It made world headlines and was banned by the US government. Her husband is said to have attempted to bring a halt to her acting career in Germany and to purchase all copies of the film. However, she later described him as extremely controlling, and wrote in her autobiography that she escaped him by disguising herself as a maid and fleeing to Paris, where she obtained a divorce in 1937. The following year she had her first serious acting role in the film *Algiers*, for which she received high praise.

While travelling to London, Hedwig met Metro-Goldwyn-Mayer studio head, Louis B. Mayer, who offered her a movie contract in Hollywood, convincing her to change her name to 'Hedy Lamarr', choosing the surname in homage to the silent film star Barbara La Marr. In 1938, Mayer took her to Hollywood and promoted her as the 'world's most beautiful woman'. Hedy became a star from the late 1930s to the 1950s, her biggest success being *Samson and Delilah* in 1949. She also went on to play a number of stage roles, including a starring one in *Sissy*, a play about Austrian royalty produced in Vienna, which won accolades from critics.

Having no formal training and being largely self-taught, Hedy worked in her spare time on various hobbies and inventions. Among those who knew of her inventiveness was aviation tycoon Howard Hughes. He actively supported her interests and placed his team of science engineers at her disposal.

Sometime during the Second World War, Hedy learned about the new technology of a radio-controlled guidance system for torpedoes, using spread spectrum and frequency-hopping technology to defeat jamming by enemy powers that could send them off course. She came up with the idea of creating a frequency-hopping signal that could not be tracked or jammed and contacted her friend, composer and pianist George Antheil, to help her develop a device for doing just that. They managed to synchronise a miniaturised player-piano mechanism with radio signals. They drafted designs for the system, which was granted a patent on 11 August 1942 (filed using her then married name, Hedy Kiesler Markey). It turned out to be technologically difficult to implement, and came at a time when the US Navy was not receptive to considering inventions coming from outside the military. However, the Navy did adopt it in the 1960s, the principles of their work being arguably incorporated into Bluetooth technology with their findings similar to methods used in legacy versions of CDMA (code-division multiple access) and Wi-Fi. An updated version of this system had been first implemented during the 1962 Cuban Missile Crisis and subsequently emerged in numerous military applications on Navy ships.

It was the 'spread spectrum' technology that Hedy helped to invent that would launch the digital communications era, forming the technical backbone that makes cellular phones, fax machines and other wireless operations possible. She therefore became a pioneer in the field of wireless communications in the USA, where she became a naturalised citizen, at age thirty-eight, on 10 April 1953. The great significance of George and Hedy's invention was not realised until decades later when, in 1997, they received the Electronic Frontier Foundation Pioneer Award and the Bulbie Gnass Spirit of Achievement Bronze Award, given to individuals whose creative lifetime achievements in the arts, sciences, business, or invention fields have significantly contributed to society. Hedy was also featured on the Science Channel and the Discovery Channel, and in 2014 she and Antheil were posthumously inducted into the National Inventors Hall of Fame. For her contribution to the motion picture industry, Hedy has a star on the Hollywood Walk of Fame.

In 1981, with her eyesight failing, Hedy retreated from public life and settled in Miami Beach, Florida. In her later years, she became increasingly reclusive. Her only means of communication with the outside world became the telephone, regularly spending up to six or seven hours a day on the phone. She had also become estranged from her son, James Lamarr Loder, when he was twelve years old and they did not speak for almost fifty years. She was married and divorced six times, to Friedrich Mandl (1933–37), Gene Markey (1939–41), John Loder (1943–47), Ernest 'Ted' Stauffer (1951–52), W. Howard Lee (1953–60) and Lewis J. Boies (1963–65), who was her own divorce lawyer. Following her sixth and final divorce, Hedy remained unmarried for the final thirty-five years of her life.

Hedy Lamarr died of heart disease in Casselberry, Florida, on 19 January 2000, aged eighty-five. Her son, Anthony Loder, spread her ashes in Austria's Vienna Woods in accordance with her last wishes. In December 2014, an 'honorary grave' was created to honour her, in Vienna's Central Cemetery, Group 33 D No. 80. She has certainly earned a place among the twentieth century's most important women inventors, with a vision for technology that was well ahead of her time.

Inge Lehmann

Born: 13 May 1888 in Østerbro,
Copenhagen, Denmark
Died: 21 February 1993
(aged 104)
Field: Seismology and
geophysics
Awards: Gordon Wood Award
(1960); the Emil Wiechert
Medal (1964); the Gold Medal
of the Danish Royal Society
of Science and Letters (1965);
William Bowie Medal (1971)

Inge Lehmann was born 13 May 1888 in Østerbro, Copenhagen, Denmark, the elder of her only sibling, Harriet. Her mother, Ida Sophie (Tørsleff) Lehmann, was a housewife, and her father the experimental psychologist Alfred Georg Ludvik Lehmann. Both parents came from prominent families. Inge received her school education at a private coeducational school *Fællesskolen* ('shared school') that was relatively new, having been founded in 1893. It was led by Hanna Adler, a wealthy woman and the aunt of the future Nobel Prize winning physicist Niels Bohr. Inge later remarked that her father and Adler were the most significant influences on her intellectual development.

Founded by a woman, a feature of the school was that boys and girls were to be treated the same, studying the same subjects and taking part in the same sports and activities. As a result, the discipline was not as rigorous as in other schools and, although Inge enjoyed her time there, she did not feel intellectually challenged.

In 1906, aged eighteen, she passed the entrance examination for Copenhagen University with a first rank mark and enrolled in a mathematics degree. Her studies were interrupted by ill health and she continued them at Newnham College in Cambridge University in 1910–11. She was exhausted from all the work she had put into her studies and took a break for several years, taking the opportunity to develop valuable computational skills in an actuarial office. She resumed studies at Copenhagen University in 1918. where she completed the *Candidata magisterii* (Master of Arts) degree in physical science and mathematics in two years.

In 1923, Inge accepted a position at Copenhagen University as an assistant to the Danish mathematician Johan Frederik Steffensen, a professor of actuarial science. Two years later, she became an assistant to the head of the Royal Danish Geodetic Institute, Niels Erik Nørlund, who gave her the job of setting up seismological observatories in Denmark and Greenland. It is said that she was 'captivated' by seismology and, in 1928, at the age of forty, she earned the *Magister Scientiarum* (Master of Science) degree in geodesy. She then accepted a position at the Geodetic Institute of Denmark as State Geodesist and Head of the Department of Seismology. Her research involved improving the coordination and analysis of measurements from Europe's seismographic observatories.

In a 1936 research article titled, 'P', Inge suggested a three-shelled model of the Earth's interior, consisting of the mantle, outer core and inner core. Scientists already knew that vibrations from earthquakes travelled all the way through the earth, some as transverse waves (S-waves) and others as longitudinal waves (P-waves). The time these waves take to travel from an earthquake's epicentre to different seismic observatories around the world reveals information about the paths the waves take. Inge was the first to interpret P-wave arrivals that surprisingly appeared in the P-wave shadow of the Earth's core as reflections from an inner core. An example of this phenomenon occurred in the strong earthquake on 17 June 1929 in Murchison that was felt all over New Zealand. Within a few years, other leading seismologists of the day – including Beno Gutenberg, Charles Richter and Harold Jeffreys – adopted this conclusion. However, it was not confirmed until 1971 when computer calculations verified it.

Inge served as the chair of the Danish Geophysical Society in 1940 and 1944, and in 1950 became the president of the European Seismological Federation. In 1952, she unsuccessfully applied for a professorship at Copenhagen University in geophysics, and the following year retired from the Geodetic Institute. She then moved to the USA for several years and collaborated with geophysicists Maurice 'Doc' Ewing and Frank Press on investigations of the Earth's crust and upper mantle. Working with American seismologist Beno Gutenberg in 1954, Inge noticed the existence of a region in the Earth's upper mantle in which P-waves and S-waves abruptly travelled faster. That region, now known as the 'Lehmann discontinuity', occurs between about 118 miles (190 km) and 150 miles (240 km) below the Earth's surface. It appears beneath continents, but not usually beneath oceans. The eminent American geophysicist Francis Birch remarked that the 'Lehmann discontinuity was discovered through exacting scrutiny of seismic records by a master of a black art for which no amount of computerization is likely to be a complete substitute'.

In 1964, Inge was awarded the Emil Wiechert Medal for outstanding scientific achievements by the German Geophysical Society, named after the founder of the society and his significance for geophysics. The following year, she was awarded the Gold Medal of the Royal Danish Royal Academy of Science and Letters and was elected a Fellow of the British Royal Society in 1969. In 1971, she was awarded the William Bowie Medal of the American Geophysical Union (AGU), its highest honour named after one of the co-founders of the AGU, for her contributions to the field of geophysics, and later received the Medal of the Seismological Society of America in 1977. In 1987, at the age of ninety-nine, she wrote her last scientific article,

'Seismology in the Days of Old'. In 1988, she attended the party held for her 100th birthday at her former workplace, the Geodetic Institute.

In 1995, the AGU created the Inge Lehmann Medal in her honour, first awarded in 1997. The medal, which bears her portrait, is awarded to researchers displaying 'outstanding contributions to the understanding of the structure, composition, and dynamics of the Earth's mantle and core'.

Inge Lehmann died at age 104 on 21 February 1993, making her one of the longest-lived scientists in history. Never marrying nor having children, she left all of her possessions to the Royal Danish Academy. Throughout her long life she was very shy and did not enjoy being in the spotlight, although she accomplished so much.

Rita Levi-Montalcini

Born: 22 April 1909 in Turin, Italy
Died: 30 December 2012 (aged 103)
Field: Neurology
Awards: EMBO Membership (1974); Louisa Gross Horwitz Prize (1983); Lasker Award (1986); Albert Lasker Award for Basic Medical Research (1986); Nobel Prize in Physiology or Medicine (with Stanley Cohen 1986); National Medal of Science (1987); Foreign Member of the Royal Society (1995); Leonardo da Vinci Award (2009)

Rita Levi-Montalcini was born on 22 April 1909 in Turin, Italy, to a wealthy Sephardic Jewish family. She and her twin sister, Paola, were the youngest of four children. Her mother, Adele Montalcini, was a painter, and her father, Adamo Levi, an electrical engineer and mathematician. Their families had moved from Asti and Casale Monferrato, respectively, to Turin at the turn of the twentieth century.

By her own account, Rita had a rewarding family life, although it was a typical early twentieth century household, where her father made most of the important decisions, including the education of his children. Although Adamo had 'a great respect for women,' he initially believed that 'a professional career would interfere with the duties of a wife and mother'. He consequently decided that his three daughters should not have a profession, nor would they attend university.

By the age of twenty, Rita realised that this did not suit her and asked her father if she could undertake a profession. In the next eight months she furthered her studies in Latin, Greek and mathematics, completing high school and entering the medical school in Turin. While there, she was inspired by the neurohistologist Giuseppe Levi and his work in the developing nervous system. In 1936, she graduated *summa cum laude* (with the highest distinction) in medicine and surgery, and then enrolled in the three-year specialisation in neurology and psychiatry. She remained at the university as Levi's assistant, but her academic career was interrupted by Benito Mussolini's *Manifesto per la Difesa della Razza* (Racial Manifesto) in 1938 and the subsequent introduction of laws excluding Jews from academic and professional careers. As a result, Rita lost her position.

During the Second World War, Rita set up a laboratory in her bedroom. where she researched the growth of nerve fibres in chicken embryos, laying the foundation for much of her later research. When the Germans invaded Italy in September 1943, her family fled south to Florence where, using false identities, they survived the Holocaust. In a corner of their shared living space, she again set up a laboratory, and during the Nazi occupation she was in contact with the liberal-socialist political Action Party, her 'close, dear friends and courageous partisans'. After the liberation of Florence, in August 1944, Rita volunteered her medical expertise for the Allied health service, working as a doctor in a centre for refugees.

In 1945, her family returned to Turin, and in September the following year she was granted a one-semester research fellowship at Washington University in St Louis, Missouri, in the laboratory of Professor Viktor Hamburger. The aim was to repeat the experiments which they had performed many years earlier in the chicken embryo. That decision, she said, was to 'change the course of [her] life'. In that first year, Hamburger offered her a research associate position that she would hold for thirty years. It was during this period that she undertook her most significant work. In 1952, Rita isolated nerve growth factor from observations of certain cancerous tissues that cause extremely rapid growth of nerve cells.

By transferring pieces of tumours to chicken embryos, Rita established a mass of cells that was full of nerve fibres. She made the surprising discovery that nerves grow everywhere, like a halo, around tumour cells, describing it to be 'like rivulets of water flowing steadily over a bed of stones'. The detection of nerve growth produced by the tumour was a revolutionary result. The nerves took over areas that would become other tissues and then entered veins in the embryo. However, nerves did not grow into the arteries, which would flow from the embryo back to the tumour. Rita concluded that the tumour itself was releasing a substance that was stimulating the growth of nerves. She was made a full professor in 1958, and in 1962 established a second laboratory in Rome, dividing her time between there and St Louis.

In 1962, Rita established the Research Centre of Neurobiology in Rome, and from 1969 to 1978 she held the position of Director of the Institute of Cell Biology of the Italian National Council of Research. Upon retirement in 1979, she became Guest Professor of this institute. In 1986, Rita was awarded the Nobel Prize in Physiology or Medicine, jointly with colleague Stanley Cohen, 'for their discoveries of growth factors', in particular the nerve growth factor or 'NGF'. In 1987, she was awarded the National Medal of Science, the highest US scientific honour.

During the 1990s, she was one of the first scientists to point out the significance of the mast cell in human pathology, and in 1993 she identified the endogenous compound palmitoylethanolamide as an important modulator of this cell.

On 1 August 2001, Rita was appointed as Senator for Life by the then President of Italy, Carlo Azeglio Ciampi, honouring her significant scientific contributions, and in 2002 she founded the European Brain Research Institute, serving as its president. In 2006, she received an honorary degree in biomedical engineering from the Polytechnic University of Turin.

In 2008, she received a PhD Honoris Causa from the Complutense University of Madrid, Spain, and in 2009 the Leonardo da Vinci Award from the European Academy of Sciences, established to honour one person a year for outstanding lifelong achievements. In 2011, McGill University, Canada, conferred on her a PhD Honoris Causa, which was presented at the Sapienza University of Rome, as she was then over 100 years old.

Rita never married nor had children, declaring in a 2006 interview, 'I never had any hesitation or regrets in this sense. My life has been enriched by excellent human relations, work and interests. I have never felt lonely.' On 22 April 2009, she became the first Nobel laureate ever to reach the age of 100 and the event was honoured with a party at Rome's City Hall.

Rita Levi-Montalcini passed away in her home in Rome on 30 December 2012, at the age of 103. The then Mayor of Rome, Gianni Alemanno, stated it was a great loss 'for all of humanity', praising her as someone who represented 'civic conscience, culture and the spirit of research of our time'.

In a tribute to her on Sky TG24 TV, an Italian news channel, the notable Italian astrophysicist Margherita Hack remarked, 'She is really someone to be admired', while Italy's then Prime Minister, Mario Monti, paid homage to her 'charismatic and tenacious' character and her lifelong efforts to 'defend the battles in which she believed'. Vatican spokesman Federico Lombardi praised Rita's civil and moral efforts, declaring her to be an 'inspiring' example for Italy and the world.

In her autobiography she wrote, 'It is imperfection – not perfection – that is the end result of the program written into that formidably complex engine that is the human brain and of the influences exerted upon us by the environment, and whoever takes care of us during the long years of our physical, psychological and intellectual development.'

Barbara Jane Huberman Liskov

Born: 7 November 1939 in Los Angeles,
California, USA
Field: Technology and computer science
Awards: Fellow of the American Academy
of Arts and Sciences (1992); Fellow of the
Association for Computing Machinery
(ACM) (1996); Society of Women Engineers
Achievement Award (1996); IEEE John von
Neumann Medal (2004); A. M. Turing Award
(2008); National Inventors Hall of Fame
(2012); Computer Pioneer Award (2018, IEEE
Computer Society)

Barbara Huberman was born on 7 November 1939 in Los Angeles,
California, the eldest of the four children of Jane (Dickhoff) Huberman
and Moses Huberman. She attended the University of California, Berkeley,
where in 1961 she earned her bachelor's degree in mathematics with a
minor in physics. She was one of only two women in her classes. She then
applied to graduate programmes in mathematics at Berkeley and Princeton,
although Princeton was not accepting female mathematics students at the
time. Barbara was accepted at Berkeley, but instead chose to move to Boston
and began working at Mitre Corporation, a non-profit organisation with
a mission of 'solving problems for a safer world'. They operate multiple
federally-funded research and development centres in the USA.

It was at Mitre that Barbara became interested in computers and
programming, working there for twelve months as a member of the
technical staff. She then accepted a programming position at Harvard
University where, in 1962–63, she worked on a language translation project
before deciding to return to postgraduate studies, applying for admission
to Berkeley, Stanford and Harvard. She settled on Stanford, and in 1963
enrolled in a master's degree, which she obtained in 1965. She then entered
the doctoral programme under the supervision of the eminent computer
and cognitive scientist John McCarthy, an expert in artificial intelligence.
Her thesis topic was 'A Program to Play Chess End Games', and in 1968
she became one of the first women to be awarded a PhD from a computer
science department in the USA. In 1970, she married Nathan ('Nate') Liskov
and the couple had a son, Moses, in 1975.

After graduating from Stanford, Barbara returned to her research work at
Mitre where she stayed until 1972, when she accepted a role at Massachusetts
Institute of Technology (MIT) as an assistant professor in the Department

of Electrical Engineering and Computer Science. She led many significant projects, including the Venus operating system – a small, low-cost and interactive timesharing system. With her students, between 1974 and 1975 she designed and implemented a programming language, CLU. It introduced many features that are still widely used today and is seen as a step in the development of object-oriented programming or OOP. She was promoted to associate professor in 1976, and to full professor in 1980. Between 1986 and 1997 she was the NEC Professor of Software Science and Engineering, and since 1997 she has been the Ford Professor of Engineering.

Between 1982 and 1988, Barbara created another programming language, Argus, in collaboration with Maurice Herlihy, Paul Johnson, Robert Scheifler and William Weihl. It was an extension of the CLU language, and utilised most of the same syntax and semantics. It was the first high-level language to support implementation of distributed programs and to demonstrate the technique of 'promise pipelining'. She also developed Thor, describing it as 'an object-oriented database system designed for use in a heterogeneous distributed environment. It provides highly-reliable and highly-available persistent storage for objects, and supports safe sharing of these objects by applications written in different programming languages.' With Jeannette Wing, Barbara developed what became known as the 'Liskov substitution principle', a particular definition of subtyping.

In 2002, Barbara was recognised as one of the top women faculty members at MIT, and by *Discover* magazine as among the top fifty faculty members in the sciences in the USA. In 2004, she won the prestigious John von Neumann Medal, awarded by the Institute of Electrical and Electronics Engineers (IEEE) in honour of the eminent mathematician John von Neumann, for 'fundamental contributions to programming languages, programming methodology, and distributed systems'. On 19 November 2005, Barbara and Donald E. Knuth were awarded Honorary Doctorates from the *Eidgenössische Technische Hochschule Zürich* (ETH Zürich), the Swiss Federal Institute of Technology, a science, technology, engineering and mathematics university. They were also featured in the ETH Zürich Distinguished Colloquium Series.

In 2007, Barbara was made the Associate Provost for Faculty Equity at MIT, a position she still held in 2019. In March 2009, Barbara received the 2008 A. M. Turing Award from the Association for Computing Machinery (ACM), an international learned society for computing – its most prestigious award and the computer science community's equivalent of the Nobel Prize. She received it for her work in the design of programming languages and software methodology that led to the development of object-oriented programming. The ACM cited her contributions to the practical and theoretical foundations of 'programming language and system design, especially related to data abstraction, fault tolerance, and distributed computing'.

In 2014, Barbara was inducted into the National Inventors Hall of Fame. In 2019, she was leading the Programming Methodology Group at MIT with a research focus in Byzantine fault tolerance and distributed computing. She is the author of three books and over 100 technical papers. Her legacy lies in her outstanding contributions to building software programs that form the infrastructure of an information-based society.

In 2011, Barbara Liskov was awarded honorary doctorates from both Northwestern University, Chicago and the University of Lugano, Switzerland. The MIT president Susan Hockfield said of her,

> Barbara is revered in the MIT community for her role as scholar, mentor and leader. Her pioneering research has made her one of the world's leading authorities on computer language and system design. In addition to her seminal scholarly contributions, Barbara has served MIT with great wisdom and judgment in several administrative roles, most recently as Associate Provost for Faculty Equity.

(Augusta) Ada Byron Lovelace

Born: 10 December 1815 in London, England
Died: 27 November 1852 (aged 36)
Field: Technology and mathematics

Augusta Ada Byron, known as Ada, was the daughter and only legitimate child of the renowned poet Lord George Gordon Byron, who left the family home when she was just five weeks old and never saw her again. His acrimonious separation from his wife, Annabella, a mathematician, meant that Ada's mother was the only significant parent in her life. She did not even see a portrait of her father until her twentieth birthday. Even so, Annabella was not very close to her daughter and Ada was mostly raised by her maternal grandmother, Judith, the Honourable Lady Milbanke. At an early age, Ada was fascinated by mathematics, music and crafts, especially model building, and these would be her lifelong passion.

Ada suffered from ill health during her early years, but at twelve became interested in flight, examining the anatomy of birds to ascertain the correct proportion between wing length and body size. This led to her writing a book, *Flyology*, on the subject. At fourteen she contracted measles, resulting in paralysis so severe she was bedridden for almost a year, after which she learned to walk again with the aid of crutches.

Ada's technological skills were becoming apparent during her teenage years and her mother arranged a series of private tutors in mathematics and science, including William Frend, William King and the celebrated Mary Somerville (*see entry*). In a letter to Ada's mother in 1832, another of her tutors, the mathematician and logician Augustus De Morgan, wrote that Ada had the capacity to be 'an original mathematical investigator, perhaps of first-rate eminence'. Upon turning eighteen, Ada was presented at court, and it was reported that 'her greatest delight was to go to the Mechanics Institute to hear the first of Dr Dionysius Lardner's lectures on the differential engine.

Miss Byron, young as she was, understood its working and saw the great beauty of the invention.'

At twenty, Ada married William, 8th Baron King, who in 1838 was made the Earl of Lovelace and Viscount Ockham. Lord Lovelace supported his wife's interests and they spent much of their time in the country where she could research and study. But it was the calculating device invented by Charles Babbage, with whom Ada became lifelong friends, which held her interest. Babbage paid her frequent visits.

In 1840, Babbage presented an invited lecture to the Prime Minister of Italy on his proposed 'Analytical Machine' (that had succeeded his 'Difference Engine'). An Italian engineer, Luigi Menabrea, transcribed the presentation into French and it was later published in the *Bibliothèque de Genève* (Library of Geneva) in October 1842. Ada was invited by Charles Wheatstone, a friend of Babbage's, to translate it into English, a task that occupied her for some nine months in meticulously explaining the difference between the two devices.

She was very particular about her writing, taking advice from Babbage only when she felt it necessary, reportedly stating, 'I cannot endure another person to meddle with my sentences.' In the end, the treatise had been expanded to three times its original size. Her work was much more than a mere translation, with one section including a technique for using the Analytical Engine to calculate a sequence of Bernoulli numbers, although the machine had not yet been built. As a result of her efforts, Ada is now largely regarded as the world's first computer programmer and her method as the first computer program. Her work 'Sketch of the Analytical Engine with notes by AAL' was published in Taylor's *Scientific Memoirs*. Not one to hold back, she praised her 'masterly' style and 'its superiority to that of the original memoir itself'.

Ada could see that the Analytical Machine had the capacity to be programmed rather than simply performing basic operations. She wrote,

[The Analytical Engine] might act upon other things besides number, were objects found whose mutual fundamental relations could be expressed by those of the abstract science of operations, and which should be also susceptible of adaptations to the action of the operating notation and mechanism of the engine ... Supposing, for instance, that the fundamental relations of pitched sounds in the science of harmony and of musical composition were susceptible of such expression and adaptations, the engine might compose elaborate and scientific pieces of music of any degree of complexity or extent.

In 1850, when in her mid-thirties, Ada developed a fascination with betting on horse racing, but it soon turned to disaster as, after finding herself deeply in debt, she reportedly 'pawned the family jewels and then implored the help of her mother in redeeming them, and in concealing the whole affair from her husband'. In 1852, after several months' illness, Ada died from uterine cancer at the age of only thirty-six. At her request, she was buried next to the father she never knew, in the Church of St Mary Magdalene in Hucknall in Nottinghamshire.

Ada was the first person to see the enormous potential of Babbage's device. Her name is honoured in the Lovelace Medal, awarded since 1998 by the British Computer Society, and since 2008 in an annual competition for women students. In 2009, the British journalist, writer and consultant Suw Charman-Anderson launched Ada Lovelace Day as an international celebration of the achievements of women in science, technology, engineering and mathematics (STEM). Its aim is to increase the profile of women in STEM, thereby creating new role models who will encourage more young women into STEM careers, as well as supporting women already working in STEM. It is now held every year on the second Tuesday of October, featuring the Ada Lovelace Day Live! 'science cabaret' event in London, in which women in STEM give short talks about their work or about other women who have inspired them. They also perform short comedy or musical interludes with a STEM focus.

Ada Lovelace was a truly remarkable woman who achieved so much in her short life. She would hold a pride of place in any list of pioneering women scientists.

Elizabeth 'Elsie' Muriel Gregory MacGill

Born: 27 March 1905 in Vancouver, Canada
Died: 4 November 1980 (aged 75)
Field: Aeronautical engineering
Awards: Centennial Medal (Canadian
Government, 1967); Order of Canada (1971);
Amelia Earhart Medal (1975); Ontario Association
of Professional Engineers Gold Medal (1979);
Canadian Aviation Hall of Fame (1983); Canadian
Science and Engineering Hall of Fame (1992);
Women in Aviation International's Pioneer Hall
of Fame (2012)

Elizabeth 'Elsie' Muriel Gregory MacGill was born in Vancouver, Canada, on 27 March 1905. She was the youngest daughter of James Henry MacGill, a prominent Vancouver lawyer, part-time journalist and Anglican deacon, and Helen (Gregory) MacGill, a journalist and British Columbia's first woman judge. Helen was the first woman in the British Empire to earn her bachelor's degree in music (1886), and by 1890 had completed bachelor's and master's degrees in mental and moral philosophy. Elsie had two older stepbrothers, Eric Herbert Gregory (b. 1891) and Frederic Philip Gregory (b. 1894), from her mother's first marriage to Frederick Charles Flesher, who had died in 1901. Elsie also had an older sister, Dr Helen 'Young Helen' MacGill (b. 1903), with whom she was so close that the family often referred to them as 'HelNelsie'.

Elsie and her sister were initially home-schooled to reflect the education that her older step-brothers received when they attended the Lord Roberts public school in Winnipeg. They had drawing lessons with the artist and writer Emily Carr, and swimming lessons with Joe Fortes (Vancouver's first official lifeguard). They later attended the public King George Secondary School, which was affiliated with McGill University. When Elsie was twelve, her mother was appointed Judge of the Juvenile Court of Vancouver. At the age of sixteen, Elsie was admitted to the applied sciences programme at the University of British Columbia. However, after only one term, the dean of the faculty asked her to leave.

In 1923, aged eighteen, Elsie enrolled in a Bachelor of Applied Sciences programme at the University of Toronto. During the summers, she honed her skills in the emerging field of aeronautical engineering by working in machine shops, where she repaired electrical motors to gain practical experience. Shortly before graduation, Elsie contracted polio and was advised that she would probably not walk again. However, with the aid of two sturdy metal canes, she did learn to walk and in 1927 went on to be the first female graduate in electrical engineering in Canada.

In 1940, Elsie remarked,

My presence in the University of Toronto's engineering classes in 1923 certainly turned a few heads. Although I never learned to fly myself, I accompanied the pilots on all test flights – even the dangerous first flight – of any aircraft I worked on.

After graduation, Elsie found a low-level position with a firm in Pontiac, Michigan, and during her time there, the company began producing aircraft. This fuelled her interest in aeronautics and she commenced part-time graduate studies in aeronautical engineering at the University of Michigan, enrolling in the autumn of 1927 in the full-time Master of Science in Engineering. It was here that she began to design aircraft and conduct research and development in the University's new aeronautics facilities. In 1929, Elsie became the first woman in North America to be awarded a master's degree in aeronautical engineering.

Elsie then enrolled in a PhD programme at MIT, financing herself by writing magazine articles about aircraft and flying. But she abandoned her studies in 1934, when she was offered employment as an Assistant Aeronautical Engineer at Fairchild Aircraft's operations in Longueuil, Quebec. In 1938, she was the first woman elected to corporate membership in the Engineering Institute of Canada (EIC) and, on 22 March the same year, presented a paper, 'Simplified Performance Calculations for Aeroplanes', to the Royal Aeronautical Society in Ottawa. It was well received and published in *The Engineering Journal*. She also participated in the Canadian Broadcasting Corporation's six-part series *The Engineer in War Time*, with a segment titled 'Aircraft Engineering in Wartime Canada'. In 1942, Elsie was elected to the position of chairman of the EIC, Lakehead Branch, after having also served as their vice chairman.

During the Second World War, Elsie worked as an aeronautical engineer and was instrumental in making Canada a driving force in aircraft construction during her years at Canadian Car and Foundry (CC & F) in Fort William, Ontario. She was the first woman in the world to hold such a position and it was there that she tested a new training aircraft, the Maple Leaf Trainer II. When the company was selected to build the Hawker Hurricane fighter aircraft for the Royal Air Force, Elsie's main role changed to streamlining operations in the production line as the factory's output rapidly grew. She also took charge of designing solutions to allow the aircraft to operate during the winter, introducing de-icing controls and a system for fitting skis for landing on snow. In 1940, she wrote and presented a paper, 'Factors Affecting Mass Production of Aeroplanes', published in *The Engineering Journal* in July that year. In 1943, the production line at

CC & F closed down, having produced over 1,400 Hurricanes. Her role in this successful enterprise earned Elsie celebrity status, to the point where the *American True Comics* series published a comic book about her in 1942, using her nickname, 'Queen of the Hurricanes'.

In 1943, Elsie married E. J. (Bill) Soulsby, the couple then moving to Toronto, where they set up an aeronautical consulting business. In 1946, she became the first woman to serve as technical adviser for the International Civil Aviation Organisation, a specialist agency of the United Nations, assisting in the drafting of International Air Worthiness regulations for the design and production of commercial aircraft. In 1947, she became the chairman of the United Nations Stress Analysis Committee.

In March 1953, she was made an honorary member of the American Society of Women Engineers who named her 'Woman Engineer of the Year', the first time that the award had been given outside the USA. Elsie was also a member of the Ontario Status of Women Committee, an affiliate of the National Action Committee on the Status of Women. In 1971, Elise was awarded the Order of Canada, the nation's second highest national honour. She remarked, 'I have received many engineering awards, but I hope I will also be remembered as an advocate for the rights of women and children'.

After a short illness, Elsie MacGill passed away at age seventy-five on 4 November 1980 while visiting her sister, Helen MacGill Hughes, in Cambridge, Massachusetts. In 1983, Elsie was posthumously inducted into Canada's Aviation Hall of Fame, and in 1992 she was a founding inductee in the Canadian Science and Engineering Hall of Fame in Ottawa.

Margaret Eliza Maltby

Born: 10 December 1860 in Bristolville, Ohio, USA
Died: 3 May 1944 (aged 83)
Field: Physics
Awards: Fellow, American Association for
the Advancement of Science (1889); Fellow,
American Physical Society (1900)

Margaret Maltby was born on 10 December 1860 on her family's farm in
Bristolville, Ohio. Nicknamed 'Minnie' by her two older sisters, she later
legally changed it to Margaret Eliza at the first opportunity. Selecting art as
her initial interest, she graduated from the private liberal arts Oberlin College
with a Bachelor of Arts in 1882, after which she spent a year studying at the
Arts Students League in New York City. She then returned to Ohio where
she was a schoolteacher for four years, during which time she developed an
interest in chemistry and physics.

Enrolling as an undergraduate student, she attended MIT in 1887, doing
so as a special (non-degree) student, as MIT did not accept female students.
During her degree, Margaret developed her lecturing skills by teaching
physics at the nearby Wellesley College between 1889 and 1893, and in
1891 she became the first woman to graduate with a Bachelor of Science
degree. For the next two years she stayed at MIT, undertaking postgraduate
work, and in 1893 won a travel scholarship to enrol in a PhD at the
University of Göttingen in Lower Saxony, Germany. Under the supervision
of Friedrich Kohlrausch, her thesis topic involved the measurement of
high electrolytic substances. Margaret became the first American woman
to earn a PhD from the university and stayed in Germany to undertake
postdoctoral work supported by a second scholarship from the Association
of College Alumni. She worked at the newly founded *Institut für
Physikalische Chemie* (Institute of Physical Chemistry) in Göttingen, under
the supervision of Hermann Nernst, a German chemist who was known
for his work in thermodynamics.

In 1896, Margaret returned to the USA, where she led the physics
department at Wellesley College, followed by a teaching role in mathematics

and physics at Lake Erie College in Ohio. It was there that a curious event took place when she suddenly resigned her post, supposedly as the result of an accident. It was reported by the physics professor and historian Katharine Sopka, and other sources, that in 1901, Margaret had adopted the orphaned son of a close friend. However, subsequent autosomal DNA tests of the son's two daughters revealed their descent from ancestral families of Margaret's mother and of her father, evident in the DNA his daughters share with numerous other descendants of those families. It is now thought that Philip Randolph Meyer was her natural son, being born in June 1897, six months after the alleged accident. Three months later, in September, she resurfaced to resume her teaching career as an instructor at Lake Erie College. It is reported that she left Philip in the care of a friend.

In 1898, Margaret travelled to Charlottenburg, Germany, to act as a research assistant to her PhD supervisor, Kohlrausch, a physicist who investigated the conductive properties of electrolytes. After twelve months, she returned to the USA to undertake research in theoretical physics with Arthur Gordon Webster, founder and president of the American Physical Society. After a year, she moved on to Barnard College in New York where, re-uniting with her son, she initially became a chemistry instructor before switching to the physics department in 1903.

For the next decade, Margaret served as an adjunct professor, in 1906 being listed in the first edition of *American Men and Women of Science* as one of the 1,000 most eminent scientists in the country. In 1913, she was appointed as an assistant professor and head of the physics department, quite a feat in a field that was almost entirely dominated by men. In her role as chair, she strongly opposed the forced resignation of a faculty member, the Canadian Harriet Brooks, when she planned to marry. At the time, the Barnard College's Dean's Rule stated, 'the College cannot afford to have women on the staff to whom the college work is secondary; the College is not willing to stamp with approval a woman to whom self-elected home duties can be secondary'. Brooks was Canada's first female nuclear physicist and the first woman at McGill University in Montreal to receive a master's degree.

Margaret became renowned for her dedication as a teacher and administrator, introducing physics courses for non-physicists, although these activities occupied so much of her schedule that it left little time for research. She was a strong proponent of women gaining entry to postgraduate and postdoctoral programmes, and between 1912 and 1927 she was on the fellowship committee of the American Association of University Women (AAUW), acting as the chair from 1913 to 1926. Two years later, the AAUW established the Margaret E. Maltby Fellowship in her honour, a scheme that was critical for maintaining many female physicists, as women were not eligible for many research fellowships.

Margaret Maltby left an enduring legacy with her dedication to the education of women, with one writing 'Professor Maltby was my mentor – a gracious lady – a friend and a counsellor. Her most memorable advice to me was not to forgo marriage for a career – which advice I followed and lived happily ever after'. In 1929 Margaret published her book, *A History of the Fellowships Awarded by the American Association of Women, 1888–1929*, containing biographies of the fellows. She had a love of music and enjoyed attending the opera and symphonies. Margaret never married, passing away on 3 May 1944, aged eighty-three, spending most of the last years of her life in the Columbia University Morningside Heights community in New York City.

Lynn Petra Alexander Margulis

Born: 5 March 1938 in Chicago, USA
Died: 22 November 2011 (aged 73)
Field: Biology
Awards: Fellow of the American Association for
the Advancement of Science (1975); Guggenheim
Fellowship (1978); National Academy of
Sciences (1983); Miescher-Ishida Prize (1986); the
*Commandeur de l'Ordre des Palmes académiques de
France* (1989); Chancellor's Medal for Distinguished
Faculty of the University of Massachusetts at
Amherst (1992); Russian Academy of Natural
Sciences (1997); Distinguished Service Award
of the American Institute of Biological Sciences
(1998); Fellow, American Academy of Arts and
Sciences (1998); William Procter Prize for Scientific
Achievement (1998); National Medal of Science
(awarded by President William J. Clinton, 1999);
Alexander von Humboldt Prize (2002–05);
Darwin-Wallace Medal (2008); NASA Public
Service Award for Astrobiology (2010)

Lynn Petra Alexander was born on 5 March 1938 in Chicago, USA, to a Jewish Zionist family. Her father was Morris Alexander, an attorney, and her mother, Leona Wise Alexander, operated a travel agency. Lynn was the eldest of four daughters. In 1952, she attended the Hyde Park Academy High School, where she was a poor, but precocious, student who frequently had to 'stand in the corner'. At the age of fifteen, she was accepted into the University of Chicago Laboratory Schools and, at nineteen in 1957, graduated with a bachelor's degree in Liberal Arts. The same year she married the twenty-three-year-old astronomer and astrophysicist Carl Sagan, who would later become quite prominent in his fields. Lynn then enrolled in a master's degree in genetics and zoology at the University of Chicago, obtaining her degree at the age of twenty-two.

Lynn then joined the University of Wisconsin, where she studied biology under the supervision of Walter Plaut, her supervisor, and cytologist Hans Ris. She graduated in 1960 with a Master of Science degree in genetics and zoology. Following this, she continued her research at the University of California, Berkeley, under the direction of zoologist Max Alfert. In 1964, before she could complete her thesis, Lynn was offered a research associateship and then a lectureship at Brandeis University in Massachusetts. Lynn and Carl were divorced in 1964. They had two sons, Dorion Sagan, who later became a popular science writer, and Jeremy Sagan, a software developer and founder of Sagan Technology.

In 1965, under the supervision of Alfert, Lynn obtained her PhD from the University of California, Berkeley with a thesis titled, 'An Unusual Pattern of Thymidine Incorporation in Euglena'. In 1966, she moved to Boston University, where she was appointed as an adjunct assistant professor, becoming an assistant professor the following year. In 1966, Lynn published her landmark paper, 'On the Origin of Mitosing Cells', a revelation in endosymbiotic theory, in the *Journal of Theoretical Biology*. Astonishingly, this work had previously been rejected by about fifteen other journals. Her theory that cell organelles, such as mitochondria and chloroplasts, were once independent bacteria, was essentially disregarded for another ten years. It only gained widespread acceptance after it was robustly substantiated through genetic evidence.

In 1967, Lynn married the crystallographer Thomas N. Margulis and adopted his family name. Their son, Zachary Margulis-Ohnuma, later became a New York City criminal defence lawyer, and their daughter, Jennifer Margulis, a teacher and author. In 1971, Lynn was promoted to associate professor and to full professor in 1977. Lynn and Thomas divorced in 1980, and she later commented, 'I quit my job as a wife twice: It's not humanly possible to be a good wife, a good mother and a first-class scientist. No one can do it – something has to go.'

Lynn was the primary modern driving force behind the significance of symbiosis (the close and often long-term interaction between two different biological species) in evolution. The prominent Canadian biologist and historian Jan Sapp declared, 'Lynn Margulis's name is as synonymous with symbiosis as Charles Darwin's is with evolution'. Of particular note was her transforming and fundamentally framing the current understanding of the evolution of cells with nuclei. The leading evolutionary biologist Ernst Mayr called her discovery 'perhaps the most important and dramatic event in the history of life', because Lynn suggested that it was the result of symbiotic mergers of bacteria. She remarked, 'Life did not take over the globe by combat, but by networking.' Lynn, with the British chemist James Lovelock, developed the 'Gaia hypothesis', a proposition that the Earth functions as a single self-regulating system. She was also the principal defender and promulgator of the five-kingdom classification of the distinguished American plant ecologist Robert Whittaker.

Lynn was elected a member of the US National Academy of Sciences in 1983, and in 1986 was promoted to professor. She was appointed a Distinguished Professor of Botany at the University of Massachusetts at Amherst in 1988, becoming a Distinguished Professor of Biology in 1993. In 1997, she transferred to the Department of Geosciences at Amherst, where she became a Distinguished Professor of Geosciences, a position she held until her death.

While Lynn was an avid evolutionist, she completely rejected the modern evolutionary synthesis. She said, 'I remember waking up one day with an epiphanous revelation: I am not a neo-Darwinist! I recalled an earlier experience, when I realized that I wasn't a humanistic Jew. Although I greatly admire Darwin's contributions and agree with most of his theoretical analysis and I am a Darwinist, I am not a neo-Darwinist.' She added, 'Natural selection eliminates and maybe maintains, but it doesn't create', instead maintaining that symbiosis was the major driver of evolutionary change.

Lynn won many awards and honours for her ground-breaking work. In 1995, she was elected as a Fellow of the World Academy of Art and Science, and in 1997 was elected to the Russian Academy of Natural Sciences. US President Bill Clinton presented her with the National Medal of Science in 1999, and in 2008 the Linnean Society of London awarded her the Darwin-Wallace Medal.

In 2009, with seven co-authors, Lynn published a position paper outlining research on the viability of round body forms of some spirochetes (these being members of the phylum *Spirochaetes*, which contains distinctive diderm bacteria). Titled, 'Syphilis, Lyme disease, & AIDS: Resurgence of "the great imitator"?', it stated that, 'Detailed research that correlates life histories of symbiotic spirochetes to changes in the immune system of associated vertebrates is sorely needed', and recommended the 'reinvestigation of the natural history of mammalian, tick-borne, and venereal transmission of spirochetes in relation to impairment of the human immune system'. It also suggested that 'the possible direct causal involvement of spirochetes and their round bodies to symptoms of immune deficiency be carefully and vigorously investigated'. In 2010, she won the NASA Public Service Award for Astrobiology and, in all, received honorary doctorates from fifteen universities.

Lynn Margulis passed away at age seventy-three on 22 November 2011 at her home in Amherst, Massachusetts, five days after suffering a haemorrhagic stroke. In accordance with her wishes, she was cremated, and her ashes scattered in her favourite research areas, near her home. She was survived by her daughter and three sons.

Mileva Marić-Einstein

Born: 29 December 1875 in Titel, Serbia
Died: 4 August 1948 (aged 72)
Field: Mathematics

Mileva Marić was born the eldest of three children in Titel, Siberia, on 29 December 1875, into a wealthy family of the Austro-Hungarian Empire (now Serbia). Her parents were Miloš Marić and Marija Ružić-Marić and soon after Mileva's birth, her father ended his military career and took a position at the court in Ruma and later in Zagreb.

In 1886, Mileva began her secondary education at a high school for girls in Novi Sad, but changed the following year to a high school in Sremska Mitrovica. In 1890, she started attending the Royal Serbian Grammar School in Šabac, but the following year her father obtained special permission to enrol her as a private student at the all-male Royal Classical High School in Zagreb. She passed the entrance exam and entered the tenth grade in 1892, gaining special permission to attend physics lectures in February 1894. Passing her final exams in September that year, her grades in mathematics and physics were the highest awarded, but almost immediately she became seriously ill, prompting her to move to Switzerland. In 1896, she passed her Matura-Exam, and enrolled for a medical degree at the University of Zürich for one semester.

In the autumn of 1896, Mileva abandoned her medical studies and transferred to the Zürich Polytechnic (later *Eidgenössische Technische Hochschule* (ETH), the Swiss science and technology university), having passed the mathematics entrance examination with an average grade of 4.25 on a scale of 1–6. She enrolled for the diploma course to teach physics and mathematics in secondary schools (section VIA) at the same time as Albert Einstein. She was the only woman in her group of six students, and only the fifth woman to enter that section, an outstanding achievement as, at the time, women were not usually admitted. Mileva and Einstein soon became close friends, and in October she travelled to Heidelberg to study at Heidelberg

University for the winter semester 1897–98, auditing lectures in physics and mathematics. She rejoined the Zürich Polytechnic in April 1898, where her courses included differential and integral calculus, descriptive and projective geometry, mechanics, theoretical physics, applied physics, experimental physics, and astronomy.

In 1901, Mileva's academic career was disrupted when she became pregnant to Einstein and, three months into the pregnancy, re-sat the diploma examination but failed for the second time without improving her grade. Their daughter, named Lieserl, was born in Novi Sad in early 1902 and the contents of one of Einstein's letters in September 1903 suggest that she was either given up for adoption or died of scarlet fever in infancy. In January 1903, Mileva and Einstein were married in Bern, Switzerland, where he had found a job at the Federal Office for Intellectual Property. In May 1904, their first son, Hans Albert, was born and the couple remained living in Bern until 1909, when Einstein got a teaching position at the University of Zürich. In July 1910, their second son, Eduard, was born there. In 1911, they moved to Prague, where Einstein held a teaching position at the Charles University and, a year later, they returned to Zürich, as Einstein had accepted a professorship there. The couple separated in 1914 and divorced in 1919.

Mileva's contribution to Einstein's early work, especially the *Annus Mirabilis Papers*, is a matter for conjecture, although there appears to be at least some evidence that she helped him in his early research. Hans Albert, later said that when his mother married Einstein, she gave up her scientific ambitions. Other evidence for her as a co-author of some of Einstein's early work, includes her telling a Serbian friend, referring to 1905, that 'we finished some important work that will make my husband world famous'. However, some historians have cast some doubt on this assertion. After suffering a breakdown and being diagnosed with schizophrenia at around age twenty, Eduard Einstein was institutionalised until his death in 1965.

Mileva Marić-Einstein died at the age of seventy-two on 4 August 1948 in Zürich. She was interred there at the Friedhof Nordheim Cemetery where a memorial gravestone was dedicated to her in June 2009. After Mileva's death, a bust in her honour was installed on the campus of the University of Novi Sad and a memorial plaque was placed on her former residence in Zürich, at Huttenstrasse 62, in her memory. In 2005, she was honoured in Zürich by the ETH and the *Gesellschaft zu Fraumünster* and a bust was placed in her high-school town, Sremska Mitrovica.

Historians have disagreed for decades as to whether Mileva Marić-Einstein was a colleague or merely a sounding board for her famous husband but, in any case, she is worthy of a place as a pioneer for women in science. One view is that she was a collaborator and even co-authored his papers, while another is that she was only an intelligent listener. The importance of her work has never been fully evaluated. Her story also highlights the plight of intellectual women of science, particularly during the first half of the twentieth century.

Antonia Caetana de Paiva Pereira Maury

Born: 21 March 1866 in Cold Spring, New York, USA
Died: 8 January 1952 (aged 85)
Field: Astronomy
Award: Annie Jump Cannon Award in Astronomy (1943)

Born in Cold Spring, New York, on 21 March 1866, Antonia Caetana de Paiva Pereira Maury was the daughter of Reverend Mytton Maury, a protestant minister, editor of a geographical magazine and a naturalist, and Virginia (Draper) Maury. Her ancestors were distinguished teachers and scientists, with her father's great-grandfather Reverend James Maury, being the teacher of three US presidents – Thomas Jefferson, James Madison, and James Monroe. She was also the granddaughter of John William Draper and the niece of Henry Draper, both prominent physicians and pioneering astronomers in celestial photography. At a young age, Antonia and her two siblings were exposed to science with her younger sister, Carlotta Maury, later becoming a geologist, stratigrapher and paleontologist.

Antonia attended Vassar College, graduating in 1887 with honours in physics, astronomy, and philosophy. While there she studied under the supervision of the eminent astronomer Maria Mitchell (*see entry*) and, in October of her senior year, published a paper, 'Hints of National Character in Language', in the journal *The Miscellany*. From there she attended the Harvard College Observatory as one of the so-called 'Harvard Computers', a group of highly-skilled women who processed astronomical data. One of her tasks was to classify northern stars using spectroscopy. This was achieved by placing a series of prisms in front of a telescope, diffracting light from the star, splitting it into a spectrum that shows the light the star absorbs and that it bounces back. From that light can be read information about the star's temperature, chemical composition and luminosity. Antonia created a system of star classification based on temperature, with divisions to describe the variation in width and definition of the spectral lines. She argued that the sharply defined spectral lines of a star 'represented a fundamental property of the stars', but her sorting system was rejected by the other 'Harvard Computers', who instead adopted a simpler sorting system devised by one of their colleagues, Annie Jump Cannon (*see entry*), who had received widespread recognition for the work.

The eminent American astronomer and director of the observatory, Edward Charles Pickering, also preferred Cannon's classification, which led to a fallout between him and Antonia when she refused to change her method. Pickering did not appreciate dissent and ruled the women under him strictly – they were referred to as 'Pickering's harem' and they were paid 25 cents per hour, half of that received by men. He told Antonia that her work was slowing down the other more valuable activities and classifications of his group. Her conflict with Pickering hindered her creativity and, as a result, she left the observatory in 1891.

Antonia returned in 1893 to continue her work, and in 1897 published an important catalogue of her classifications, 'Spectra of Bright Stars Photographed with the 11-inch Draper Telescope as Part of the Henry Draper Memorial'. The conclusions were based on her examination of 4,800 photographs and provided a detailed analysis of 681 bright northern stars, her catalogue becoming the first observatory publication credited to a woman. As part of this work, she noticed periodic doubling of some lines in the spectrum of ζ Ursae Majoris (Mizar A), leading to the publication of the first spectroscopic binary orbit. She wrote to Pickering, 'I worked out the theory at the cost of much thought and elaborate comparison and I think that I should have full credit for my theory of the relations of the star spectra'. It turned out that there was a difference between very bright red stars (giants) and very faint ones (dwarf stars) – and only Antonia's classification scheme, with its three divisions, captured the distinction. It was not until 1922 that the International Astronomical Union agreed, incorporating her work into its classification system, which had been based on Cannon's.

For the remainder of her life, Antonia worked only sporadically at Harvard. From 1896 to 1918, she taught physics and chemistry at the Castle School (Miss C. E. Mason's suburban school for girls) in Tarrytown-on-Hudson, New York. In 1908, she returned for a while to work at Harvard, where she continued to research spectroscopic binary stars. Over many years she investigated the complex spectroscopic binary *Beta Lyrae*, examining nearly 300 spectra of this star. She returned to Harvard College Observatory in 1918 as an adjunct professor, and with the death of Pickering the following year, she found herself better able to work with his successor, Harlow Shapley.

In 1933, Antonia published two studies of three extremely complex binaries in *Annals of Harvard College Observatory*. She continued to examine the puzzling spectra of *Beta Lyrae* well beyond her official retirement from Harvard in 1935. At the age of seventy-seven, she received the prestigious Annie Jump Cannon (*see entry*) Award of the American Astronomical Society in 1943.

When she retired, Antonia pursued a number of interests, including ornithology, along the lines of her father, a noted naturalist and geographer.

She was a supporter of naturalist and conservationist causes, fighting to save western sequoia forests when they were endangered by the cutting down of trees during wartime and she remained a member of the American Astronomical Society, the Royal Astronomical Society, and the National Audubon Society. For three years, she also served as curator of the Draper Museum in Hastings-on-Hudson, New York, where both her grandfather and her uncle had built observatories.

Antonia Caetana de Paiva Pereira Maury passed away in Dobbs Ferry, New York, on 8 January 1952 at age eighty-five. She never married. The 1978 revised *M. K. Spectral Atlas for Stars Earlier Than the Sun* was dedicated to 'Antonia C. Maury, Master Morphologist of Stellar Spectra'. One of its authors, William Wilson Morgan, described her 1897 work as 'the most remarkable phenomenological investigation in modern astronomy'.

Maria Göppert Mayer

Born: 28 June 1906 in Kattowitz, German
Empire (now Katowice, Poland)
Died: 20 February 1972 (aged 65)
Field: Physics
Award: Nobel Prize in Physics (1963)

An only child, Maria Göppert was born on 28 June 1906 in Kattowitz, a city in Prussia, now Katowice, Poland. Her parents, Friedrich Göppert and Maria (Wolff) Göppert, moved with Maria to Göttingen in 1910 when Friedrich was appointed as the professor of paediatrics at the University of Göttingen. Maria later admitted being closer to her father than her mother, explaining that 'my father was more interesting; he was, after all, a scientist'.

Maria was educated at the only school in Göttingen that prepared girls for the university entrance examination (*abitur*), the *Frauenstudium*, run by suffragettes who aimed to prepare girls for university. Although the school shut during the severe inflation in Germany in the early 1920s, the teachers continued to give instructions to the pupils.

In the spring of 1924, Maria entered the University of Göttingen, where she planned to become a mathematician as there was a perceived shortage of women mathematics teachers for girls' schools. This led to an influx of women studying mathematics at a time of high unemployment, and there was even an outstanding female professor of mathematics at Göttingen, Emmy Noether (*see entry*), but most of the girls were only interested in qualifying for their teaching certificates.

Before long, Maria became more interested in physics, as this was an era when quantum mechanics was in its early stages, an area she found exciting and attractive, and she decided to undertake a PhD in the discipline. Her 1930 doctoral dissertation, 'On Elemental Processes with Two Quantum Jumps', examined the theory of possible two-photon absorption by atoms. It was described nine years later by the eminent Hungarian American theoretical physicist, engineer and mathematician Eugene Paul Wigner (a 1963 Nobel Prize winner) as a 'masterpiece of clarity and concreteness'. Her thesis examiners were three Nobel Prize winners, Max Born (1954), James

Franck (1925) and Adolf Otto Reinhold Windaus (1928). At the time, there seemed little chance that the theory she expounded could be verified through experimentation. But, in 1961, the development of lasers allowed the detection of two-photon-excited fluorescence in a europium-doped crystal. In honour of her fundamental contribution to this area, the unit for the two-photon absorption cross section is named the 'Göppert-Mayer (GM) unit'.

On 19 January 1930, Maria married Joseph Edward Mayer, an American Rockefeller fellow and one of James Franck's assistants, whom she had met when Joseph had boarded with her family. The couple moved to the USA, where he had been offered a position as associate professor of chemistry at Johns Hopkins University. This was the period of the Great Depression and the university would not consider hiring the wife of a professor. Instead she was given a role as an assistant in the Physics Department working with German correspondence. For this she received only a very small salary, but it gave her a place to work and access to the facilities. She also taught a few courses and published a paper on double beta decay in 1935.

In 1937, Joseph was fired, an act he attributed to the hatred of women on the part of the dean of physical sciences, which he thought was provoked by Maria's presence in the laboratory. Joseph then accepted a position as chairman of the physics department at Columbia University, part of the arrangement being that Maria would have an office there but receive no salary. Two years later, the renowned physicists Harold Urey and Enrico Fermi joined the department and the latter asked her to investigate the valence shell of the undiscovered transuranic elements. Using the Thomas-Fermi model, Maria correctly predicted that they would form a new series similar to the rare earth elements.

In December 1941, Maria took up her first paid professional position as a part-time science teacher at Sarah Lawrence College and, in the spring of 1942, with the USA heavily involved in the Second World War, she joined the Manhattan Project, a research and development undertaking that ultimately produced the first nuclear weapons. Urey offered her a part-time research post with Columbia University's Substitute Alloy Materials (SAM) Laboratories, to find a means of separating the fissile uranium-235 isotope in natural uranium. Maria researched the chemical and thermodynamic properties of uranium hexafluoride, investigating the possibility of separating isotopes by photochemical reactions, a technique that proved impractical at the time. However, the development of lasers later opened the possibility of separation of isotopes by laser excitation.

Maria's friend, the Hungarian American theoretical physicist Edward Teller arranged a position for Maria at Columbia with the Opacity Project, in which she researched the properties of matter and radiation in conditions of extreme heat for use in thermonuclear weapons. The aim was to develop Teller's 'Super' bomb, the wartime programme for the development of such weapons. In February 1945, as Maria's husband was sent to the Pacific War, she decided to leave her children in New York and join Teller's group at the Los Alamos Laboratory. When Joseph returned from the Pacific earlier than expected, they returned to New York together in July 1945.

In February 1946, Joseph became a professor in the Chemistry Department at the new Institute for Nuclear Studies at the University of Chicago, and Maria became a voluntary associate professor of physics at the school. When Teller also accepted a position there, she was able to continue her Opacity Project work with him. When the nearby Argonne National Laboratory opened on 1 July 1946, Maria was also offered a part-time job as a senior physicist in the theoretical physics division. It was there that she programmed the Aberdeen Proving Ground's Electronic Numerical Integrator and Computer (ENIAC), the first electronic general-purpose computer, to solve criticality problems for a liquid metal cooled reactor using the Monte Carlo method.

During her time at Chicago and Argonne in the late 1940s, Maria developed a mathematical model for the structure of nuclear shells, her model explaining why certain numbers of nucleons in an atomic nucleus result in particularly stable configurations. These numbers are what physicist Eugene Wigner called 'magic numbers' – 2, 8, 20, 28, 50, 82, and 126. She also correctly postulated that the nucleus is a series of closed shells and pairs of neutrons and that protons tend to couple together.

In 1960, at the age of fifty-four, Maria was appointed a full professor of physics at the University of California, San Diego, and, although suffering a stroke shortly after arriving there, she continued to teach and conduct research for a number of years. In 1963, Maria and the German nuclear physicist Johannes Hans Daniel Jensen were awarded a half-share of the Nobel Prize 'for their discoveries concerning nuclear shell structure'. The other half was awarded to Eugene Wigner. In 1965, she was elected a Fellow of the American Academy of Arts and Sciences.

On 20 February 1972, Maria Göppert Mayer passed away in San Diego, California, aged sixty-five. She had been left comatose after a heart attack the previous year. She was buried at El Camino Memorial Park in San Diego, and her memory is honoured in several ways. The American Physical Society established the Maria Göppert Mayer Award to honour young female physicists at the beginning of their careers. The University of California, San Diego, hosts an annual Maria Göppert Mayer symposium to bring together female researchers to discuss current science, while their physics department is housed in Mayer Hall, which is named after Maria and her husband Joseph. On Venus, the crater Göppert Mayer, with a diameter of about 21 miles (34 km), is named after her, and in 2011 Maria was included in the third release of the collection of US postage stamps honouring American Scientists. Maria's papers are held in the Geisel Library at the University of California, San Diego.

(Eleanor) Barbara McClintock

Born: 16 June 1902 in Hartford, Connecticut, USA
Died: 2 September 1992 (aged 90)
Field: Cytogeneticist
Awards: National Medal of Science (1970); Thomas Hunt Morgan Medal (1981); Louisa Gross Horwitz Prize (1982); Nobel Prize in Physiology or Medicine (1983)

Barbara McClintock was born on 16 June 1902 in Hartford, Connecticut, USA. Although she was baptised 'Eleanor' McClintock, her parents soon started calling her Barbara, the name she stayed with all her life. She was the third of four children (the others being Marjorie, born 1898, Mignon, born 1900 and Malcolm, born 1903) to Sara (Handy) McClintock, a housewife, artist and poet, who was descended from an old American *Mayflower* family, and father Thomas Henry McClintock, a homeopathic physician. To relieve the financial burden on her parents while her father established his medical practice, from the age of three until she began school Barbara mostly lived with an aunt and uncle in Massachusetts. Although close to her father, Barbara was described as a solitary and independent child who enjoyed being alone, while having an uneasy relationship with her mother.

Barbara returned to her parents in Hartford to begin school, and in 1908 the whole family moved to Brooklyn, New York. She completed her secondary education there at Erasmus Hall High School, graduating in early 1919, developing along the way a love for science. Her mother did not want Barbara to go to university as she considered this would make her unmarriageable, but her father prevailed and so, at age seventeen, Barbara enrolled at Cornell University in Ithaca, New York. After leaving home. she seemed happier and became less of a loner as an undergraduate, even socialising with fellow students, joining a jazz band and being elected as president of the women's freshman class.

In the autumn of 1921, Barbara attended the only course in genetics open to undergraduate students at Cornell University. There were relatively few students enrolled and most were interested in pursuing agriculture as a profession, although she had an eye on genetics that had not yet then received general acceptance. It was only twenty-one years since the rediscovery of

(Eleanor) Barbara McClintock

Mendel's principles of heredity. At the completion of the genetics course, Barbara was invited to take part in the only other genetics programme given at Cornell, although it was supposed to be for graduate students. This set the path for Barbara's future studies, and she also enrolled in a cytology course that discussed the structure of chromosomes and their behaviours at mitosis and meiosis (stages of cell division).

After graduating with a Bachelor of Science in 1923, Barbara enrolled in a postgraduate programme and set her sights firmly on the field of cytogenetics. Her Master of Science degree in botany was awarded in 1925 and a PhD in the same field in 1927, both at Cornell. During this time, she also worked as a research assistant for the Cornell botanist Lowell Fitz Randolph and later for Lester Whyland Sharp.

Following her doctorate, Barbara accepted a role as an instructor in the botany department at Cornell, continuing her work in plant cytogenetics, involving microscopes to investigate plant genetics at the cellular level – particularly studying chromosomes. While traditional genetics involved breeding successive generations of an organism and observing differences visible to the naked eye, cytogeneticists also correlate their observations with changes taking place within cells. Barbara was instrumental in assembling a group that studied the new field of cytogenetics in maize. This group brought together plant breeders and cytologists, including Marcus Rhoades, the 1958 Nobel laureate George Beadle, and Harriet Baldwin Creighton who was a graduate student to whom Barbara was a mentor.

In 1931, Barbara and Harriet provided a ground-breaking discovery that the recombination of genes linked on a chromosome requires the physical exchange of segments of the chromosome with its homologous partner, obtaining a corn plant which had a knob on the 9th chromosome. They also showed that crossing-over occurs in sister chromatids as well as homologous chromosomes.

In 1938, Barbara produced a cytogenetic analysis of the centromere, describing the organisation and function of the centromere as well as the fact that it can divide. Her cytogenetic research focused on developing ways to visualise and characterise maize chromosomes, influencing a generation of students. Barbara also developed a technique using carmine staining to visualise maize chromosomes, and showed for the first time the morphology of the ten maize chromosomes. This discovery was made because she observed cells from the microspore, as opposed to the root tip. By studying the morphology of the chromosomes, Barbara was able to link specific chromosome groups of traits that were inherited together.

Barbara's research results and publications resulted in her being awarded several postdoctoral fellowships from the National Research Council, the associated funding allowing her to continue to study genetics at Cornell, the University of Missouri (where she expanded her research on the effect of X-rays on maize cytogenetics) and the California Institute of Technology, where she worked with E. G. Anderson. During the summers of 1931 and 1932, she worked at Missouri with geneticist Lewis Stadler, who introduced her to the use of X-rays as a mutagen (a physical or chemical agent that changes the genetic material). Exposure to X-rays can increase the rate of mutation above the natural background level, making it a powerful research tool for genetics. Barbara also received a fellowship from the Guggenheim Foundation that made possible six months of training in Germany during 1933 and 1934.

In December 1941, Barbara was offered a research position in the Department of Genetics of the Carnegie Institution of Washington, in the Cold Spring Harbor Laboratory. She then became a permanent member of the faculty, a position she held until 1967. On her retirement, she was honoured by being made a Distinguished Service Member. This allowed her to continue working with graduate students and colleagues in the Cold Spring Harbor Laboratory as *scientist emerita*.

In 1947, Barbara received the Achievement Award from the American Association of University Women. In 1959, was elected a Fellow of the American Academy of Arts and Sciences. In 1967, she was awarded the Kimber Genetics Award; three years later, she was presented with the National Medal of Science by US President Richard Nixon, the first female recipient of this award. In 1973 Cold Spring Harbor named a building in her honour, and in 1978 she received the Louis and Bert Freedman Foundation Award and the Lewis S. Rosenstiel Award for Distinguished Work in Basic Medical Research. In 1981, Barbara became the first recipient of the MacArthur Foundation Grant and was awarded the Albert Lasker Award for Basic Medical Research and the Wolf Prize in Medicine. She was also the recipient of the Thomas Hunt Morgan Medal of the Genetics Society of America, awarded in recognition of lifetime contributions to the field of genetics, named after the 1933 Nobel Prize winner. In 1982, she was awarded the Louisa Gross Horwitz Prize from Columbia University for her research in the 'evolution of genetic information and the control of its expression', awarded for outstanding contributions in basic research in the field of biology or chemistry.

Barbara's most outstanding award was the Nobel Prize for Physiology or Medicine in 1983, the first woman to win the prize in that category unshared, for discovering 'mobile genetic elements', and the third woman to win an unshared Nobel science prize, after Marie Curie (*see entry*) and Dorothy Hodgkin (*see entry*), both for chemistry. Following this honour, she became a key leader and researcher in the field at Cold Spring Harbor Laboratory on Long Island, New York. Her reputation was established in the 1930s after she had doggedly carried out her work, ahead of her time. When she was asked in 1943 about the long delay in recognition for her discoveries, she replied, 'If you know you're right, you don't care. You know that, sooner or later, it will come out in the wash.'

Barbara McClintock passed away in Huntington, New York, on 2 September 1992, at the age of ninety. She never married or had children.

Elise 'Lise' Meitner

Born: 7 November 1878 in Vienna, Austria
Died: 27 October 1968 (aged 89)
Field: Physics
Awards: Lieben Prize (1925); Max Planck
Medal (1949); Otto Hahn Prize (1955);
Fellowship of the Royal Society (ForMemRS)
(1955); Wilhelm Exner Medal (1960); Enrico
Fermi Award (1966)

The third of eight children, Elise Meitner was born on 7 November 1878 into an upper-middle-class Jewish family in Vienna, Austria. Her father, Philipp Meitner, was one of the first Jewish lawyers in Austria, as well as a chess master. She shortened her name from Elise to 'Lise' and, as an adult, converted to Christianity, following Lutheranism, and was baptised in 1908.

The earliest evidence of Lise's research interests occurred in 1887 when she was eight, as she kept a notebook of her records underneath her pillow. She was particularly interested in mathematics and science, and first studied colours of an oil slick, thin films, and reflected light. At the time when she completed high school, women were not allowed to attend public institutions of higher education, but Lise studied physics at a private educational institution, in part because of the support of her parents. She completed her course in 1901 with an *externe Matura* (matriculation) examination at the *Akademisches Gymnasium* (academic high school). She then enrolled in postgraduate studies at the University of Vienna, where she was undecided whether to study mathematics or physics, and so attended multiple lectures in both areas of study. While examining a beam of alpha particles, in her experiments with collimators and metal foil, Lise found that scattering increased with the atomic mass of the metal atoms, a result which later led Ernest Rutherford to examine the nuclear atom. In 1905, she became only the second woman to obtain a PhD in physics at the university, graduating *summa cum laude* with a dissertation on 'heat conduction in an inhomogeneous body'.

After receiving her doctorate, Lise rejected an offer to work in a gas lamp factory but, encouraged and supported by her father, she went to Friedrich Wilhelm's University in Berlin where Max Planck, the renowned physicist, allowed her to attend his lectures. Before this, Planck had not allowed women to sit in on his classes. After one year of attending his lectures, Lise became his assistant. She also worked with chemist Otto Hahn, together discovering several new isotopes. In 1909, Lise and Hahn published two papers on beta-radiation, as well as discovering and developing a physical separation method known as 'radioactive recoil', in which a daughter nucleus is forcefully ejected from its matrix as it recoils at the moment of decay.

In 1912, the research group known as 'Hahn-Meitner' moved to the newly-founded Kaiser Wilhelm Institute (KWI) in Berlin. There Lise worked without salary as a 'guest' in Hahn's Department of Radiochemistry. It was not until 1913, at the age of thirty-five and following an offer of an appointment in Prague as an associate professor, that Lise obtained a permanent position at KWI. Hahn and Lise were a formidable team, with her knowledge of physics and his knowledge of chemistry, and in 1918 while studying radioactivity, they discovered the element 'protactinium'. In 1922, Lise discovered the cause of the emission of electrons from surfaces of atoms with 'signature' energies known as the 'Auger effect', named after the French scientist Pierre Victor Auger who independently discovered the effect two years later. In 1926, Lise became the first woman in Germany to assume a post of full professor in physics, at the University of Berlin.

In 1938, after Austria was annexed by Germany, and because of her Jewish heritage, Lise was forced to flee. What she did not know at the time was that her escape from Berlin had been arranged by the Danish physicist Niels Bohr and the international physics community. With Bohr's help, she reached Sweden, where she was able to continue her work at the Manne Siegbahn Institute of Physics in Stockholm. Lise received little support, however, due in part to Siegbahn's prejudice against women in science. She encouraged Hahn to continue the research they had started in Berlin, concerning a strange 'bursting' that happened to uranium, which he wrote up in a paper submitted in December 1938, without crediting Lise's contributions. A new round of experiments with her nephew, the physicist Otto Frisch, provided Lise her own evidence for nuclear 'fission', a term coined by Frisch, and between January and March 1939 they wrote a series of articles on the nuclear fission of uranium.

In a biographical entry on Lise, Patricia Rife wrote,

> Set against the backdrop of war, intrigue, and prejudices against women in gaining acceptance/admission to a scientific career (Meitner, like her French colleague Madame Curie [(*see entry*)], was often the only woman in many famous physics circles throughout the early twentieth-century), Meitner's story becomes all the more ironic. Lise Meitner and Albert Einstein were among the few scientists who did not work on weapons research during World War II.

Even though she never worked on atomic bomb research herself, she was given 'celebrity treatment' in the USA. At a dinner for the Women's Press Club honouring her in 1945, it is reported that US President Harry Truman said, 'So you're the little lady who got us into all of this!'

In 1944, Hahn was awarded the Nobel Prize for Chemistry for his research into fission, but Lise was ignored, partly because Hahn downplayed her role after she left Germany. Her omission was only partly rectified in 1966, when Hahn, Meitner and Strassman were awarded the Enrico Fermi Award, an award of $50,000 USD, with a certificate signed by the US President and the Secretary of Energy, honouring scientists of international standing for their lifetime achievement in the development, use or production of energy.

Lise became a Swedish citizen in 1949, and in 1960 retired to Cambridge, England, living close to her nephew Otto Frisch. She received many honorary doctorates from universities in the USA and Europe. She travelled a great deal in her seventies and eighties, encouraging women students to 'remember that science can bring both joy and satisfaction to your life'.

In 1964, she suffered a heart attack, during a strenuous trip to the USA, from which she spent several months recovering. Affected by atherosclerosis, she passed away in her sleep at Cambridge on 27 October 1968, at the age of eighty-nine. She was buried in the village of Bramley in Hampshire at St James parish church, near her younger brother Walter, who had died in 1964. Her nephew, Otto Frisch, composed the inscription on her headstone that read, 'Lise Meitner: a physicist who never lost her humanity'.

In 1992, element 109, the heaviest known element in the universe, was named Meitnerium (Mt) in her honour. Lise Meitner is regarded as one of the most significant women scientists of the twentieth century and was praised by Albert Einstein as the 'German Marie Curie' (*see entry*). She never married.

Maud Leonora Menten

Born: 20 March 1879 in Port Lambton, Ontario, Canada
Died: 26 July 1960 (aged 81)
Field: Biochemistry
Award: Canadian Medical Hall of Fame (1998)

Maud Leonora Menten was born on 20 March 1879 in Port Lambton, Ontario, Canada. Her father, Captain William Menten, piloted boats across the Fraser River to Chilliwack, a centre of trade for agricultural products. The family moved to Harrison Mills, where her mother, Emma, worked as a postmistress as the family also owned and operated a hotel and a general store. After completing high school, Maud enrolled at the University of Toronto completing a bachelor's degree in 1904 and a master's degree in physiology in 1907, with a thesis on the distribution of chloride compounds in nerve cells. During her graduate studies she worked in the university's physiology laboratory as a demonstrator.

Showing much promise, in 1907 Maud was appointed a Fellow at the Rockefeller Institute for Medical Research (now Rockefeller University) in New York City, where she studied the effect of radium bromide on cancerous tumours in rats. After a year at the Institute, Maud worked as an intern at the New York Infirmary for Women and Children before returning to Canada a year later to begin studies at the University of Toronto. In 1911, Maud became one of the first Canadian women to receive the degree of Doctor of Medicine.

In 1912, Maud moved to Berlin where she worked with George Crile, a renowned surgeon. Together they worked on the control of acid-base balance during anaesthesia. Maud also communicated with the German biochemist, physical chemist and physician Leonor Michaelis, one of the world-leading experts in pH and buffers. The pair co-authored a paper in *Biochemische Zeitschrift* (Biochemical Journal), which showed that the rate of an enzyme-catalysed reaction is proportional to the amount of the enzyme-substrate complex. In recognition of their findings, the relationship between reaction rate and enzyme-substrate concentration became known as the 'Michaelis-Menten' equation. After studying with Michaelis in Germany, Maud enrolled for a PhD in medicine at the University of Chicago, rather than in Canada, where women were unable to formally undertake such research. As part of her studies, Maud travelled to other countries, including Germany, to complete her thesis, 'The Alkalinity of the Blood in Malignancy and Other Pathological Conditions; Together with Observations on the Relation of the Alkalinity of the Blood to Barometric Pressure'. She was awarded her PhD in 1916.

In 1923, Maud was employed as an assistant professor at the University of Pittsburgh in the School of Medicine and was later promoted to associate professor and head of Pathology at the Children's Hospital of Pittsburgh. In 1948, at the age of sixty-nine, she was made a full professor of Pathology. One of her inventions was the azo-dye coupling reaction for alkaline phosphatase, which is still used today in histochemistry (the staining of cells with chemicals such as dyes, enabling microscopic visualisation and quantification of specific cell components). Maud characterised bacterial toxins from *B paratyphosus*, *Streptococcus scarlatina* and *Salmonella ssp* that were used in a successful immunisation programme against scarlet fever in Pittsburgh in the 1930s and 1940s.

In 1944, Maud also conducted the first electrophoretic separation of blood haemoglobin proteins and researched the properties of haemoglobin, the regulation of blood sugar level, and kidney function. In addition, she investigated the mobility of proteins in the presence of electric fields (electrophoresis). Her results provided crucial information on differences in size and mobility of haemoglobin molecules. She also used histochemical approaches to study glycogen and nucleic acids in bone marrow.

In 1950, after retiring from the University of Pittsburgh, Maud returned to Canada where she continued to do cancer research at the British Columbia Medical Research Institute. In 1955, she retired due to poor health. She was a prolific researcher, authoring or co-writing over seventy research papers. The American freelance science writer Rebecca Skloot described her as a 'pocket dynamo', who wore 'Paris hats, blue dresses with stained-glass hues, and Buster Brown shoes'. Maud was somewhat of an adventurer, driving a Model T Ford through the University of Pittsburgh area for over thirty years, although reportedly not well, and people kept out of her path. She played the clarinet, was a talented artist, climbed mountains, went on an Arctic expedition and studied astronomy. Maud also spoke several languages, including Russian, French, German, Italian and at least one Native American language.

On 17 July 1960, Maud passed away at the age of eighty-one, in Leamington, Ontario, retaining her Canadian citizenship throughout her life. On her death, colleagues Aaron H. Stock and Anna-Mary Carpenter honoured her in an obituary in the prestigious journal *Nature*, writing,

> Menten was untiring in her efforts on behalf of sick children. She was an inspiring teacher who stimulated medical students, resident

physicians and research associates to their best efforts. She will long be remembered by her associates for her keen mind, for a certain dignity of manner, for unobtrusive modesty, for her wit, and above all for her enthusiasm for research.

In 1998, Maud Menten was posthumously inducted into the Canadian Medical Hall of Fame and was also honoured by the University of Toronto with a plaque. At the University of Pittsburgh there is the named Menten Chair in Experimental Pathology and the annual Dr Maud L. Menten Memorial Lecture Series is hosted in her honour. In 2015, her birthplace of Port Lambton installed a commemorative bronze plaque in her memory. She did not marry and had no children.

Maria Sibylla Merian

Born: 2 April 1647 in Frankfurt am
Main, Germany
Died: 13 January 1717 (aged 69)
Field: Naturalist and ecologist

Maria Sibylla Merian was born on 2 April 1647 in Frankfurt am Main,
Germany. Her father was the Swiss engraver and publisher Matthäus
Merian the Elder, who married her mother, his second wife, Johanna
Sybilla Heyne, in 1646. When Maria was born the following year, she
was his ninth child. Her father passed away in 1650, and a year later her
mother remarried, to Jacob Marrel, a flower and still life painter who
encouraged Maria to paint and draw. At age thirteen she painted her first
pictures of insects and plants from specimens she had captured, and from
an early age she had access to many books about natural history. Details
of her life as a young girl were provided in the foreword to her 1705 book,
Metamorphosis insectorum Surinamensium (Life cycles of the insects of
Suriname), where she wrote,

> I spent my time investigating insects. At the beginning, I started with silk
> worms in my home town of Frankfurt. I realized that other caterpillars
> produced beautiful butterflies or moths, and that silkworms did the
> same. This led me to collect all the caterpillars I could find in order to
> see how they changed.

In 1665, at the age of eighteen, Maria married Johann Andreas Graff. In
1668, their daughter, Johanna Helena, was born in Nuremburg, where the
couple had moved. It was there that Maria painted watercolours, mainly
flowers. She had a fascination with the caterpillars she found on plants,
allowing them to pupate and emerge as butterflies.

In 1675, Johann published the first of Maria's two-volume series on
caterpillars, the second volume appearing in 1683. Each volume contained
fifty plates, engraved and etched by Maria, including a description of

caterpillar life cycles. She also documented evidence on the process of metamorphosis and the plant hosts of 186 European insect species.

The first of a three-part volume of books, each with twelve plates, known as a *Blumenbuch* (Book of Flowers) of natural illustrations also appeared in 1675. The initial volume was titled *Florum Fasciculus Primus*; the second volume, from 1677, was called *Florum Fasciculus Alter*; and in 1680, the third volume, *Florum Fasciculus Tertius*, appeared.

In 1678, her second daughter, Dorothea Maria, was born. The family had moved to Frankfurt am Main, but her marriage was reportedly not a happy one.

In 1680, Maria published a second edition of her *Blumenbuch*, titled *Neues Blumenbuch* (New Book of Flowers), containing all three volumes together. Like the first edition, the second was sold without a binding. Only six complete copies of this work are known to have survived to this day.

When her stepfather died in 1681, Maria moved in with her mother and, in 1685, travelled with her mother, her husband and their children to Friesland, where her half-brother, Caspar Merian, had lived since 1677. She was attracted to the religious Labadists community there, a Protestant sect, which had settled on the grounds of a stately home, Walt(h)a Castle, at Wieuwerd in Friesland. However, Johann was not allowed in (although he later came back twice), as the Labadists declared his marriage to Maria invalid. Their divorce was finally made official in 1692. Maria stayed with the group for about five years, with the Labadists being engaged in a number of occupations, including printing, farming and milling. At its peak, the community numbered around 600, with visitors arriving from England, Italy, Poland and elsewhere.

In 1690, Maria moved with her daughters to Amsterdam and it was there, in the Vijzelstraat, that she started her first private studio. A number of watercolours painted either by her, or one of her daughters, still remain to this day. Johanna Helena had become a very skilled artist in her own right, as did Dorothea Maria, who later followed suit. Meanwhile, Maria's fame had spread, and she mingled with prominent residents of Amsterdam who were studying nature and collecting curiosities. These included the renowned Nicolaes and Jonas Witsen, the botanist and anatomist Frederik Ruysch and later the pharmacist Albertus Seba.

In 1699, Maria travelled to the Dutch colony of Suriname to examine tropical insects. In 1705, she published *Metamorphosis insectorum Surinamensium* (Life cycles of the insects of Suriname) in both a coloured and an uncoloured copy. On her own initiative, a Latin and a Dutch edition of this work containing sixty plates was published and Maria paid the production costs herself. As there were few colour images printed before 1700, this volume has been credited with influencing a range of naturalist illustrators. As a result of Maria's careful observations and documentation of the metamorphosis of the butterfly, the eminent naturalist Sir David Attenborough considered her to be among the most significant contributors to the field of entomology. Maria was no doubt a leading entomologist of her time, discovering many new facts about insect life through her research.

In 1715, Maria Merian suffered a stroke, but continued her work despite being partially paralysed. She passed away on 13 January 1717 at age sixty-nine in Amsterdam, the Netherlands, and was buried four days later at the *Leidse Kerkhof* cemetery. After her death, Maria became even more renowned, with various birds, animals and insects named after her. These include two butterflies: a common postman butterfly (*Heliconius melpomene meriana*) and a split-banded owlet butterfly (*Opsiphanes cassina merianae*). The Cuban sphinx moth has been named *Erinnyis merianae*, and a bug with no common name has been named *Plisthenes merianae*. A genus of mantises has also been named *Sibylla*.

The bird-eating spider *Avicularia merianae* was named in her honour, and in 2017 the spider *Metellina merianae* followed suit. The Argentine black-and-white tegu lizard has been named *Salvator merianae* and a cane toad was named *Rhinella merianae*. A snail with no common name was named *Coquandiella meriana* and the Madagascan African stonechat bird was called *Saxicola torquatus sibilla*. A genus of exotic flowering plants was named *Meriana* and a bulbil bugle-lily is known as *Watsonia meriana*.

In 2005, a modern research vessel named *Maria S. Merian* was launched at Warnemünde, Germany, and on 2 April 2013 she was honoured with a Google Doodle to mark the 366th anniversary of her birth. In 2016, her *Metamorphosis insectorum Surinamensium* was re-published with updated scientific descriptions.

Three hundred years after her death, on 24 March 2017, the Lloyd Library and Museum in Cincinnati, Ohio, hosted 'Off the Page', an exhibit rendering many of her illustrations as 3D sculptures with preserved insects, plants, and taxidermy specimens. Three months later, a symposium was held in her honour in Amsterdam. Maria Merian is regarded as one of the greatest botanical illustrators and the world's first ecologist. She was indeed a pioneering woman of science.

Maryam Mirzakhani

Born: 3 May 1977 in Tehran, Iran
Died: 14 July 2017 (aged 40)
Field: Mathematics
Awards: Blumenthal Award (2009); Satter Prize (2013); Clay Research Award (2014); Fields Medal (2014)

Maryam Mirzakhani was born on 12 May 1977 in Tehran, Iran. Her father, Ahmad, was an electrical engineer, and while Maryam initially planned on becoming a writer it was mathematics that became her passion. She attended the all-girls Tehran Farzanegan School as part of the National Organisation for Development of Exceptional Talents (NODET). The principal was undeterred by the fact that no girl had ever competed for Iran's International Mathematical Olympiad team and Maryam first gained international recognition during the 1994 and 1995 competitions. In 1994 she earned a gold medal, and in 1995 was awarded a perfect score and another gold medal. In doing so, Maryam became the first Iranian student to achieve a perfect score and to win two gold medals.

After obtaining her Bachelor of Science in mathematics in 1999 from the Sharif University of Technology in Tehran, Maryam enrolled in the graduate school at Harvard University, where she was supervised by Curtis McMullen, the winner of the prestigious Fields Medal in mathematics in 1998. At Harvard, Maryam was renowned for her determination and relentless questioning, despite difficulties with the English language. In 2004, Maryam obtained her PhD, the same year becoming a research fellow at the Clay Mathematics Institute and an assistant professor of mathematics at Princeton University. In 2009, she became a professor at Stanford University.

Maryam made a number of contributions to the theory of moduli spaces of Riemann surfaces. She obtained a new proof for the formula discovered by Edward Witten and Maxim Kontsevich on the intersection numbers of tautological classes on moduli space, as well as an asymptotic formula for the growth of the number of simple closed geodesics on a compact hyperbolic surface. In doing so, she generalised the theorem of the three geodesics for spherical surfaces.

In 2008, Maryam married Jan Vondrák, a Czech theoretical computer scientist and applied mathematician. They had a daughter named Anahita.

In 2014, with Alex Eskin and input from Amir Mohammadi, Maryam proved that complex geodesics and their closures in moduli space are surprisingly regular, rather than irregular or fractal. The closures of complex geodesics are algebraic objects defined in terms of polynomials and therefore have certain rigidity properties, which is analogous to a celebrated result that mathematician Marina Ratner of the University of California had arrived at during the 1990s. In the same year, Maryam was awarded the prestigious Fields Medal for 'outstanding contributions to the dynamics and geometry of Riemann surfaces and their moduli spaces'. Awarded by the International Mathematical Union, the Fields Medal is regarded as the 'mathematician's Nobel Prize' and one of the highest honours that a mathematician can receive. As at 2019, Maryam was the only woman to win the award.

The award was made in Seoul at the International Congress of Mathematicians on 13 August 2014. However, the year before, Maryam had been diagnosed with breast cancer and while she had recovered sufficiently from the rigours of her treatment to attend and accept the medal, she was not able to deliver the invited lecture for medal recipients. At the time of the award, the internationally renowned mathematician Jordan Ellenberg explained her research to a popular audience:

> [Her] work expertly blends dynamics with geometry. Among other things, she studies billiards. But now, in a move very characteristic of modern mathematics, it gets kind of meta: She considers not just one billiard table, but the universe of all possible billiard tables. And the kind of dynamics she studies doesn't directly concern the motion of the billiards on the table, but instead a transformation of the billiard table itself, which is changing its shape in a rule-governed way; if you like, the table itself moves like a strange planet around the universe of all possible tables ... This isn't the kind of thing you do to win at pool, but it's the kind of thing you do to win a Fields Medal. And it's what you need to do in order to expose the dynamics at the heart of geometry; for there's no question that they're there.

By 2016, the cancer had spread; on 14 July 2017, at the age of forty, Maryam died at Stanford Hospital in Stanford, California. President Hassan Rouhani of Iran and other officials published condolence messages and praised her scientific achievements. Rouhani said that 'the unprecedented brilliance of this creative scientist and modest human being, who made Iran's name resonate in the world's scientific forums, was a turning point in showing the great will of Iranian women and young people on the path towards reaching the peaks of glory and in various international arenas.'

On her death, the prominent Australian mathematician Professor Nalini Joshi (*see entry*) wrote, 'I know that in the world somewhere, now or in the future, there is a little girl or a woman, brave, persistent, driven to discovery through relentless questioning, who is deeply talented, who may win a Fields Medal again. The bell is still ringing in my heart.'

Maryam Mirzakhani was survived by her husband, Jan, who, in 2019, was an associate professor at Stanford University, and their daughter, Anahita, who once referred to her mother's work as 'painting' because of the doodles and drawings that marked her process of working on proofs and problems.

On 2 February 2018, Satellogic, a high-resolution Earth observation imaging and analytics company, launched a ÑuSat type micro-satellite named in her honour. Asteroid 321357 Mirzakhani was named in her memory with the official naming citation being published by the Minor Planet Center (MPC 108698). She also has an Erdős number of 3.

Maria Mitchell

Born: 1 August 1818 in Nantucket,
Massachusetts, USA
Died: 28 June 1889 (aged 70)
Field: Astronomy
Awards: King of Denmark's
Cometary Prize Medal (1848); Fellow
of the American Association for the
Advancement of Science (1848)

Maria Mitchell was born on 1 August 1818 in Nantucket, Massachusetts, the third of ten children, to a large Quaker family. Her mother was Lydia (Coleman) Mitchell and her father, William Mitchell, as a young man was a cooper, then conducted his own school and was the principal officer in a bank from 1836 to 1861. As was the case with other Quakers, Maria's parents were very supportive of education for all their offspring, including the girls. They insisted that the girls receive an education equal to the boys.

By all accounts, Maria was a quiet child who enjoyed reading and was diligent in her schoolwork, attending Elizabeth Gardener's small school when quite young, then moving to the North Grammar School in 1827, where her father was principal. When she reached the age of eleven, Maria's father established his own school on Howard Street, and she became his teaching assistant. William had a passion for astronomy and the family owned a small telescope which he used to teach Maria the finer points of astronomy to the extent that when was aged just twelve, she aided him in calculating the exact moment of an annular eclipse. When her father's school closed, the now fifteen-year-old Maria attended Unitarian Minister Cyrus Peirce's 'school for young ladies' for the next two years.

After completing her schooling, Maria worked at Peirce's school as a teaching assistant before opening her own school in 1835, a controversial move as she allowed non-white children to attend at a time when there was segregation at the local public school. Twelve months later she accepted a position as a librarian at the Nantucket Athenaeum, continuing her astronomical observations all the while.

Her father's connections meant that Maria was able to meet some of the country's most eminent scientists. She also took every chance to go to the roof of the house to survey the heavens. Although some comets had been discovered by others, it was still viewed as a significant event and King Frederick VI of Denmark offered a gold medal prize for the discovery of each new 'telescopic comet', one too faint to be seen with the naked eye

Two months after turning twenty-nine, at 10.30 p.m. on the evening of 1 October 1847, Maria went to the roof of the Pacific National Bank to 'sweep the sky' with the family's telescope. She noticed a small blurry streak that, although invisible to the naked eye, was clear in the telescope. Suspecting immediately that it may well be a comet, she ran to tell her father, who was inclined to announce her discovery right away. But Maria asked him to wait until she could be certain and continued to observe it and recording its position. Determined that his daughter's discovery be recognised, on 3 October William wrote to his friend and colleague William C. Bond, who was the director of the Observatory at Harvard College (now University) in Cambridge, Massachusetts, announcing the discovery. The president of Harvard, Edward Everett, then wrote to William Mitchell, asking whether he was aware that Maria could claim a medal from the King of Denmark for her new comet.

As it happened, several others, including Father de Vico in Rome, had also independently observed the comet on 3 October and reported it to European authorities, but Maria received priority and she was recognised for the feat with the medal from the new King Christian VIII of Denmark. This brought her immediate fame and honours, the following year being the first woman elected to membership of the American Academy of Arts and Sciences and becoming a Fellow of the American Association for the Advancement of Science. The only women to have previously discovered a comet were the renowned astronomers Caroline Herschel (*see entry*) and Maria Margarethe Kirch (*see entry*). Maria Mitchell's comet was originally designated as Comet 1847 VI but was later referred to as C/1847 T1.

Maria was frequently mystified by all the attention she received, an entry in her diary after one scientific meeting reading,

It is really amusing to find one's self lionized in a city where one has visited quietly for years; to see the doors of fashionable mansions open wide to receive you, which never opened before. One does enjoy acting the part of greatness for a while! I was tired after three days of it, and glad to take the cars and run away.

She was not only struck by the knowledge she gained from her observations, but also the beauty of the sky. She recorded this observation in her journal,

Feb. 12, 1855 ... I swept around for comets about an hour, and then I amused myself with noticing the varieties of colour. I wonder that I have so long been insensible to this charm in the skies, the tints of the different stars are so delicate in their variety ... What a pity that some of our manufacturers shouldn't be able to steal the secret of dyestuffs from the stars.

In 1865, Maria was appointed a professor of astronomy at Vassar College, making her the first such female professor in the USA and was also appointed director of the College Observatory. She began recording sunspots in 1868 but, from 1873, she and her students made daily photographic records, thereby obtaining more accuracy. These were the first photographs of the sun and for the total solar eclipse. In July 1878, she travelled 2,200 miles (3,500 km) with five assistants to Denver where they used a 4 inch (10 cm) telescope to make their observations.

Maria never married, but she remained close to her family throughout her life. After retiring from Vassar College on Christmas Day in 1888, she returned to live in Lynn, Massachusetts, where she had previously lived with her father after the death of her mother in 1861. On 28 June 1889, Maria passed away at age seventy and was buried in Lot 411 in Prospect Hill Cemetery, Nantucket.

The Maria Mitchell Observatory in Nantucket is named in her honour and is part of the Maria Mitchell Association (formed after her death) located there. She is the namesake of the SS *Maria Mitchell,* a cargo ship built in the Second World War. A crater on the moon is called after her and her name appears on the front of the Boston Public Library. She was awarded two honorary doctorates, one from Hanover College in Indiana in 1853 and the other from Columbia University in 1887. She also received an honorary PhD from Rutgers Female College in 1870.

As the first US woman astronomer and an advocate for women, Maria Mitchell paved the way for others. In addition to her scientific work, she was also active in opposing slavery and in advocating for women's rights.

Marcia Neugebauer

Born: 27 September 1932 in New York City, New York, USA
Field: Space physics
Awards: California Woman Scientist of the Year (Museum of Science and Industry, 1967); Exceptional Scientific Achievement Award (NASA); Outstanding Leadership Medal (NASA); Distinguished Service Medal (NASA); Women in Technology International Hall of Fame (1997); William Kaula Award (2004); Arctowski Medal (National Academy of Sciences, 2010); George Ellery Hale Prize (Solar Physics Division of the American Astronomical Society, 2010)

Marcia Neugebauer was born on 27 September 1932 on Manhattan Island, New York City, New York, USA – 'Fifth generation born on Manhattan Island', as she later said. In 1946, her family moved to Vermont, where she attended a small private high school in Manchester with only 180 pupils. It was assumed that Marcia would attend the all-girls Vassar College, because her mother and grandmother has gone there, but a chance meeting while on holiday in Florida with a girl from Rochester, New York, who was going to Cornell University, convinced her to do likewise.

Marcia enrolled at Cornell, selecting physics as her major with a minor study area of philosophy that included logic, history of science and semantics in her senior year. She later remarked,

> I majored in physics as an undergraduate at Cornell. I am not sure why. Perhaps it was because my father had taught me to use a slide rule which made high school physics a lot easier and more interesting than it might have been.

Two of her classmates in physics, Sheldon Glashow and Steve Weinberg, were later joint winners, with Abdus Salam, of the Nobel Prize in Physics in 1979. Marcia received a Bachelor of Arts in physics with honours from Cornell in 1954, followed by a Master of Science in physics from the University of Illinois in Urbana in 1956. There she worked in the laboratory of the physicist David Lazarus, where she measured diffusion in metals. But it was at Cornell that she met her husband, the physicist Gerry Neugebauer. The couple were married in 1956 and later had two daughters, Carol and Lee. In 1956, Marcia also began her role as Senior Research Scientist at the Jet Propulsion Laboratory (JPL), which in 1958 was transferred to NASA (the National Aeronautics and Space Administration). She was to remain with JPL for forty years.

In her research on solar wind, Marcia developed analytical techniques vital to understanding the flow of energetic particles from the sun and its impact on Earth. Marcia developed analytical instruments that orbited Earth, some of which were set up on the Moon by the Apollo astronauts, while others flew by Halley's Comet on the European Giotto mission. She was a primary investigator of the Mariner Mark II plasma analyser, which made the first extensive measurements of the solar wind and discovery of its properties. With an increase in solar activity, the influence of solar wind on orbital and terrestrial electronics and power systems will become stronger. As communications systems become increasingly vital to all levels of businesses worldwide, an understanding of solar wind will be essential to adjusting these technologies accordingly. Her pioneering research yielded the first direct measurements of solar wind and uncovered its physics and interaction with comets.

During her long career with NASA, she was the study scientist for many space missions, while holding several management positions at Caltech's Jet Propulsion Laboratory in Pasadena, California. These included manager of the Physics and Space Physics sections, manager of the Mariner Mark II study team, and project scientist for Rangers I and II and the Comet Rendezvous Asteroid Flyby mission. In 1967 the Museum of Science and Industry named Marcia 'California Woman Scientist of the Year'. This was one of the awards that she said meant most to her.

Marcia received many awards from NASA, including the Exceptional Scientific Achievement Award, the Outstanding Leadership Medal, and the Distinguished Service Medal (the highest award given by NASA). She also served as president of the American Geophysical Union 1994–96 and was Editor-in-Chief of its journal, *Reviews of Geophysics*. Marcia also chaired the National Academy of Sciences' Committee on Solar and Space Physics.

In 1997, Marcia was inducted into the Women in Technology International Hall of Fame and the following year was awarded an honorary Doctorate of Physics by the University of New Hampshire. In 2004, she was awarded the William Kaula Award, the citation reading,

> Marcia Neugebauer is one of the pioneers of the Space Age. She started work at Jet Propulsion Laboratory in June 1956 and contributed directly to the first identification and studies of the solar wind using some of the first space missions. Later Marcia brought her enthusiasm, thoroughness, and broad impact to the American Geophysical Union, including working for the AGU publications programme as editor-in-chief of Reviews of Geophysics and then serving as president of the AGU.

Marcia was employed at JPL until her retirement in 1996, with occasional visiting appointments at Cambridge University, Caltech, and the University of California, Los Angeles (UCLA). In 2002, after a forty-five-year career at JPL, she moved to Tucson to continue her research at the University of Arizona while a resident of the Academy Village, where she became the president of the Arizona Senior Academy.

In 2010, Marcia was awarded the Arctowski Medal from the National Academy of Sciences, a distinguished award presented to honour outstanding contributions to the study of solar physics and solar-terrestrial relationships. In the same year, she was awarded the Hale Prize of the Solar Physics Division of the American Astronomical Society for outstanding contributions to the field of solar astronomy.

On 26 September 2014, Marcia's husband, Gerry, passed away in Tucson.

Marcia Neugebauer has written more than 200 scientific publications, is the editor of six books, and has served as president of the American Geophysical Union. In an article, 'Pioneers of space physics: A career in the solar wind', published in the *Journal of Geophysical Research* in 1997, Marcia wrote,

> It is difficult to think of myself as one of the 'Pioneers of Space Physics'. I certainly did not enter the field because of any strong pioneering spirit. It was simply a matter of being at the right place at the right time. I feel very fortunate to have been able to participate in what may have been the golden age of space physics. My generation may have cleaned up most of the easy stuff, but the next generation still has many exciting and important tasks to tackle.

Ida Tacke Noddack

Born: 25 February 1896 in Rhine Province,
Germany
Died: 24 September 1978 (aged 82)
Field: Physics and chemistry
Awards: Liebig Medal of the German
Chemical Society (1931); Scheele Medal of the
Swedish Chemical Society (1934)

Ida Tacke was born on 25 February 1896 in Lackhausen (today Wesel) in Rhine Province, Germany. Her father, Adalbert Tacke, was a varnish producer. In 1915, Ida attended the *Technische Hochschule* (technical high school) in Berlin, now the *Technische Universität* (technical university), just six years after women were allowed to study in all of Berlin's universities. Only nine out of the eighty-five students in her class studied chemistry. Ida obtained her *Diplom Ingenieur* (engineering degree) in 1919, for which she won first prize in chemistry and metallurgy. Her specialty was the field of higher aliphatic fatty acid anhydrides. She was one of the first generation of women in Germany to study chemistry, the percentage of women who did so increasing from 3 per cent before the First World War to 35 per cent during the war.

Ida immediately began postgraduate research on the organic chemistry of fatty acids, achieving a master's degree in 1919 and the degree *Doktor Ingenieur* (Doctor of Engineering) in 1921. She gained a position in the chemistry laboratory of the Berlin turbine factory of *Allgemeine Elektricitäts-Gesellschaft* (German Electricity Company, AEG), a company associated with General Electric in the USA. Located in the Moabit district of Berlin, the factory where Ida worked was designed in the shape of a turbine by the leading architect Peter Behrens.

Since 1922, Ida had been in contact with Walter Noddack, a researcher in the University of Berlin's Physical Chemistry Department. He was working in the chemistry division of the prestigious German *Physikalisch-Technische Reichsanstalt* (Imperial Physical Technical Institute, PTR), a government laboratory in Berlin with a specific interest in searching for the remaining missing elements of the Periodic Table. In 1924, Ida resigned her position at AEG to work full-time as an unpaid collaborator with Walter. Their PTR group concentrated on elements 43 and 75 that were predicted by Russian chemist and inventor Dmitri Mendeleev to be part of the Periodic Table, located in the same column as manganese ores. However, Walter and Ida considered that these new elements could be found in the minerals of manganese's horizontal neighbours in the Periodic Table and not in the same column.

In 1925, following intense analytical work, Walter and Ida announced the discovery of elements 75 (co-authored by the German scientist Otto Berg) and 43, respectively named 'rhenium', in deference to Ida's birthplace and the River Rhine, and 'masurium', honouring Walter's East German background. The following year, Ida and Walter were married. Other scientists who had been searching for the same elements disagreed with their findings, and so the couple undertook the mammoth job of surveying around 1,800 ores and meteorites to obtain sufficient quantities of the new elements. However, they succeeded only in the case of rhenium, with masurium proving to be a far more difficult proposition. They secured sponsorship from the industrial manufacturing company Siemens, which constructed a laboratory especially for their research, as the company was interested in using rhenium instead of tungsten in the filaments of electric lamps.

Walter and Ida decided to patent their work and, in 1929, were granted a German patent for the rhenium coating of lamp filaments. They were also awarded a British patent for the use of rhenium as a catalyst for oxidation processes. Three more patents followed in 1931 and 1932 in the USA for filaments for incandescent lamps and vacuum tubes, rhenium concentrates and the use of metallic rhenium as an electric glower for incandescent lamps. For these, in 1931, they were awarded jointly the German Chemical Society's esteemed Liebig Medal and, for their discovery of rhenium and masurium, they were unsuccessfully nominated for the Nobel Prize in 1932, 1933, 1935 and 1937. In 1934 they received the prominent Scheele Medal of the Swedish Chemical Society, awarded to 'a particularly prominent and internationally renowned pharmaceutical scientist', and were awarded a German patent for rhenium concentrates in the same year.

In 1934, Ida published a paper in which she criticised Enrico Fermi's alleged discovery of element 93 as the product of nuclear fusion by the bombardment of uranium with neutrons. Instead, she maintained that his experiments suggested nuclear fission and leading German scientists of the day scoffed at her notion, even ridiculing it. She remarked, 'When heavy nuclei are bombarded by neutrons, it would be reasonable to conceive that they break down into numerous large fragments which are isotopes of known elements but are not neighbours of the bombarded elements.'

The following year, the couple moved to Freiburg University where Walter was appointed full professor of Physical Chemistry, a position formerly occupied by a Jewish scientist, Georg Hevesy. Ida and Walter were unsuccessful in proving the existence of the element masurium by means of X-ray spectroscopic devices. In 1937, element 43 was isolated by Emilio Segre and Carlo Perrier in a cyclotron and was named 'technetium' due to its artificial origins. It is known today that it is nuclear and decomposes very rapidly, but this was still a mystery in their time and so masurium remained contentious until 1945.

Ida Noddack was the first to appreciate the concept of nuclear fission. She maintained that, when atoms are bombarded by protons or α-particles, the nuclear reactions that occur involve the emission of an electron, a proton, or a helium nucleus and the mass of the bombarded atom suffers little change. However, when atoms are bombarded by neutrons, which carry no electric charge, different types of nuclear reaction from those previously known would take place. It was not until 1939 that the prominent physicist Lise Meitner (*see entry*) at last explained the results of the work as fission.

Ida passed away on 24 September 1978, aged eighty-two, in Bad Neuenahr-Ahrweiler, Germany. She had spent her last years in Wohnstift Augustinum, an apartment building for elderly people in Bad Neuenahr on the Rhine near Bonn. Walter predeceased her, dying on 7 December 1960 aged sixty-seven. They had no children.

Amalie 'Emmy' Noether

Born: 23 March 1882 in Erlangen, Germany
Died: 14 April 1935 (aged 53)
Field: Mathematics and physics
Award: Ackermann-Teubner Memorial Award (1932)

Amalie Noether came from a Jewish family of eminent scientists, with her father, Max, being a research mathematician and full professor at the University of Erlangen. Her mother, Amalia, was a talented musician. At an early age, Amalie adopted her middle name 'Emmy' and, after completing high school in Erlangen, decided to study mathematics at university. However, two years earlier the University of Erlangen had re-confirmed its policy that admitting female students would 'overthrow all academic order'. As a result, Emmy and another woman were permitted to attend lectures as auditors only. After two years she transferred to the University of Göttingen, but during her first semester there learned that Erlangen had changed its policy and gave women the same rights as men in taking exams and matriculating. As a result, she returned in 1904, staying there to complete her PhD, 'On the Formation of the Forming System of the ternary biquadratic form', involving algebraic invariants, under the algebraist and long-time family friend Paul Gordon. She graduated *summa cum laude*.

As there was essentially no employment for women with PhDs, between 1908 and 1915 Emmy worked without pay at the Mathematical Institute in Erlangen. In 1918, she published a pivotal paper on the inverse Galois problem. Instead of determining the Galois group of transformations of a given field and its extension, she pondered whether, given a field and a group, it is always possible to find an extension of the field that has the given group as its Galois group. It later became known as 'Noether's problem'. Soon after this, the renowned mathematician David Hilbert invited Emmy to lecture in Göttingen but, as powerful as he was, even he was unable to organise a university appointment for her. However, in 1922, he was able to arrange for her a title of 'unofficial associate professor' where she also secretly took up swimming at a men-only pool.

It was during this period that Emmy investigated relative dependency, devising mathematical formulations for several concepts of Einstein's general theory of relativity. Of particular interest to her was the relationship between his theory and invariant theory (a branch of abstract algebra dealing with actions of groups on algebraic varieties, such as vector spaces, from the point of view of their effect on functions). She continued to work on differential invariants (invariants for the action of a Lie group on a space that involves the derivatives of graphs of functions in the space, a fundamental part of projective differential geometry) throughout the First World War.

She remained in Göttingen for eleven years, although during this period she held visiting professorships at Moscow in 1928, where she taught classes in abstract algebra and algebraic geometry as well as working with leading topologists, and Frankfurt in 1930. She never used lesson plans for her classes, but instead used the time for spontaneous discussion to solve difficult mathematical problems. Although this technique was not to everyone's liking, it was nevertheless effective enough that the class notes from some of her students formed the basis for textbooks.

Her attention then turned to an investigation of the theory of ideals, rings and chains, being influenced by the work of the German mathematician (Julius) Richard Dedekind. It was then that Emmy changed the face of algebra with her co-authored paper (with Werner Schmeidler) '*Moduln in nichtkommutativen Bereichen, insbesondere aus Differential und Differenzenausdrücken*' (Modules in non-commutative areas, especially from differential and difference expressions), that appeared in the journal *Mathematische Zeitschrift* (Mathematical Journal). This work heralded her new ideas.

In 1915, Emmy had proved a theorem that stated, 'every differentiable symmetry of the action of a physical system has a corresponding conservation law' and eventually published it in 1918. It has since become known as 'Noether's Theorem' and is used in theoretical physics and the calculus of variations. Her results were mainly in the area of algebraic structures, a knowledge of which provides insight into the optimal way in which computers may be designed, computation can be performed and how data can be optimally stored. Her theorem was described by the physicist Peter G. Bergmann as 'one of the cornerstones of work in general relativity as well as in certain aspects of elementary particles'.

Among the relationships to follow from the theorem, the most remarkable was one that linked time and energy. It showed that a symmetry of time, such as whether you throw a ball in the air today, or make the same toss next week, will have no effect on the trajectory of the ball. This directly relates to the Law of Conservation of Energy that energy can neither be created nor destroyed and can only be transformed from one form to another. In 1924, Dutch mathematician Bartel Leendert van der Waerden joined her research group and promoted her ideas between 1924 and 1928, leading to the foundation for the second volume of his textbook *Moderne Algebra* in 1931. In all, Emmy supervised sixteen doctoral students.

Her work was original and ground-breaking and her interest was now peaked by linear transformations as applied to commutative number fields. Her two seminal works, both appearing in *Mathematische Zeitschrift*, were '*Hyperkomplexe Größen und Darstellungstheorie*' (Large Hypercomplex and Representation Theory) (1929) and '*Nichtkommutative Algebra*' (Noncommutative Algebra) (1933).

In April 1933, with the rise of the Nazi regime, Emmy and other Jewish faculty members were given a directive that read, 'I hereby withdraw from you the right to teach at the University of Göttingen.' However, she continued to teach in secret from her apartment. German mathematician Hermann Weyl wrote fondly of Emmy in this period:

A stormy time of struggle like this one we spent in Göttingen in the summer of 1933 draws people closely together; thus I have a vivid recollection of these months. Emmy Noether – her courage, her frankness, her unconcern about her own fate, her conciliatory spirit – was in the midst of all the hatred and meanness, despair and sorrow surrounding us, a moral solace.

Emmy managed to escape Germany by accepting an invitation to undertake lecturing and research at Bryn Mawr College, a women's liberal arts institution in Pennsylvania, USA, commencing in the autumn of 1933. Emmy also undertook research at the Institute of Advanced Study in Princeton, New Jersey, but in April 1935 underwent surgery for an ovarian cyst, passing away four days later after the operation.

On 2 January 1935, three months before her death, the American mathematician Norbert Wiener, who established the science of cybernetics, wrote that Emmy was 'the greatest woman mathematician who has ever lived; and the greatest woman scientist of any sort now living, and a scholar at least on the plane of Madame Curie [(*see entry*)]'. And, on 1 May 1935, after her passing, Albert Einstein wrote to the *New York Times*:

In the judgment of the most competent living mathematicians, Fräulein Noether was the most significant creative mathematical genius thus far produced since the higher education of women began. In the realm of algebra, in which the most gifted mathematicians have been busy for centuries, she discovered methods which have proved of enormous importance in the development of the present-day younger generation of mathematicians.

Nearly three decades after her death, at an exhibition at the 1964 World's Fair devoted to Modern Mathematicians, Emmy Noether was the only woman represented among the notable mathematicians of the modern world.

Christiane 'Janni' Nüsslein-Volhard

Born: 20 October 1942 in Magdeburg, Germany
Field: Biology
Awards: Gottfried Wilhelm Prize (1986); Louisa Gross Horwitz Prize (1992); Albert Lasker Award for Basic Medical Research (1991); Nobel Prize in Physiology or Medicine (1995)

The second of five children, Christiane ('Janni') Volhard was born on 20 October 1942 in Magdeburg, Germany. Her mother was Brigite Volhard, a musician and painter, and her father was Franz Volhard, a professor of medicine in Frankfurt and a heart and kidney specialist. The family lived in a flat in the south of Frankfurt. An avid reader as a child, Christiane was especially interested in animals and plants, and by the age of twelve was convinced that she wanted to be a biologist. As a small child she spent a number of vacations on a farm in a little village, which had been the refuge of her grandparents in the last year of the Second World War. Christiane recalled that, in the post-war period, when it was hard to purchase things, her parents made toys and books for their children, who also learned to sew and make things for themselves. Both her parents were good musicians and painted, 'so we kids did that too', she recounted. Christiane learnt to play the flute.

Christiane was the only one of her family with a lasting interest in the sciences and was supported in this by her parents, at the same time learning much from her high school teachers. She said that she was often 'intensely interested in things, obsessed by ideas and projects in many areas'. However, she rarely did her homework, admitting to being 'lazy', a verdict shared by her teachers, and finished high school with only average results, barely scraping by in English. Her father had passed away on 26 February 1962, the day of the exam. She then did a one-month course as a nurse in a hospital, an experience that intensified her desire to become a doctor.

Christiane enrolled at Johann Wolfgang Goethe University in Frankfurt, but found the subjects quite dull, apart from botany and later physics. She married while studying in Frankfurt, but the marriage was short-lived. She completed her bachelor's degree in 1964, studying biology, physics and chemistry. In mid-1964, she discovered that a new curriculum in biochemistry was being introduced at the Eberhard Karl University of Tübingen, the only one of its kind in Germany. She then completed a diploma in biochemistry in 1968, when she said she got her 'first real training in a laboratory', with people such as the chemist Heinz Schaller, with whom she then embarked on doctoral studies. He taught her 'to think in quantitative terms, yields, completeness of reactions', being 'an excellent experimenter'. Her thesis project involved researching protein-DNA interaction. After that, she said, she 'got bored' and became interested in new methods for DNA sequencing and began to study the use of genetics in researching developmental problems. This took Christiane to Basel in 1975 to work with the Swiss developmental biologist Walter Gehring at the *Biozentrum*, the Centre for Molecular Life Sciences, of the University of Basel.

Soon after she started her postdoctoral work in Basel, most people in the laboratory began to work on recombinant DNA and molecular biology, with the aim of cloning developmentally interesting genes. It was there that she met Eric Wieschaus, who was finishing his PhD at the time. With a ellowship from the research funding organisation *Deutsche Forschungsgemeinschaft* (DFG) in 1977, Christiane spent a year working in Freiburg in the laboratory of Klaus Sander, the eminent insect embryologist, who had been the first person to describe gradients in the insect egg. He had done experiments in which he translocated a symbiont ball localised to the posterior pole in a leaf hopper embryo and in doing so changed the polarity and pattern over large distances of the egg. In Freiburg, Christiane and Sander did a fate map for the larval cuticle with Margit Schardin, using laser ablations of *Drosophila* (fruit fly) blastoderm cells. This experiment was important as it showed that the primordia of individual segments in the blastoderm stage were no more than three cells wide. It also led to a very detailed examination and description of the segmental pattern of the *Drosophila* larva, which they later used in their screens. She continued her work on *dorsal*, discovering the recessive phenotype and interpreted the phenotype postulating a gradient determining the dorsoventral axis. She said, 'I loved working with flies. They fascinated me, and followed me around in my dreams.'

In 1978, Christiane accepted a job at the new European Molecular Biology Laboratory in Heidelberg and Eric Wieschaus was hired at the same time. They began work together analysing *Drosophila*, leading to a landmark paper in the prestigious journal *Nature*, in 1980. In 1981, Christiane moved back to Tübingen to continue her work on *Drosophila*, and in 1985 she took up a position as director at the Max Planck *Institut für Entwicklungsbiologie* (Institute of Developmental Biology) – a position she held until 2014. In 1986, she received the Gottfried Wilhelm Leibniz Prize of the *Deutsche Forschungsgemeinschaft* (German Research Foundation), which is the highest honour awarded in Germany for research.

In 1991, Christiane won the Albert Lasker Award for Basic Medical Research for her work involving the discovery of genes that control development in animals and humans, and the demonstration of morphogen gradients in the fly embryo. Awarded by the Lasker Foundation for outstanding discovery, contribution and achievement in the field of medicine and human physiology, the award frequently precedes a Nobel Prize in Medicine – almost half of the winners have gone on to win one. This was the case for Christiane, who was awarded the Nobel Prize in Physiology or Medicine in 1995, with Eric Wieschaus and Edward B. Lewis, for research on the genetic control of embryonic development. At the time she led a research group at the Max Planck Institute that focused on pattern formation, growth and cell migration in the zebrafish, a new vertebrate model organism. In her Nobel Banquet Speech on 10 December 1995, she said,

> The three of us have worked on the development of the small and totally harmless fruit fly, *Drosophila*. This animal has been extremely cooperative in our hands – and has revealed to us some of its innermost secrets and tricks for developing from a single-celled egg to a complex living being of great beauty and harmony. ... None of us expected that our work would be so successful or that our findings would ever have relevance to medicine.

In 1997, Christiane was recognised in the German civil honours list with the award of the *Pour le Mérite* for Sciences and Arts, and in 2005 with the Grand Merit Cross with Star and Sash of the Federal Republic of Germany. After being vice-chancellor of the Order of *Pour le Mérite* from 2009 to 2013, she was made chancellor in 2013. She was secretary general of the European Molecular Biology Organization (EMBO), which promotes excellence in the life sciences, until 2009, and a member of many scientific councils (including *Gesellschaft Deutscher Naturforscher und Ärzte* (GDNÄ – Society of German naturalists and doctors) and the National Ethics Council of Germany (*Nationaler Ethikrat*). In order to support women with children in science, she founded the Christiane Nüsslein-Volhard Foundation in 2004.

Christiane Nüsslein-Volhard has been awarded honorary degrees from Yale, Harvard, Princeton, Rockefeller, Utrecht, University College London, Oxford (June 2005), Sheffield, St Andrews (June 2011), Freiburg, Munich and Bath. The asteroid 15811 Nüsslein-Volhard is named in her honour.

In a 2017 interview with *The Node*, she said, 'I love music – I play the flute and sing. I have taken lessons for the last ten years and love to sing German Kunstlieder, and I sometimes give concerts to my friends.' In 2019 she received the Schiller Prize of the City of Marbach, endowed with 10,000 euros. This is awarded every two years, on 10 November, Friedrich Schiller's birthday, to personalities who are committed to the poet's tradition of thought in their life or work. In 2023, Christiane was residing in Bebenhausen, Germany.

Muriel Wheldale Onslow

Born: 31 March 1880 in Birmingham, England
Died: 19 May 1932 (aged 52)
Field: Biochemistry

Muriel Wheldale was born on 31 March 1880 in Birmingham, England, the only child of the barrister John Wheldale and Fannie (Heyward) Wheldale. She was educated there at King Edward's High School, an institution that was once the most prominent girls' schools for science teaching. In 1900, she enrolled in Newnham College, Cambridge, attaining a first class in both parts of the Natural Sciences Tripos with botany as her principal subject. However, she did not graduate from Cambridge as it did not award degrees to women until 1948.

In 1903, Muriel joined the genetics laboratory of the English biologist William Bateson at Cambridge, where she began her study focusing on the interaction of factors and of the inheritance of petal colour in *Antirrhinum* (snapdragons). Bateson was the first person to use the term 'genetics' to describe the study of heredity and made the ideas of Gregor Mendel popular following their rediscovery in 1900. Muriel and Bateson carried out a series of breeding experiments in various plant and animal species during the next seven years.

Muriel held a research position with a Bathurst studentship in 1904, and was an assistant lecturer in botany in Newnham College from 1906 to 1908. Her teaching and research stimulated the students' interest in the botanical aspects of biochemistry, and her classes in plant biochemistry were much valued in the teaching of advanced botany.

Muriel investigated the theory of 'multiple-allelism', a situation in which more than two alleles affect a phenotype, explaining the inheritance patterns of fur colour in mice. She examined inheritance patterns in snapdragons, in which the inheritance of flower colours also violated Mendel's laws. This research led to the first description of the phenomenon of 'epistasis' (the interaction of genes that are not alleles). In her 1907 paper, 'The Inheritance of Flower Colour in *Antirrhinum majus*', she described certain genes' influences on other genes in production of flower colour.

In 1909, Muriel won a Newnham College fellowship and then pursued the study of genetics in its biochemical aspect. The same anthocyanin pigments she had researched genetically, she now worked on chemically, her studies in this field culminating in the publication in 1916 of her celebrated book *The Anthocyanin Pigments of Plants* (which she revised in 1925). This book established Muriel's reputation both in England and abroad, as she was among the first to visualise and attempt to obtain a chemical interpretation of genetic data.

From 1911 to 1914, Muriel was part of the Bateson genetics group at the John Innes Horticultural Institution in Merton, Surrey. In 1914, she returned to Cambridge, joining the biochemistry laboratory of the English biochemist Frederick Gowland Hopkins, who was later a joint Nobel Prize in Physiology or Medicine in 1929 for the discovery of vitamins. It was at Merton that she investigated the biochemical aspects of petal colour, the genetics of which she had revealed in her work with Bateson. In combining genetics and biochemistry, Muriel became one of the first biochemical geneticists and soon after became one of the first women members of the Biochemical Club in Cambridge University.

In Hopkins' laboratory, Muriel also worked on the problem of anthocyanin metabolism and chemical structure. She discovered apigenin, and luteolin in 1914 and 1915, respectively. Both apigenin and luteolin are types of flavones, precursors of a number of yellow pigments. In 1917, she worked for the Food Investigation Board. In 1919, she married a colleague, the biochemist Victor Alexander Herbert Huia Onslow, the second son of the 4th Earl of Onslow. Victor, a paraplegic, had recently entered the field of chemical genetics and the couple had common research interests. They collaborated closely in research from 1917 until his death, focusing on the problem of colour and iridescence of insect scales.

From 1919, Muriel investigated the oxidase systems of the higher plants and, in 1920, published her second book, *Practical Plant Biochemistry*. From 1922 she worked at the Cambridge Low Temperature Station, directing the chemical section of the team's work there in progress on the changes involved in the ripening of fruit. Her results were outlined in the Annual Reports of the Food Investigation Board. Victor passed away in that year, and in 1924 Muriel wrote a memoir outlining his achievements as a man of courage and great mental strength. (His successes in the face of incredible difficulties were largely due to the encouragement and assistance he received from Muriel.)

In 1926, Muriel was appointed to a university lectureship in biochemistry at Cambridge, making her one of the first women to achieve that role in the university. She was renowned as an inspiring teacher with a gift for clear exposition. In 1931, she published the first volume of her textbook *Principles of Plant Biochemistry*, written at her house in Norfolk, one of her two favourite places, the other being the Balkans, where she spent her holidays. She had planned a second volume that did not appear due to her illness and death.

Muriel's followers included Rose Scott-Moncrieff who went on to identify, in about 1930, the first crystalline form of primulin, this being the first crystalline anthocyanin pigment ever identified. Muriel and Scott-Moncrieff have both been credited with founding biochemical genetics, each having their own claim to the honour.

On 19 May 1932, Muriel Onslow passed away in Cambridge at the age of fifty-two, at the height of her professional success. Her understanding of the genetics of pigment formation enabled her to do cutting-edge work in biochemistry. Newnham College honours her memory in a prize and a research fellowship. In 2010, the Royal Institution of Great Britain staged a play, *Blooming Snapdragons*, about four early twentieth-century women biochemists, one of whom was Muriel.

Ruby Violet Payne-Scott

Born: 28 May 1912 in Grafton, NSW, Australia
Died: 25 May 1981 (aged 68)
Field: Physics

Ruby Violet Payne-Scott was born on 28 May 1912 in Grafton, New South Wales, the elder child of Cyril Hermann Payne-Scott, a London-born accountant, and his Sydney-born wife Amy Sarah (Neale) Payne-Scott. Ruby attended Penrith Public Primary School from 1921 to 1924 and then Cleveland Street Girls' High School in Sydney from 1925 to 1926. She completed her secondary schooling at Sydney Girls' High School, with honours in both mathematics and botany.

Ruby then enrolled at the University of Sydney, where she graduated with first-class honours in mathematics and physics in 1933. Her outstanding results won her the Norbert Quirk prize for mathematics and, jointly with R. H. Healey, the Deas Thomson and Walter Burfitt scholarships for physics.

These were the years of the Great Depression, with employment difficult to secure, but Ruby found work at the University of Sydney as a physicist with the cancer research committee. It was there that she concentrated on radiation, a recently discovered treatment for cancer. She completed a Master of Science degree in 1936 with a thesis on the wave-length distribution of the scattered radiation in a medium traversed by a beam of X- or gamma rays. From 1936 to 1938, she conducted research with William H. Love at the University of Sydney's Cancer Research Laboratory, establishing that the magnetism of the earth had little of nor effect on the vital processes of beings living on the Earth, through experiments involving chicken embryos being subjected to magnetic fields up to 5,000 times more powerful than that on Earth.

When this research project concluded, and she was unable to find further work in her field, Ruby obtained a Diploma of Education in 1938 and taught at Woodlands Glenelg Church of England Girls' Grammar School, Adelaide.

In 2004, Sydney University academic Dr Claire Hooker said of Ruby that,

> At that point in her life, Ruby may well have been the story of many other brilliant women of her era. She may have disappeared into teaching and we may not have heard anything of her again. But I guess she loved physics and was looking for a way back in and she applied to Australian Wireless Amalgamated (AWA), an enormous company in those days that ran all the wireless services in Australia, it was the major hirer of physicists at that time.

The following year, Ruby was appointed librarian with Amalgamated Wireless (Australasia) Ltd (AWA) in Sydney, the first woman hired in any kind of research capacity, but 'she quickly turned the word librarian into a whole lot more', editing their journal and undertaking research, turning her 'full-time "librarian job" [into] a full-time physicists research job'.

In 1941, Ruby and other young engineers from AWA familiar with research on receivers and transmission, were hired by the division of radiophysics of the Council for Scientific and Industrial Research (later known as the Commonwealth Scientific and Industrial Research Organisation or CSIRO). Their task was to conduct research into radar which was hailed as a new, secret defence weapon. About three months after her appointment, the head of the division wrote a note about Ruby as a probationary employee, saying, 'Well, she's a bit loud and we don't think she's quite what we want and she may be a bit unstable, but we'll let her continue and see how she works out.'

Ruby's wartime research on small-signal visibility on radar displays and the accurate measurement of receiver noise factors brought her into contact with the group leader, the research physicist Joseph Lade Pawsey. Both were interested in reports of extra-terrestrial radio signals and they conducted what is thought to be the first radio astronomy experiment in the southern hemisphere, from the grounds of Sydney University in 1944. This was followed the year later by some of the key early solar radio astronomy observations, and between 1945 and 1947 Ruby discovered three of the five categories of solar bursts originating in the solar corona, making major contributions to the techniques of radio astronomy.

On 8 September 1944, at the District Registrar's office in Ashfield, Sydney, Ruby secretly married telephone mechanic William (Bill) Holman Hall. The CSIRO management officially heard of this in 1950, causing them to invoke a rule of the Australian Public Service (that remained until 1966) which required women to resign upon marriage. As a result, Ruby lost her permanent position and became a temporary employee, also losing her pension rights, a loss of status that she indignantly protested. A note was placed on her staff file by the Secretary of the CSIRO that no 'disciplinary action of any kind' was to be applied, because they wanted to keep Ruby on staff. During the Second World War, she had engaged in top secret work investigating radar and became an expert on the detection of aircraft using PPI (Plan Position Indicator) displays.

Ruby Violet Payne-Scott

In her private life, she was an atheist and an avid advocate for women's rights. She confronted inequality and injustice wherever she came across it, including the reduction in the rate of pay for women after the end of the war, to the old 'female rate', whereas women had been paid equal wages during the last stage of the war. Her style of dress was unconventional for a woman, including wearing shorts to work. When called in for an interview about this, she said, 'Well, this is absurd. We're climbing up ladders, up on aerials every day. I'm not going up on a ladder with a skirt on. The shorts are much better attire for us.' She also protested against a rule that allowed men to smoke, but not women, by reportedly going into the discussion of the issue smoking a cigarette. The Australian Security Intelligence Organisation (ASIO) was interested in her and kept a substantial file of thirty-nine pages on her activities as a 'person of interest' from 1948 to 1959, which was revealed in 2004 when a file was uncovered at the National Archives of Australia, although greatly redacted. A CSIRO informant, whose name was withheld, accused her of being a communist and also wrongly alleged that Ruby had been dismissed for failing to give notification of her marriage.

At the end of the Second World War, Ruby and Pawsey, with others from the radiophysics division, formed one of only two teams of scientists in the world to use survey work to investigate 'cosmic static' that emanated from the sun, radio nebulae and other astronomical objects. In consequence, Australia became a world leader in radio astronomy, with Ruby playing a central role with other pioneers, such as Bernard Mills and John Bolton. Her research focused on solar noise, particularly its correlation with sunspot activity, enabling her to more fully investigate the structure and character of the new types of non-thermal emissions from the solar corona. Ruby also played a pivotal role in the discovery of Type I, II and III bursts, the latter two relating to plasma emission processes.

On 26 January 1946, Ruby played a major role in the first-ever radio astronomical interferometer observation, when the sea-cliff interferometer was used to determine the position and angular size of a solar burst. But her most significant contribution to radio astronomy was to demonstrate, with Pawsey and Lindsay McCready, that the distribution of radio brightness across the sky could be treated mathematically as a two-dimensional sum of an infinite series of simple waveforms of varying frequency known as the 'Fourier components' of the distribution, and that therefore the components could be computed by performing a 'Fourier transform'. This work, published in the *Proceedings of the Royal Society of London* on 12 August 1947, was recognised as the mathematical foundation of future research in radio astronomy. In the same year she published, with co-authors J. G. Bolton and D. E. Yabsley, 'Relative times of arrival of bursts of solar noise on different radio frequencies', in the prestigious journal *Nature*.

Ruby left the CSIRO and radio astronomy in 1951. She was pregnant for the second time, having miscarried a few years earlier. There was no maternity leave, so she had no choice but to resign. The CEO of CSIRO wrote to her on the matter:

This event must be giving you a great deal of pleasure but I can well imagine that you regret having to leave off research, at least for the time being. Unfortunately we cannot give a married woman leave without pay, but I can assure you that I at least would be very pleased to see you return to Radiophysics in due course. I hope that the event comes off successfully.

At this time, Ruby had been receiving one of the highest salaries of the scientific staff who were not in administration and had been promoted to Senior Research Officer Grade I. Dr Claire Hooker observed, 'Thus this 39-year-old mother's research career came to an end.' She also changed her name to Ruby Hall.

Ruby and Bill had two children, Peter Gavin Hall, a mathematician who later worked in theoretical statistics and probability theory as a professor at the University of Melbourne, and Fiona Margaret Hall, one of Australia's more prominent artists. Ruby and Bill lived in Oatley, in Sydney, and Ruby stayed home to raise her two children until 1963. Then, in her early fifties, Ruby began her teaching career as a mathematics and science teacher at Danebank Anglican School for Girls in Hurstville, retiring in 1974, aged sixty-two.

Ruby Payne-Scott passed away in the Sydney suburb of Mortdale, New South Wales, on 25 May 1981, three days before her sixty-ninth birthday. In the final years of her life she suffered from presenile dementia.

In 2008, in Ruby's honour, CSIRO established the Payne-Scott Awards, designed as career development awards to assist staff returning from extended leave for parenting, family duties, sick or carer's leave. The awards provide funding for travel, courses, conferences and other professional development to assist the awardee to integrate their professional networks following an absence. In 2013, the Australian Institute of Physics (AIP), marking its fiftieth anniversary, initiated an award to recognise outstanding physics research by an early career researcher, naming it the Ruby Payne-Scott Early Career Research Award. The AIP Executive decided to name the awards after Ruby, as 'one of Australia's most outstanding physicists' because she made 'her pioneering contributions to radio-astronomy in Australia during the time (1941–51) she was an early career researcher'.

Rózsa Péter

Born: 17 February 1905 in Budapest, Hungary
Died: 16 February 1977 (aged 71)
Field: Mathematics
Awards: Kossuth Prize (1951); Manó Beke Prize (János Bolyai Mathematical Society, 1953); Silver State Prize (1970); Gold State Prize (1973)

Rózsa Politzer was born in Budapest, Hungary, on 17 February 1905 and was raised in a country torn by war and civil unrest. She attended the Maria Terezia Girls' School until 1922 before enrolling at Pázmány Péter University (in 1950 renamed Eötvös Loránd University). Rózsa intended to major in chemistry, but her interest soon turned to mathematics and she studied with world-famous mathematicians, including Lipót Fejér and József Kürschák.

After graduating in 1927, Rózsa was unable to find a permanent teaching position, although she had passed the qualifying exams to be a mathematics teacher. Due to the effects of the Great Depression, many university graduates could not find work and so she began private tutoring, as well as doing some high school teaching as a substitute teacher. At this time, Rózsa also began her graduate studies, initially concentrating her research on number theory. Before long, she found that her results had already been independently proven by the work of Robert Carmichael and L. E. Dickson, so she abandoned mathematics to focus on poetry. However, she was convinced to return to mathematics by her friend and long-time collaborator, the Hungarian mathematician László Kalmár, who suggested she research the work of Kurt Gödel on the theory of incompleteness. Rózsa prepared her own, different proofs from Gödel's work on the recursive functions. These involved a branch of mathematical logic, computer science and the theory of computation, which later included the study of computable functions and Turing degrees.

In 1932, she presented a paper, '*Rekursive Funktionen*' (Recursive Functions), at the International Congress of Mathematicians in Zürich, Switzerland, proposing for the first time that such functions be studied as a separate subfield of mathematics. More papers followed, and she received her PhD *summa cum laude* in 1935. The following year, Rózsa presented her paper *'Über rekursive Funktionen der zweiten Stufe'* (About recursive functions of the second stage) to the International Congress of Mathematicians in Oslo and her papers formed the foundation of the modern field of recursive function theory as a separate area of mathematical research, as she suggested. The German mathematician Walter Felscher spoke of his admiration of Rózsa's work, declaring that 'it may well be said that she forged, with her bare hands, the theory of primitive recursive functions into existence'. In 1937, Rózsa became a contributing editor of the *Journal of Symbolic Logic*.

As a result of the Fascist laws passed in 1939, Rózsa was forbidden to teach, due to her Jewish background, and was briefly confined to the Jewish ghetto in Budapest. In the 1930s, Rózsa changed her German-style name of 'Politzer' to a Hungarian one, 'Péter'. She continued working during the war years and, in 1943, wrote and printed a book, *Playing with Infinity: Mathematical Explorations and Excursions*, a discussion of ideas in number theory and logic for the lay person. A review of the book described it as 'a perfect compromise between rigour and clarity'. Unfortunately, many of the copies were destroyed by bombing and the book was not distributed until the war had ended. Originally published in Hungarian, her book has since been translated into English and at least fourteen other languages. Rózsa lost her brother, a number of friends and fellow mathematicians to fascism, something she outlined in the foreword of later editions of the book.

After the end of the Second World War, she obtained her first regular teaching position at the Budapest Teachers' Training College. In 1951 she published a book, *Recursive Functions*, which went through many editions. In 1952, Rózsa was the first Hungarian woman to be made an Academic Doctor of Mathematics and continued to apply recursive function theory to computers from the mid-1950s. She was committed to improving mathematical education in Hungary, writing school textbooks and working on improving the curriculum. In 1951 she was awarded the Kossuth Prize by the Hungarian government, and in 1953 she received the Manó Beke Prize by the János Bolyai Mathematical Society, awarded since 1950 to people prominent in furthering mathematical education.

When the Teachers' Training College was closed in 1955, Rózsa became a professor at Eötvös Loránd University. In 1959, she presented a major paper, *'Über die Verallgemeinerung der Theorie der rekursiven Funktionen für abstrakte Mengen geeigneter Struktur als Definitionsbereiche'* (On the generalisation of the theory of recursive functions for abstract sets of suitable structure as domains of definition), to the International Symposium in Warsaw (later published in two parts in 1961 and 1962). Rózsa won the Silver State Prize in 1970 and the Gold State Prize in 1973, both high national civilian honours. In the latter year she also became the first woman to be elected to the Hungarian Academy of Sciences.

In 1975, Rózsa retired from Eötvös Loránd University. Her popularity among students led to her being known as 'Aunt Rózsa', and she was an avid supporter of increased opportunities in mathematics for girls and young women. The following year she published her final book, *Recursive Functions in Computer Theory*, which was originally published in Hungarian. This was only the second Hungarian mathematical book to be published in the Soviet Union because its subject matter was considered indispensable to the theory of computers. In 1981 it was translated into English.

In her lectures to general audiences, which were often titled 'Mathematics is Beautiful', Rózsa declared that 'no other field can offer, to such an extent as mathematics, the joy of discovery, which is perhaps the greatest human joy'.

Rózsa Péter died of cancer on 16 February 1977, the day before her seventy-second birthday. In her eulogy, her student Ferenc Genzwein recalled that she taught 'that facts are only good for bursting open the wrappings of the mind and spirit' in the 'endless search for truth'. Her legacy is such that she is now referred to as the 'founding mother of recursion theory'.

Agnes Luise Wilhelmine Pockels

Born: 14 February 1862 in Venice, Italy
Died: 21 November 1935 (aged 73)
Field: Chemistry
Award: Laura Leonard Award from the
Colloid Society (with Henri Devaux) (1931)

Agnes Luise Wilhelmine Pockels was born on 14 February 1862 in Venice, Italy, at a time when the city was under Austrian rule and her father served in the Royal Austrian Army. When he fell ill with malaria in 1871, the family moved to Brunswick (Braunschweig) in Lower Saxony, which was part of the newly-formed German Empire. It was there that Agnes attended the Municipal High School for Girls. She said later in life that she had 'a passionate interest in natural science, especially physics, and would have liked to have studied them', but the school had little in the way of science on offer. Although she wanted to explore physics after completing high school, women were not allowed to enter the universities.

Her younger brother, Friedrich 'Fritz', enrolled at the University of Göttingen and graduated with a degree in physics, later becoming a professor of physics. By sending letters to Agnes and giving her access to his textbooks, he helped her learn advanced physics from her home in Brunswick. She also read *Naturwissenschaftliche Rundschau*, the German equivalent of *New Scientist*. Despite studying from home, due to a lack of equipment Agnes was not able to perform the experiments that her brother could. However, it seemed that her daily activities running the family household provided opportunities to implement what she had studied. It has been said that she became interested in the effect of impurities on the surface tension of liquids while washing up in her kitchen.

As Agnes was the only daughter, and unmarried, managing the household and caring for her unwell parents fell to her and she continued to live there throughout her life. As she looked after the house, Agnes had many opportunities to use soaps, oils and other products, and to see their effect on water. Her sister-in-law observed that, 'what millions of women see every day without pleasure and are anxious to clean away, i.e., the greasy washing-up water, encouraged this girl to make observations and eventually to ... scientific investigation.'

She was able to measure the surface tension of water by devising an apparatus known as the 'slide trough', a key instrument in the emerging discipline of 'surface science'. Irving Langmuir, an American chemist, used an improved version of her slide trough to make further discoveries on the properties of surface molecules, which earned him a Nobel Prize in chemistry in 1932. Her trough was also the forerunner of the 'Langmuir-Blodgett trough', developed later by Langmuir and physicist Katharine Blodgett (*see entry*). This is a laboratory apparatus that is used to compress monolayers of molecules on the surface of a given sub-phase (usually water) and measures surface phenomena due to this compression.

After two years of experimenting with her trough, Agnes received a letter from her brother advising her of a publication by Lord Rayleigh (John William Strutt) in which he investigated the properties of a thin layer of oil on the surface of water. As this was similar research to her own, Agnes sent him a letter, in German, with the results of her experiments,

> My lord, will you kindly excuse my venturing to trouble you with a German letter on a scientific subject? Having heard of the fruitful researches carried on by you last year on the hitherto little understood properties of water surfaces, I thought it might interest you to know of my own observations on the subject.

She said he could keep her results for himself, but he was so impressed that he had the letter translated and submitted it to the prestigious journal *Nature*, under her name, as well as another article of his own. He wrote to the journal stating, 'I shall be obliged if you can find space for the accompanying translation of an interesting letter which I have received from a German lady, who with very homely appliances has arrived at valuable results respecting the behaviour of contaminated water surfaces.' *Nature* published the letter under the heading 'Surface Tension' and this heralded the commencement of Agnes's career and reputation in studying surface films. However, her personal circumstances remained unchanged. She declared, 'Like a soldier, I stand firm at my post caring for my aged parents.'

Agnes continued her kitchen experiments and, with Rayleigh's encouragement, published several more papers on surface science, mostly in German journals. However, the declining health of her parents left her less time for study, and with the death of her brother in 1913, and the onset of the First World War, she retreated into private life.

After the war, Agnes stopped performing experiments and maintained only irregular contact with scientists in her field. Although she never held a formal appointment, her published papers eventually led to her recognition as a pioneer in the new field of surface science.

In 1931, with Henri Devaux, Agnes was awarded the Laura R. Leonard Prize of the German Colloid Society for 'Quantitative Investigation of the

Properties of Surface Layers and Surface Films', the first woman to win the award. The following year, she received an honorary doctorate from the Carolina Wilhelmina University of Brunswick, in recognition of her seventieth birthday.

Agnes's sister-in-law said that, in her later years,

> She led a quiet life as 'Auntie Agnes', like many other middle-aged women in Brunswick. She had many acquaintances, and two puzzle-solving societies met in her home … She herself always lived simply, and kept her thought to herself without saying much. The information about her special scientific knowledge was now only noised abroad in whispers.

Agnes Pockels, who never married, passed away 21 November 1935 in Brunswick. Shortly before her death, the eminent biologist Sir William Bate Hardy FRS wrote,

> I think I may say without exaggeration that the immense advances in the knowledge of the structure and properties of this fourth state of matter, which have been made during this century, are based upon the simple experimental principles introduced by Miss Pockels.

Helen Rhoda Arnold Quinn

Born: 19 May 1943 in Melbourne, Australia
Field: Physics
Awards: Dirac Medal of the ICTP (2000);
Honorary Officer of the Order of Australia
(2005); Oskar Klein Medal (2008); J. J. Sakurai
Prize (2013); Karl Taylor Compton Medal
(2016); Benjamin Franklin Medal (2018)

Helen Rhoda Arnold was born on 19 May 1943 in Melbourne, Australia, later attending Tintern Church of England Girls' Grammar School, in Ringwood East, Victoria, surrounded by 50 acres (20 hectares) of bushland. Though neither of her parents had university degrees, her father, Ted Adamson Arnold, was a self-educated engineer and inventor. Her mother, Helen Ruth (Down) Arnold, worked as a dietician and later as a home economics teacher. After completing her secondary schooling in 1959, Helen enrolled at the University of Melbourne with the support of the Australian Weather Bureau, where she spent two years studying general science with the intention of becoming a meteorologist.

Then, in 1962, she and her family emigrated to the USA, where she had to select a major that would not force her to start her degree all over again. She chose to study physics at Stanford University in California. She was not deterred by the fact that there were no women in the faculty at that time, and hardly any female students in the Stanford Physics Department. Helen's passion for mathematics had found an outlet in physics, assisted by the proximity of the Stanford Linear Accelerator Laboratory (now the Stanford Linear Accelerator Center or SLAC), which was on the cutting edge of theoretical and experimental research. She was so captivated that she enrolled in graduate school there, being one of only two per cent of those enrolled in PhD programmes in the US at the time who were women.

In 1966, she married Daniel James Quinn, a fellow physics graduate student. They had two children.

She completed her PhD in 1967 under the supervision of James Daniel 'BJ' Bjorken. Helen later remarked, 'Really, the beginning was the fact that particle physics was bubbling at that time at Stanford, and that's where I got hooked on it.'

After finishing her PhD, Helen travelled to Germany for postdoctoral study at the *Deutsches Elektronen-Synchrotron* (DESY) laboratory in Hamburg, a national research centre that operates particle accelerators to investigate the structure of matter. She then moved to Boston in the USA. There she taught high school for a brief period, before being offered a position at Harvard University. It was there that she collaborated with theoretical physicists Steven Weinberg (a future 1979 Nobel Prize winner in Physics) and Howard Georgi, working on 'grand unified theories', also known as 'GUTS'. These models attempted to bring together the three forces described by quantum physics: electromagnetism, which holds together atoms, and the weak and strong forces, which govern nuclear structure.

During a leave of absence from Harvard, Helen returned to Stanford where she joined forces with physicist and a particle theorist Robert Peccei. The pair had frequent discussions with Weinberg and Gerard 't Hooft (winner of the 1999 Nobel Prize in Physics with M. J. G. Veltman), both of whom were visiting Stanford at that time. Their research involved an attempt at understanding the strong force which governs the structure of particles such as protons. This theory is known as 'quantum chromodynamics' (QCD), in reference to the colour charge of quarks, which is analogous to electric charge. One difficulty was that QCD predicted certain outcomes that disagreed with experiments, including an electrical property of neutrons.

Working together, Helen and Peccei figured out that there would not be an issue if one type of quark had no mass. She said, 'That led me to think, well, in the very early universe when it's hot ... quarks are massless.' By adding a new symmetry, once quarks acquired their masses from the Higgs field the QCD issue could be resolved. In 1977, Helen and Peccei published their results (now known as the 'Peccei-Quinn theory'). As soon as their paper came out, Steven Weinberg and Frank Wilczek independently realised the theory also made a prediction of the existence of a light particle, the 'axion'. If axions exist (they are yet to be observed) and have low mass within a specific range, they are a possible component of cold dark matter, which permeates space and helps to bind galaxies together.

Helen then collaborated with experimentalists and other theorists at SLAC to understand the interactions involving the bottom quark (the fifth quark to be detected and a member of the third and heaviest pair of quarks). Their hope was that studying particles containing bottom quarks could reveal differences in the laws of physics for matter and antimatter that might possibly explain why there is a lot more matter than antimatter in the cosmos. Though such differences were indeed observed, they were not large enough to explain that puzzle, which remains a challenge to particle physics today.

In 2000, Helen was awarded, with two others, the Dirac Medal of the International Centre for Theoretical Physics, in Trieste, Italy, 'for pioneering contributions to the quest for a unified theory of quarks and leptons and of the strong, weak and electromagnetic interactions'. She was the first woman to receive this award. In 2001, Helen was elected as president of the American Physical Society (APS) for the year 2004, becoming only the fourth woman to be elected to that position since the society's establishment in 1899. In 2005, she undertook an Australian Institute of Physics lecture tour in celebration of the International Year of Physics. In the same year, she was appointed an Honorary Officer of the Order of Australia (AO) 'for service to scientific research in the field of theoretical physics and to education'. (The award was honorary because she was no longer an Australian citizen.) In 2008, she was the first woman to receive the Oskar Klein Medal from the Royal Swedish Academy of Sciences.

In 2009, Helen began a five-year term as chair of the Board on Science Education of the National Academy of Sciences and the following year she retired, and Stanford University granted her the title of Professor Emerita of SLAC.

In 2013, Helen won the J. J. Sakurai Prize for Theoretical Particle Physics from the American Physical Society (jointly with Roberto Peccei): 'For their proposal of the elegant mechanism to resolve the famous problem of strong-CP violation which, in turn, led to the invention of axions, a subject of intense experimental and theoretical investigation for more than three decades.'

Between 2015 and 2018, Helen Quinn was a member of the Board (*Comisión Gestora*) of the National University for Education (UNAE) in Ecuador. In 2018, she was awarded the Benjamin Franklin Medal in Physics from the Franklin Institute, awarded for outstanding achievements in science, engineering and industry.

In January 2019, Helen became chair of the board of the Concord Consortium, a nonprofit organisation dedicated to creating innovative educational technology for STEM learning. In the same year, she delivered the J. Robert Oppenheimer Lecture at the University of California, Berkeley, with the topic 'Teaching for Learning: What I have learned from learning research'. In 2021 she was awarded the Harvey Prize, an annual Israeli award for breakthroughs in science and technology

About physics, Helen said, 'It's really a lot of fun ... That's one reason to reach out to high school students. From what they get now, they'd never believe it was so interesting!'

Mina Spiegel Rees

Born: 2 August 1902 in Cleveland, Ohio, USA
Died: 25 October 1997 (aged 95)
Field: Mathematics
Awards: Award for Distinguished Service to Mathematics from the Mathematical Association of America (1962); Achievement Award by the American Association of University Women (1965); King's Medal for Service in the Cause of Freedom (UK) (1948); Public Welfare Medal from the National Academy of Sciences (1983)

Mina Spiegel Rees was born on 25 October 1902 in Cleveland, Ohio, USA, the fifth and youngest child of Moses and Alice Louise (Stackhouse) Rees. When she was two years of age, the family moved from Cleveland to New York, where Mina received her primary education in the city's public schools. She attended Hunter College High School, an institution for gifted girls, and was the top student. Mina then attended Hunter College where she majored in mathematics. At the end of her first year, she was asked if she would like to teach a course called 'transit' that dealt with surveying. She agreed, and that summer took a course on the subject at Columbia University. For the remainder of her undergraduate years she taught this trigonometric laboratory course, being paid half the salary of a regular instructor.

In 1923, Mina was invited to join *Phi Beta Kappa*, America's oldest academic honour society, and graduated *summa cum laude* even though she had held down a part-time teaching position in the mathematics department. After graduation, she taught at the Hunter College High School while also being enrolled for a master's degree at Columbia University. In an interview for *Mathematical People*, Mina said that she heard, unofficially, that 'the Columbia mathematics department was really not interested in having women candidates for PhDs' so she settled for her Master of Arts degree in 1925. She then accepted a position as a mathematics teacher at Hunter College but, in 1929, took sabbatical leave to enrol at the University of Chicago, with her sights set on obtaining a doctorate.

Mina was aware of the abstract algebraist Leonard Eugene Dickson's work on associative algebras and hoped to write a thesis in the field. Although Dickson now worked in a different area, he made an exception for Mina and supervised her thesis, 'Division algebras associated with an equation whose group has four generators', for which she was awarded a PhD in 1931. Her thesis was published in the *American Journal of Mathematics* in 1932.

Mina returned to her position as an assistant professor at Hunter College and, in 1940, was promoted to the rank of associate professor. During the Second World War, she again took a leave of absence from Hunter to work as a technical aide and executive assistant with the Applied Mathematics Panel in the Office of Scientific Research and Development. This group brought together mathematicians from many disciplines to work on military problems. For her contributions during the war, she was awarded the King's Medal for Service in the Cause of Freedom (UK), presented in recognition of wartime civilian services by foreign nationals, and the President's Certificate of Merit (USA). In 1946, Mina was invited by the US Navy to become head of the mathematics branch of the Office of Naval Research to support scientific and mathematical research in Washington DC. In 1948, Mina wrote that the Office of Naval Research was committed 'primarily to the support of fundamental research in the sciences, as contrasted with development, or with applications of known scientific results – the types of activity in which scientists were largely engaged during the war.' During 1952–53 Mina served as their deputy science director, working on projects that included fluid flow to solve antisubmarine problems, hydrofoil designs and early rocketry. In 1953, the council of the American Mathematical Society acknowledged the significance of Mina's role in a resolution adopted by its Council:

> Under her guidance, basic research in general, and especially in mathematics, received the most intelligent and wholehearted support. No greater wisdom and foresight could have been displayed and the whole post-war development of mathematical research in the United States owes an immeasurable debt to the pioneer work of the Office of Naval Research and to the alert, vigorous and farsighted policy conducted by Miss Rees.

The Institute of Mathematical Statistics adopted a similar resolution the same year. That year, 1953, Mina returned to Hunter College where she was appointed Dean of the Faculty, a position she held for the next eight years. In 1955, she married the physician Leopold Brahdy.

In 1961, Mina was appointed professor and the first Dean of Graduate Studies in the newly established City University of New York (CUNY), later becoming provost of the graduate division for 1968–69 and president of the Graduate School and University Center from 1969 until her retirement in 1972.

In 1962, Mina received the first Award for Distinguished Service to Mathematics from the Mathematical Association of America. It was given 'for outstanding service to mathematics, other than mathematical research' and for 'contributions [that] influence significantly the field of mathematics or mathematical education on a national scale'.

In 1971, she became the first woman president of the American Association for the Advancement of Science. The following year she was chair of the Executive Committee of the Board of Directors. In 1983, Mina was awarded the National Academy of Sciences Public Welfare Medal 'in recognition of distinguished contributions in the application of science to the public welfare'. She was also awarded at least eighteen honorary doctorates, and in 1985 the library at the Graduate School and University Center of CUNY was dedicated as the Mina Rees Library.

Mina passed away aged ninety-five on 25 October 1997 at the Mary Manning Walsh Home in Manhattan. She had lived on the Upper East Side. Her husband had died in 1977.

Uta Merzbach, the German American historian of mathematics who became the first curator of mathematical instruments at the Smithsonian Institution, said of her,

> Mina Rees was eminently rational. Her devotion to reason helped her formulate goals clearly and allocate resources judiciously in accordance with these goals. ...
>
> Mina Rees was eminently intelligent. She comprehended quickly, communicated effectively and thought creatively. Her ability to attach realisable pieces of basic research to mission-oriented applications of mathematics did much to develop a broadened base of support for mathematicians' work.
>
> Mina Rees was eminently civilised. Her diplomatic skills were considerable; her conversational technique bespoke her broad knowledge base as well as her wide interest in mathematical and non-mathematical topics. Experience and reflection led her to a balanced outlook on teaching and research, the arts and sciences, long-range and short-range planning and obligations of the professional and the private life.

Ellen Henrietta Swallow Richards

Born: 3 December 1842 in Dunstable,
Massachusetts, USA
Died: 30 March 1911 (aged 68)
Field: Chemistry
Award: National Women's Hall of Fame
(post., 1993)

Ellen Henrietta Swallow was born on 3 December 1842 in Dunstable, Massachusetts, USA. She was the only child of Peter Swallow and Fanny Gould (Taylor) Swallow, both of whom came from established families of modest means who believed in the value of education. Ellen was homeschooled in her early years, and in 1859, when the family moved to Westford, at the age of seventeen she attended Westford Academy. There she studied mathematics, composition and Latin. Ellen's proficiency in Latin allowed her to study French and German as well, and her flair for languages led to her being in demand as a tutor, which generated some income to support her further studies.

Ellen left the Westford Academy in March 1862 and contracted measles two months later, significantly disrupting her preparations to begin teaching. In the spring of 1863, the family moved to Littleton, Massachusetts, where her father had just purchased a larger store and expanded his business. In June 1864, twenty-one-year-old Ellen took a teaching position, but spent 1865 managing the family store and taking care of her ill mother. During the winter of 1865/66, Ellen studied and attended lectures in Worcester, Massachusetts. By the time she had reached her mid-twenties, she had saved enough to enter one of the new women's colleges, and in September 1868 she was admitted to Vassar College. Attracted to astronomy, as a pupil of Maria Mitchell (*see entry*), she graduated in 1870 with a bachelor's degree. She then focused on chemistry, earning a Master of Arts degree with a thesis on the amount of vanadium in iron ore, performing numerous experiments in mineralogy. These included the discovery of an insoluble residue of the rare mineral samarskite. On 10 December 1870, after some discussion and a vote, the Faculty of the Institute of Technology in Boston recommended to the Corporation her admission as a non-degree student in chemistry. As a result, she became the first woman admitted to Massachusetts Institute of Technology (MIT), where she was able to continue her studies, 'it being understood that her admission did not establish a precedent for the general admission of females'.

In 1873, Ellen received a Bachelor of Science degree from MIT for her thesis, 'Notes on Some Sulpharsenites and Sulphantimonites from Colorado'. She then became an unpaid chemistry lecturer at MIT for the next five years, to 1878, when she continued her studies and was set to be awarded its first advanced degree. However, MIT was reluctant to grant this honour to a woman and did not award its first doctorate to a woman until 1886.

On 4 June 1875, Ellen married Robert Hallowell Richards, chairman of the Mining Engineering Department at MIT, with whom she had worked in the mineralogy laboratory. They lived in Jamaica Plain, Massachusetts, in a 'mutually sympathetic and hospitable home life'. She continued her association with MIT, volunteering her services and contributing $1,000 USD annually to the 'Women's Laboratory', established to afford better opportunities for the scientific education of women and providing a programme in which her students, mainly schoolteachers, who lacked training in laboratory work, could perform chemical experiments and learn mineralogy. Ellen was behind the finding of financial support for the laboratory, and she was the 'guiding hand in its management, and hers the leading spirit'.

Ellen acted as a consulting chemist for the Massachusetts State Board of Health from 1872 to 1875, and in the 1880s became interested in issues of sanitation, especially air and water quality. She undertook a series of water tests on 40,000 samples. These led to the so-called 'Richards' Normal Chlorine Map', which predicted the pollution of inland water in the State of Massachusetts. Her map also plotted the chloride concentrations in waters of the state. The survey work she undertook was later described as a 'classic of its kind', and its success was attributed to her 'enthusiasm, energy, experience and insight'.

As a result of Ellen's efforts, Massachusetts established the first water quality standards in the USA, leading to the establishment of the first modern sewage treatment plant. In 1879, she was recognised by the American Institute of Mining and Metallurgical and Petroleum Engineers (now the American Institute of Mining Engineers, AIME) as their first female member.

From 1884 until her death, Ellen was an instructor at the newly founded MIT laboratory of sanitary chemistry at the Lawrence Experiment Station, the first if its kind in the USA. From 1887 to 1897, she held the position as the official water analyst of Massachusetts, as well as serving as a nutrition expert for the US Department of Agriculture.

Ellen also applied her scientific knowledge to the home and family nutrition, feeling that all women should be educated in the sciences. She wrote to her parents, 'Perhaps the fact that I am not a radical and that I do not scorn womanly duties but claim it as a privilege to clean up and sort of supervise the room and sew things is winning me stronger allies than anything else.' She also demonstrated model kitchens, devised curricula and organised conferences.

In 1882, Ellen wrote a book about science for use in the home, *The Chemistry of Cooking and Cleaning: A Manual for House-keepers*. Her next book, in 1885, *Food Materials and Their Adulterations*, led to the passing of the first Pure Food and Drug Act in Massachusetts. She also applied these principles to her own home, improving air quality there by moving from coal heating and cooking oil to gas. She and her husband installed fans to pull air from the home to the outside to create better air quality inside. The couple also used chemicals to test the water quality of the property's well, to ensure that waste water was not contaminating their drinking water.

Ellen derived the term 'euthenics', meaning 'the science of better living' and was the first to use the term in her 1905 book, *The Cost of Shelter*. In 1910, her book *Euthenics: the science of controllable environment* refined the definition to 'the betterment of living conditions, through conscious endeavour, for the purpose of securing efficient human beings'.

Ellen served on the board of trustees of Vassar College for many years and was granted an honorary Doctor of Science degree in 1910. She died of heart disease in Jamaica Plain at the age of sixty-eight on 30 March 1911, survived by her husband Robert. She was buried at Christ Church Cemetery, Gardiner, Maine.

A memorial tribute to her in MIT's alumni publication in 1911 concluded in saying,

> A powerful leader, a wise teacher, a tireless worker, of sane and kindly judgment, Mrs Richards has taught and inspired thousands to carry forward the movements which she has inaugurated. Her associates and co-laborers necessarily mourn their loss and miss her leadership, but they will best express their appreciation of her life and its far-reaching influence by increased activity in behalf of those phase of human progress and betterment for which she sacrificed herself so freely.

In 1992, the Ellen Swallow Richards House was designated a National Historic Landmark. The following year, she was posthumously inducted into the National Women's Hall of Fame. In her honour, MIT designated a room in the main building for the use of female students and, in 1973, on the occasion of the 100th anniversary of her graduation, established the 'Ellen Swallow Richards professorship for distinguished female faculty members'.

In 2011, Ellen Swallow Richards was listed as number eight on the MIT150 list of the top 150 innovators and ideas from MIT and is also commemorated on the Boston Women's Heritage Trail. Swallow Union Elementary School in her hometown of Dunstable, Massachusetts, is named in her honour. In 2013, the AIME established an award in her honour, the

Ellen Swallow Richards Diversity Award. In announcing the award, the president of the Minerals, Metals and Materials Society said that

> Ellen Richards was a pioneer in diversity who quietly and, in her own way, overcame seemingly insurmountable barriers to become an accomplished, professional member of the engineering establishment. We want to make sure that members of today's professional community are aware of their colleagues who, like Ellen Richards, are role models in their perseverance and courage in the face of adversity. By celebrating their achievements, we hope to provide inspiration and affirmation to those professionals who are addressing challenges of their own, as well as those who want to become involved in efforts to advance the cause of diversity and inclusion within our field.

Mary Ellen Estill Rudin

Born: 7 December 1924 in Hillsboro, Texas, USA
Died: 18 March 2013 (aged 88)
Field: Mathematics

Mary Ellen Estill was born on 7 December 1924 in the small town of Hillsboro, Texas, USA. Her family was of a middle-class, Presbyterian background. Her father, Joe Jefferson Estill, was a civil engineer and her mother, Irene (Shook) Estill, was a high school English teacher before marriage. Mary Ellen had one sibling, Joe Jefferson Estill, Jr, who was ten years her junior.

Her father was employed on road building projects around the small, isolated town of Leakey, Texas, and the family lived there. To reach Leakey in the 1920s was quite difficult, the only route being a 50-mile (80-km) dirt road through a canyon that forded the Frio River seven times. Mary Ellen spoke of her childhood in Leakey,

> We had few toys. There was no movie house in town. We listened to the radio. But our games were very elaborate and purely in the imagination. I think actually that that is something that contributes to making a mathematician – having time to think and being in the habit of imagining all sorts of complicated things.

She attended the Leakey school, the only school of any kind within 60 miles (96 km) of the town. It had ten grades altogether, each grade had only about five students. There were two full-time and one part-time teacher for the whole school.

After completing her secondary education in 1941, Mary Ellen enrolled at the University of Texas, where she developed her interest in mathematics. Thriving under the guidance of the research mathematician Professor Robert Lee Moore, his 'Moore Method' of encouraging his students to undertake original research inspired her to become a mathematician. At the University of Texas she completed her Bachelor of Arts degree in 1944, and her PhD in 1949, her thesis providing a counterexample to one of 'Moore's axioms'.

In 1950, Mary Ellen began teaching at Duke University where she started her publishing career after completing her doctoral thesis, commencing with the paper, 'Concerning abstract spaces', which was published in the *Duke Mathematical Journal*. It examined the implications, and relations to various alternatives, of an axiom system for point set theory proposed by Moore in 1932. In 1951, she published her work that examined spaces satisfying a subset of Moore's axioms in 'Separation in non-separable spaces' and, the following year, another paper, 'A primitive dispersion set of the plane', provided a positive solution to an unsolved problem contained in R. L. Wilder's 1949 book *Topology of Manifolds*. In 1952, Mary Ellen's paper, 'Concerning a problem of Souslin's', also continued her examination of the implications of Moore's axiom systems.

Mary Ellen found Duke very different from Texas, where Moore's students had been isolated from all contact with books, other mathematicians, or other mathematical ideas apart from their own. Mary Ellen declared,

> I'm a child of Moore. I was always conscious of being manoeuvred by him. I hated being manoeuvred. But part of his technique of teaching was to build your ability to withstand pressure from outside. So he manoeuvred you in order to build your confidence. He built your confidence that you could do anything. I have that total confidence to this day.

She very much enjoyed reading and listening to and talking to other mathematicians about mathematics. She considered that her opportunities at Duke were so wonderful that she stated, 'It still seems impossible that anyone would pay me for doing this.'

In 1953, Mary Ellen married the eminent mathematician Walter Rudin, and they subsequently had four children: Catherine (b. 1954), Eleanor (b. 1955), Jefferson (b. 1961) and Charles Michael (b. 1964). She left Duke in 1953 when she and Walter relocated to the University of Rochester where, from 1953 to 1958, she worked as a temporary part-time associate professor. This title 'admitted her presence there as a mathematician, avoided any nepotism problem, and an elementary course taught'. It also enabled her access to the library and to the mathematical community of the university. The couple then moved permanently to Madison, Wisconsin, where they both accepted teaching positions at the University of Wisconsin. For Mary Ellen, however, it was still only as a temporary, part-time lecturer, but this enabled her to balance her role as a mother with her professional life, of which she said, pragmatically:

> I was a mathematician, and I always thought of myself as a mathematician. I always had all the goodies that go with being a mathematician. I had graduate students. I had seminars. I had colleagues who loved me. I never had committees. I did lots of mathematics, but I did it because I wanted to do it and enjoyed doing it, not because it would further my career.

Mary Ellen Estill Rudin

Mary Ellen was renowned for her ability to solve complex problems. In 1963, her solution to one of the Dutch Prize Problems was recognised by the Mathematical Society of the Netherlands in its prize *Nieuw Archief voor Wiskunde* (New Archive for Mathematics).

Mary Ellen served as a lecturer until 1971 when she was promoted to full professor. This promotion, from the bottom to the top of the academic ladder in one step, she attributed to the 'guilt feelings in the mathematics department' – so much so that they did not ask her, but simply presented her with full professorship. In 1981, she became the first holder the Grace Chisholm Young Professorship at Wisconsin, where she remained for the rest of her career. She has also held visiting professorships in New Zealand, Mexico and China.

Mary Ellen enjoyed a very successful career in mathematics and revelled in her full family life, once stating that she 'never minded doing mathematics lying on the sofa in the middle of the living room with the children climbing all over [her]'. Her research mainly centred upon set-theoretic topology, with an emphasis on the construction of counter examples. She published about seventy research papers on this topic and was an excellent teacher, supervising many PhD students throughout her career. Mary Ellen was awarded a number of research grants as well as being involved in numerous mathematical associations and societies, including the Mathematical Association of America, the Association for Women in Mathematics, the Association for Symbolic Logic and the American Mathematical Society (AMS). She received at least four honorary doctorates for her outstanding work, and in 1984 presented the Emmy Noether (*see entry*) Lecture.

During 1980–81, Mary Ellen was vice president of the AMS, also serving on several mathematical boards, including the Committee of the National Academy of Science for Eastern Europe, the National Committee for Mathematics of the Board of Mathematical Science of the National Research Council, and the editorial board of Topology and Its Applications.

In a 1988 interview, Mary Ellen Rudin explained her mathematical interests: 'From the beginning, it was the set-theoretic aspects of topology which interested me most. I liked finite and infinite combinatorics. ... I'm basically a problem solver.' She passed away aged eighty-eight on 18 March 2013, her husband Walter having died on 20 May 2010.

In 2015, a collection of memories of her was published by the AMS. American mathematician Judith Roitman concluded her contribution in saying,

> Mary Ellen was, of course a woman mathematician at a time when there were few women mathematicians. She belonged ... to what she called the housewives' generation: women who did substantial mathematics outside the academy, with only occasional ad hoc positions. I think of those women as exhibiting enormous strength of character. I think they thought of themselves as simply doing mathematics. As F. Burton Jones wrote in his Festschrift volume, 'Wherever Mary Ellen was there was some mathematics.'

Hazel Marguerite Schmoll

Born: 23 August 1890 in McAllaster, Kansas, USA
Died: 31 January 1990 (aged 99)
Field: Botany
Award: Colorado Women's Hall of Fame (1985)

Hazel Marguerite Schmoll was born in a sod cabin in McAllaster, Kansas, on 23 August 1890, to William and Amelia Schmoll. When Hazel was two years old, the family moved to Ward, Colorado, where her father set up a livery stable. She demonstrated an early interest in wildflowers, spending much of her time exploring the area on horseback, collecting specimens and picking berries. William opened his livery and purveyor's business, eventually owning sixteen horses that he used to transport goods and tourists to and from Ward. Hazel recalled riding with her father in the buggy in which he ferried tourists who stayed at the cabins at Stapps Lake. After a few years, William was the owner of extensive property in the Ward region, which Hazel later inherited. Her parents were active in town politics, serving on the town council, board of education, and William as mayor. She was living what she described as an 'ideal childhood', riding her horse with her small cocker spaniel throughout the mountains. The town children would scour the 'mine dumps' for rich specimens from the ore that had been sorted and sell them at the depot to the tourists.

In 1900, when Hazel was ten, there was a great fire in Ward, resulting in most of the business area burning down. Locals saved the Ward School (now the Post Office and Town Hall) and the Ward Congregational Church. Her family lost everything, including their home and livery stable, but the horses survived. Many people left Ward after the fire, and the population never recovered, although her father rebuilt his livery business.

Hazel's mother supported her having a sound education and the family bought property in Boulder so that Hazel, who had finished eighth grade, could attend the Preparatory School (later named Boulder High School) there. After completing high school, Hazel enrolled in a biology degree at the University of Colorado, graduating in 1913. Between 1913 and 1917, she taught at the prestigious Vassar College in Poughkeepsie, New York, initially in the biology department and then in the botany department. She was the first University of Colorado graduate to be hired by Vassar and was active in promoting the cause of women's suffrage on campus. She also attended a summer 'bug' programme of the University of Michigan on Lake Michigan at Bayview, near Petoskey, Michigan. To continue teaching at Vassar required a higher degree, and so she enrolled in a master's degree in botany under the supervision of the ecological pioneer and botanist Henry Chandler Cowles at the University of Chicago.

She graduated in 1919, the first woman to obtain a doctorate in botany at the University of Chicago. Then Hazel looked for short-term employment, finding a position involving mounting and cataloguing the botanical collections of the Canadian American botanist Alice Eastwood (*see entry*) and botany teacher Ellsworth Bethel. She later conducted the first systematic study of plant life in the south-western part of the state, a project that would later form the basis of her doctoral thesis. In addition to her job, Hazel educated the public about plant life in the Rocky Mountains and in 1920–21 served as assistant curator for the State Bureau of Mines. In 1925, she was a leading lobbyist for efforts to pass legislation protecting the state flower, *Aquilegia coerulea* or blue columbine. As it turned out, her short-term turned into six years, and in 1925 she travelled to Europe to visit botanical gardens and learn German.

On returning to the USA, Hazel enrolled in doctoral studies at the University of Chicago, supporting herself with various occupations, ranging from cleaning houses to rewriting a high school biology textbook. She also worked at the Field Museum of Natural History and was employed as a substitute professor at a local junior college. In 1930, Hazel was named botanist and museum curator for the State of Colorado and she subsequently catalogued and mounted a display of 10,000 species of native plants.

As a member of the Board of Directors of the Colorado Mountain Club, Hazel personally lobbied the state legislature. She is regarded as the driving force behind legislation establishing the lavender columbine as the Colorado State Flower and providing fines for destroying or picking the flower. In 1932, she became the first woman to obtain a PhD in botany from the University of Chicago, with a thesis on the vegetation of the Chimney Rock area of south-western Colorado.

As these were the years of the Great Depression, Hazel was unable to find a permanent job as a scientist and so, in 1938, she built Rangeview Ranch outside Ward, adjacent to Rocky Mountain National Park. It was at first a children's camp and then a guest ranch, with Hazel acting as a nature guide for guests well into her seventies. For the rest of the year, she divided her time between the ranch in the summers and a house in Ward. In 1985, she was elected to the Colorado Women's Hall of Fame where she was named a 'Woman of Consequence' and listed in 'Women of Boulder County', Women of the West Museum.

Hazel Schmoll passed away on 31 January 1990 at the age of ninety-nine, after poor health forced her to leave Ward and enter a Denver nursing home. For at least three decades she was the oldest person in Ward, and her extensive notebooks were given to the Boulder Historical Society. She was buried in Green Mountain Cemetery, Boulder, Boulder County, Colorado. Much of her property was donated for conservation purposes, with Rangeview Ranch being donated to the Christian Science Church for use as a retreat and conference centre. To honour her legacy, the University of Colorado, Boulder, established the Hazel Schmoll Research Fellowship in Colorado Botany, with the aim of continuing her work of research and education, with an emphasis on field botany, including systematics and ecology of Colorado plants. It is open to faculty, staff and students and the first recipients, in 2002, were Rosemary Sherriff and Eric DeChaine.

In 1945, a rare and threatened species of milkvetch (also known as locoweed), *Astragalus schmolliae* or 'Schmoll's milkvetch', was named after her. It grows only on Chapin Mesa in Mesa Verde National Park.

Beatrice ('Tilly') Shilling

Born: 8 March 1909 in Waterlooville,
Hampshire, England
Died: 18 November 1990 (aged 81)
Field: Aeronautical Engineering
Awards: Order of the British Empire
(1947)

Beatrice Shilling was born on 8 March 1909 in Waterlooville, Hampshire, England, into in a middle class family. Her father was a butcher. Although she was often referred to as 'Tilly', it was reported that she did not like the name and was rarely called that to her face. She was obsessed with engines from an early age. In a later interview with *The Woman Engineer* journal, Beatrice recalled: 'As a child I played with Meccano. I spent my pocket money on penknives, an adjustable spanner, a glue pot and other simple hand tools.'

At age fourteen, Beatrice purchased a motorbike. She often tinkered with it, fuelling her desire to become an engineer. For the next three years after completing secondary school, she worked for an electrical engineering company where she installed wiring and generators. Her employer, the engineer and women's advocate Margaret Partridge, encouraged her to study electrical engineering at the University of Manchester (then called the Victoria University of Manchester). In 1932, Beatrice graduated with a bachelor's degree, one of only two female graduates, and, as there were so few women represented in the field, the student record cards did not have an option for a woman to be documented.

Beatrice continued her studies at the university to obtain a master's degree in mechanical engineering the following year, specialising in the examination of piston temperatures of high-speed diesel engines. Although it was difficult to find employment during the Great Depression, she secured a position as a research assistant to Professor G. F. Mucklow at the University of Birmingham.

Beatrice gained a reputation as a fearless motorcycle racer during the 1930s, improving on existing vehicle designs. In one instance, she enhanced her trademark Norton M30 motorcycle by adding a supercharger to the engine. She then rode it to race around the famous Brooklands motor circuit, near Weybridge in Surrey, at speeds of 106 miles (171 km) per hour. By doing so, she soon became a skilled competitor who could outperform professionals such as the record-holding Noel Pope, and she was awarded the Brooklands Gold Star for being the fastest woman on the track.

In 1936, Beatrice was recruited as a scientific officer by the Royal Aircraft Establishment (RAE), the research and development agency of the Royal Air Force (RAF) in Farnborough, Hampshire, which became a leading specialist in aircraft carburettors. It was there that her 'pragmatism came up hard against bureaucratic niceties'.

Writing an instruction leaflet for the maintenance of the Bristol Pegasus engine, the matter of cooling valve seats for insertion in the cylinder head was considered. Knowing that RAF depots had no cooling equipment she advised that maintenance officers should stop an ice-cream van, get some sacks of dry-ice (solid carbon dioxide), mix with meths and make a cooling solution. 'I had a bit of fun with that,' she later said, dismissing the furore it had created. However, her abilities were apparent and, propriety aside, she was transferred to the Engine Department. There she built a reputation as a 'no nonsense type' with a very good 'hands-on approach'.

In 1938, Beatrice married a work colleague, George Naylor, but only agreed to do so after he also had been awarded the Brooklands Gold Star for lapping the circuit at over 100 mph (161 kph). She retained the surname Shilling, rather than taking George's. The workshop personnel gave her a wedding present of a set of stocks and dies, because she borrowed theirs so often, 'she might as well have her own set'.

Throughout the Second World War, Beatrice worked on many projects for the RAE. In 1940, during the Battle of France and Battle of Britain, RAF pilots discovered a serious problem in fighter planes with Rolls-Royce Merlin engines, including the Hurricane and Spitfire. The issue occurred when the plane went nose-down to begin a dive, as the resulting negative G-force would flood the engine's carburettor, causing the engine to stall. German fighters did not have the same problem as they used fuel-injection engines. This meant they could evade a pursuing RAF fighter by flying a negative-g manoeuvre, knowing that the RAF plane could not follow. In March 1941, Beatrice led a small team that designed a RAE restrictor to solve this problem. It was a brass thimble with a hole in the middle (later versions used a flat washer), which could be fitted into the engine's carburettor without taking the aircraft out of service. The restrictor limited maximum fuel flow and prevented flooding. Her mechanism was extremely popular with pilots, who affectionately named it 'Shilling's Penny', 'Miss Shilling's orifice', or simply the 'Tilly orifice'. It remained in use as a stopgap to help prevent engine stall in the Merlin engines of the Hawker Hurricane and Supermarine Spitfire fighters, until the introduction of the pressure carburettor in 1943.

The chief engineer, Keith 'Mad Dog' Maddock, at Hangar 42, a Second World War aircraft hangar which is now used to reconstruct old Spitfires, declared,

Spitfires had to typically go into a half-roll to keep the fuel in the bottom of the tank, and dive on. By which time [your enemy] has gone – and that is the problem – you have lost your opportunity. What Tilly Shilling came up with was the idea of putting a baffle in there – a diaphragm to stop fuel surge. Her RAE restrictor was a war-winning modification, without which we would have suffered ... defeat. Beatrice Shilling helped us to win World War Two – of that there is no doubt.

After the war, Beatrice worked on a variety of projects, including the British medium-range ballistic Blue Streak missile and the effect of a wet runway upon the braking of a plane. In 1947 she was appointed an OBE in recognition of her work during the war. She and her husband became interested in racing cars which they tuned and modified extensively in their home workshop and then began racing them. Once again, it was said, Tilly 'was a force to be reckoned with'. Between 1959 and 1962 they raced an Austin-Healey Sebring Sprite, often at Goodwood Members' Meetings, where they managed a number of third places and even one win.

In 1967, Beatrice was brought in to help the prominent American racing driver Dan Gurney solve overheating problems with his Eagle Mk1 Formula One racing car. She continued to work for the RAE until her retirement in 1969 at the age of sixty, having risen to a senior post in the Mechanical Engineering Department. During her retirement ceremony she was described as being 'as outspoken with authority as she was helpful to junior staff'.

In 1957, she was awarded the Kenneth Lightfoot Medal from the Institute of Refrigeration. In 1970, she received an honorary doctorate from the University of Surrey. She was also a Chartered Engineer, a peer-reviewed qualification. Her memberships included the Institution of Mechanical Engineers and the Women's Engineering Society. Her retirement years were said to have 'few dull moments'. She was a keen pistol shot, and 'while she remained active she was to be seen screaming around town in what was probably the fastest Triumph Dolomite Sprint in the country'. She and her husband continued to race motorbikes and then cars until illness intervened.

Beatrice Shilling passed away on 18 November 1990 at the age of eighty-one, 'having packed more into her life than most, forever immortalised by the pilots she has saved'. The Director of Operations at the University of Manchester's science and engineering faculty, Rachel Brealey, remarked, 'Beatrice Shilling is such an inspiration to our students and we are delighted to be celebrating this woman who made such a significant impact to engineering and responded so brilliantly to the technical challenges of her time.'

In 2011, the UK and Ireland Weatherspoon's chain of public houses and hotels opened a pub in Farnborough, in north-east Hampshire, named The Tilly Shilling in her honour. A collection of her racing badges and trophies was bought by the Brooklands Museum in 2015.

Michelle Yvonne Simmons

Born: 14 July 1967 in London, England
Field: Quantum Physics
Awards: Australian Academy of Science
Pawsey Medal (2005); NSW Scientist of the
Year (2011); Royal Society of New South
Wales Walter Burfitt Prize (2013); Australian
Research Council Laureate Fellow (2013);
Thomas Ranken Lyle Medal (2015); Foresight
Institute Feynman Prize in Nanotechnology
(2016); L'Oréal-UNESCO Award for Women in
Science (2017); Australian of the Year (2018)

Michelle Yvonne Simmons was born on 14 July 1967 in London, England. In a 2016 interview with science journalist Elizabeth Finkel, for *Cosmos* magazine, she told a story about how, as an eight-year-old, she sat silently, week after week, watching her father, a high-ranking policeman in London, play chess with her elder brother. One day she asked her father if she could play and he reluctantly agreed – she checkmated him. Her mother was a bank manager, and her grandparents included diplomats and members of the military. She described her family as 'take-responsibility kind of people', with her father constantly telling her, 'Don't take the easy route. Do the most challenging thing.'

Michelle attended Durham University in England, where she obtained a double degree in physics and chemistry, later being awarded a PhD in 1992 with her thesis 'The characterisation of CdTe-based epitaxial solar cell structures fabricated by MOVPE', under the supervision of physicist Andrew W. Brinkman. She was appointed to the postdoctoral position of Research Fellow alongside Michael Pepper (later knighted in 2006) in quantum electronics at the Cavendish Laboratory in Cambridge, UK, where she gained an international reputation for her work in the discovery of the '0.7 feature' and metallic behaviour in 2D Gas hole systems. In 1999, Michelle was awarded a QEII Fellowship which took her to Australia, where she was a founding member of the Centre of Excellence for Quantum Computer Technology at the University of New South Wales. She said that she was drawn to the problem of quantum computing 'because it combines fundamental understanding of how the world works at the atomic-scale with my experience of engineering and creating new electronic devices. It's also an important problem. Ultimately, if we're successful, we'll create a new computing technology that is highly beneficial for humanity.'

In 2000, Michelle established a large research group dedicated to the fabrication of atomic scale devices in silicon and germanium, using the atomic precision of scanning tunnelling microscopy. Her group is the only one in the world with the ability to create atomically precise devices in silicon. They were also the first team ever to develop a working 'perfect' single-atom transistor and the narrowest conducting doped wires in silicon. This centre aims to implement quantum processors able to run error-corrected algorithms and transfer information across networks with absolute security. The centre has also developed technologies for manipulating matter and light at the level of individual atoms and photons, with the highest fidelity, longest coherence time qubits in the solid state, the world's longest-lived quantum memory, and the ability to run small-scale algorithms on photonic qubits. This new technology aims to provide a strategic advantage in a world where information and information security are of paramount importance.

In 2017, Michelle delivered the Australia Day Address for New South Wales (NSW), emphasising the importance of setting high expectations for students and making it clear that Australians need to set the bar high and tell their students they expect them to jump over it. She remarked, 'It is better to do the things that have the greatest reward; things that are hard, not easy', and was critical of the changing standards in physics education in the NSW secondary curriculum. Michelle noted that the curriculum now had a 'feminised nature' and said it was a 'disaster' to try to make physics more appealing to girls by substituting rigorous mathematical problem-solving with qualitative responses.

Michelle is a prolific researcher, with over 370 peer-reviewed journal papers that have amassed over 6,000 citations. She has also written five book chapters and in 2002 co-authored a book on nanotechnology, *Nanotechnology: Basic Science and Emerging Technologies*. She has filed four patents and delivered over 100 invited and plenary presentations at international conferences. Her scientific vision has enabled the creation of the world's smallest precision transistor, a vital component of a future quantum computer.

Michelle has been awarded two prestigious Australian professorial research fellowships, Federation Fellowships, and in 2019 was an Australian Research Council Laureate Fellow, awarded to world-class research leaders. In 2005, she was awarded the Pawsey Medal by the Australian Academy of Science, also becoming one of the youngest elected Fellows of this Academy in 2006. The following year she became an Australian citizen. In 2012 she was named NSW Scientist of the Year by the NSW Government Office of the Chief Scientist, and in 2014 became an elected Fellow of the American Academy of Arts and Sciences.

In a 2012 interview with the Australian Broadcasting Corporation's *Science* programme, Michelle said, 'I love to chill out with my family and friends, planning expeditions and keeping fit. But the thing that brings me the most joy is my funny husband (Thomas Barlow) and three adorable children.'

In 2015, Michelle was awarded the Thomas Ranken Lyle Medal of the Australian Academy of Science, awarded to a mathematician or physicist for outstanding research accomplishments. The citation for the award read,

> Professor Simmons has pioneered a radical new technology for creating atomic-scale devices producing the first ever electronic devices in silicon where individual atoms are placed with atomic precision and shown to dictate device behaviour. Her ground-breaking achievements have opened a new frontier of research in computing and electronics globally. They have provided a platform for redesigning conventional transistors at the atomic-scale and for developing a silicon-based quantum computer: a powerful new form of computing with the potential to transform information processing.

In the same year, Michelle also received the Eureka Prize for Leadership in Science. Awarded by the Commonwealth Scientific and Industrial Research Organisation (CSIRO), the prize of $10,000 AUD is awarded to an Australian individual who has demonstrated an outstanding role and impact on science and 'celebrates the role and impact of great leadership in science'. In 2016 Michelle was awarded the Foresight Institute Feynman Prize in Nanotechnology, and in 2017 she was named as the Asia-Pacific Laureate in the L'Oréal-UNESCO Awards for Women in Science 'for her pioneering contributions to quantum and atomic electronics, constructing atomic transistors en route to quantum computers'. In 2019, she was a Scientia Professor of Quantum Physics in the Faculty of Science at the University of New South Wales and Editor-in-Chief of *npj Quantum Information*, an academic journal publishing articles in the emerging field of quantum information science.

On 25 January 2018, Michelle Simmons was named as the 2018 Australian of the Year for her work and dedication to quantum information science. In receiving this honour she remarked,

> Throughout my career, I found people often underestimate female scientists. In some ways for me that has been great, it has meant I have flown under the radar and have been able to get on with things. I'm also conscious when a person starts to believe in what others think of them, that can become a self-fulfilling prophecy. That is why I feel it is important not to be defined by other people's expectations of who you are and what you might be.

On 10 June 2019, Michelle was appointed an Officer of the Order of Australia (AO) in the Queen's Birthday Honours in recognition of her 'distinguished service to science education as a leader in quantum and atomic electronics and as a role model'. In 2021, she was named a Fellow of the American Physical Society, and in 2022 was awarded the Royal Society's Bakerian Medal and Lecture for 'seminal contributions to our understanding of nature at the atomic-scale by creating a sequence of world-first quantum electronic devices in which individual atoms control device behaviour'.

Mary Fairfax Greig Somerville

Born: 26 December 1780 in Jedburgh, Scotland
Died: 29 November 1872 (aged 91)
Field: Astronomy and Science writing
Award: Patron's medal (1869)

Mary Fairfax was born on 26 December 1780 in Jedburgh, Scotland. Her father was Lieutenant (later Vice Admiral Sir) William George Fairfax and her mother was his second wife, Margaret (Charters) Fairfax, whose father was the Solicitor of Customs for Scotland. Mary was born at the house of her maternal aunt, Martha, and her husband, Dr Thomas Somerville, the Scottish minister, antiquarian and amateur scientist. She was the second of four surviving children (three of her siblings had died in infancy) and was said to be very close to her oldest brother, Sam. She grew up in Burntisland, in Fife. The family was not wealthy as her father's naval pay was relatively low while he rose through the ranks. The family income was supplemented by her mother growing vegetables, maintaining an orchard and keeping cows for milk. Mary spent one year at Miss Primrose's boarding school in Musselburgh, and otherwise was largely self-educated. Her mother taught her to read the Bible and Calvinist catechisms and, when she was not busy, doing household tasks. Mary often explored among the birds and flowers in the garden. Her aunt's husband, Dr Somerville, assisted her in learning Latin.

If the weather was bad, Mary busied herself with reading the books in her father's library, including Shakespeare. She later recalled, 'These occupied a great part of my time; besides, I had to sew my sampler, working the alphabet from A to Z, as well as the ten numbers, on canvas'. Her aunt Janet, who lived with the family, reportedly told her mother, 'I wonder you let Mary waste her time in reading, she never sews more than if she were a man'. As a result, Mary was sent to the village school to learn plain needlework, declaring that she was annoyed that her turn for reading 'was so much disapproved of', and thought it 'unjust that women should have been given a desire for knowledge if it were wrong to acquire it'.

When aged thirteen, Mary was sent to writing school in Edinburgh during the winter months, improving her writing skills and studying the common rules of arithmetic. She was under the care of Lady Burchan, making her first appearance at a ball, her first dancing partner being the Earl of Minto. During the summer, she would accompany her aunt Martha and Dr Somerville to Burntisland, where she read elementary books on algebra and geometry. Mary learned to play the piano, as well as Greek, so that she could read Xenophon and Herodotus. In Edinburgh she attended the academy of the Scottish portrait and landscape painter Alexander Nasmyth, who had opened it for ladies. He advised another student to study Euclid's *Elements* and *The Elements of Navigation* by John Robertson, to gain a foundation in perspective, astronomy and mechanical science and she took the opportunity as she felt the books would help her. Mary subsequently rose early to play the piano, paint during the day and stay up late to read her books.

In 1804, Mary married a distant cousin, Samuel Greig, who was a captain in the Russian navy and the Russian consul in London. They had two children, one of whom, Woronzow Greig, later became a barrister and scientist. Although she continued to study mathematics, Mary later wrote, 'Although my husband did not prevent me from studying, I met with no sympathy whatever from him, as he had a very low opinion of the capacity of my sex.' After he died in 1807, her status as a widow with a small inheritance meant that Mary was again free to concentrate on her mathematical studies. She returned to Burntisland with her children. Her family connections gave her access to Edinburgh social and intellectual circles, and through such connections she obtained mathematics instruction by correspondence from the mathematician William Wallace and his brother. In 1812, Mary remarried, this time to another cousin, William Somerville, an army doctor, who relished his wife's educational accomplishments. The Wallaces helped Mary to become familiar with the work in advanced mathematical analysis being pioneered in France.

Mary now began to study botany and geology and, in 1816, the couple relocated to London, where they became friends with eminent scientists, including the astronomers Sir William and Caroline Herschel (*see entry*), metallurgist William Hyde Wollaston, physicist Thomas Young, and mathematician Charles Babbage. Her connections expanded during a visit to the Continent in 1817. In 1826, Mary published her first scientific paper, 'On the Magnetizing Power of the More Refrangible Solar Rays', and the next year was asked by the lawyer Henry Brougham, an influential figure in London educational circles, to prepare a condensed version of Laplace's five-volume work *Traité de mécanique céleste* (Celestial Mechanics, 1798–1827), which gave a complete mechanical interpretation of the solar system. The publication was for the for Society for the Diffusion of Useful Knowledge, founded by Brougham, with the object of publishing information to people who were unable to obtain formal teaching. It took Mary three years to complete, but Brougham thought the work too lengthy and lost interest.

However, astronomer Sir John Herschel rated her book, *The Mechanism of the Heavens*, so highly that he recommended it to another publisher, who published it in 1831. Her introduction, summarising the current state of astronomical knowledge for the general reader, was published separately in 1832 as *A Preliminary Dissertation to the Mechanism of the Heavens,* a volume that received much praise from British mathematicians and astronomers. In 1837, Cambridge University adopted *The Mechanism of the Heavens* as a textbook in advanced mechanics.

In 1834, Mary's next book, *The Connection of the Physical Sciences,* summarised astronomy, physics, geography and meteorology, with nine subsequent editions appearing over the years. In the third edition in 1836, Mary wrote that difficulties in calculating the position of Uranus could suggest the existence of an undiscovered planet, prompting the British astronomer John Couch Adams to begin the calculations that led to the discovery of the planet Neptune.

In 1834, Mary was made a member of the Royal Irish Academy; the next year, she and Caroline Herschel (*see entry*) were the first women to be made honorary members of the Royal Astronomical Society. The same year, on the recommendation of Prime Minister Sir Robert Peel, Mary received a Civil List pension of £200 per year (two years later raised to £300). She and her husband went to Italy in 1838 because of his ill-health and she remained there for the rest of her life. In 1840, the Royal Society commissioned a marble bust of Mary from sculptor Francis Chantrey.

Mary's next book, *Physical Geography* (2 vols, 1848), was the first textbook on the subject in English and her most popular and influential work, running to six editions. In 1869, she received the Patron's Medal of the Royal Geographical Society for *Physical Geography.* Her final book, *On Molecular and Microscopic Science* (2 vols, 1869), did not enjoy the success of her previous works. Its science was by then largely out of date, but it was received with kind deference to Mary, who was then eighty-nine years old.

When the philosopher and economist John Stuart Mill organised a massive petition of 1,521 names to give women the right to vote, delivering it to the House of Commons on 7 June 1866, he had Mary put her signature the first on the list.

On 29 November 1872, Mary passed away in Naples, aged ninety-one, and was buried there in the English Cemetery. Her husband, William, had predeceased her in 1860. The *Morning Post* declared in her obituary that 'whatever difficulty we might experience in the middle of the nineteenth century in choosing a king of science, there could be no question whatever as to the queen of science'. Her autobiography, *Personal Recollections, from Early Life to Old Age* (1873), was edited by her daughter Martha and published the year after her death. It consisted of reminiscences written during her old age.

Mary Somerville received many honours before and after her death. Somerville Square in Burntisland is named after her family and marks the site of their home, while Somerville College, Oxford, bears her name, as does Somerville House, a high school for girls in Brisbane, Australia. One of the Committee Rooms of the Scottish Parliament in Edinburgh has been named after her. Somerville Island, a small island in Barrow Strait, Nunavut, was a tribute to her by Sir William Edward Parry in 1819. 5771 Somerville (1987 ST1) is a main-belt asteroid discovered on 21 September 1987 by E. Bowell at Lowell Observatory Flagstaff, Arizona, and named after her. Somerville Crater is a small lunar crater in the eastern part of the Moon. Scottish banknotes bearing her image were issued in the second half of 2017. There are 10,000 items involving her life in the Somerville Collection of the Bodleian Library at the University of Oxford. In 1829, Sir David Brewster, inventor of the kaleidoscope, wrote that Mary was 'certainly the most extraordinary woman in Europe – a mathematician of the very first rank with all the gentleness of a woman'. The science historian Mary R. S. Creese said of her,

> Although she was not among those nineteenth-century women who contributed to original work in science, Mary's long sustained and immensely successful scientific writing was unquestionably outstanding. Perhaps no woman of science until Marie Curie [(*see entry*)] was as widely recognised in her own time. Her books were remarkably influential; not only did they bring scientific knowledge in a broad range of fields to a wide audience, but thanks to her exceptional talents for analysis, organisation, and presentation, they provided definition and shape for an impressive spread of scientific work.

'Janet' Jane Ann Ionn Taylor

Born: 13 May 1804 in Wolsingham, Durham,
England
Died: 26 January 1870 (aged 65)
Field: Mathematics and navigation
Awards: Gold Medal from the King of The
Netherlands (1836); Gold Medal from the
King of Prussia (1837); Gold Medal from
Pope Gregory XVI (1843)

Jane Ann Ionn was born on 13 May 1804 in Wolsingham, Durham, England, the fourth child of the Reverend Peter Ionn and Jane Deighton, the daughter of a country gentleman. Peter Ionn was the curate and schoolmaster at the Free Grammar School in Wolsingham, where he taught, among many other things, the subject of navigation. When Jane Ann's mother died in May 1811, two months after giving birth to her sixth child, Frederick, Jane Ann was six years old.

When she was aged nine, a scholarship became available for girls aged fourteen and over to attend the Royal School for Embroidering Females, which had been established at Ampthill in Bedfordshire, under the patronage of Queen Charlotte. With the backing of the Member of Parliament for Durham City, Michael Angelo Taylor, and notwithstanding her young age, Jane Ann was accepted as a pupil, and stayed there until the death of the Queen on 17 November 1818, when the school was closed. Then aged seventeen, Jane Ann obtained a position as governess to the family of the Vicar of Kimbolton, the Reverend John Huntley. Kimbolton was a small town of only 1,500 people and only about 25 miles (40 km) north of Ampthill.

On 2 May 1821, her father Peter died suddenly, when Jane Ann was sixteen years old. She then went to live with her brother Mathew Ionn, who had opened a linen draper's shop at 44 Oxford Street in London. Jane Ann assisted in running the business, but was still fascinated by the mathematics of navigation. She knew she needed a husband to be 'respectable' and so, on 30 January 1830, now aged twenty-nine, Jane Ann married George Taylor Jane, aged forty-one. George was a widower with three children, a former naval man during the Napoleonic wars and now a publican. He was also a 'Dissenter', brought up outside the Church of England. On their marriage,

George changed his name to 'George Taylor' and she became 'Mrs Janet Taylor' for the rest of her life. The couple set up residence at 6 East Street, Red Lion Square, near Oxford Street in London.

Janet became a prodigious author of nautical treatises and textbooks, born of a particular fascination in measuring longitude by the lunar distance method. She conducted her own Nautical Academy in Minories in the East End of the City, not far from the Tower of London. She was a sub-agent for Admiralty charts, ran a manufacturing business for nautical instruments, many of which she designed herself, and embarked on the business of compass adjusting at the height of the controversies generated by magnetic deviation and distortions on iron ships.

Her singular contribution to the maritime community and to the welfare of mariners was recognised in her being awarded gold medals (the highest award possible) by the King of Prussia, the King of The Netherlands and even by the Pope, culminating in the eventual award of a Civil List pension in her country, with the support of the Bishop of Durham. That her role was significant is also evident in that she was the only woman included among 2,200 entries in the encyclopaedic work *The Mathematical Practitioners of Hanoverian England 1714–1840* by Professor Eva G. R. Taylor, and notwithstanding that Janet's contribution continued well after the 1840s, beyond the period considered by Professor Taylor.

Janet's entry into publishing began in 1833, at the age of twenty-nine, and she went on to produce a number of major works of importance to the maritime community, most of which went into many editions. There were seven editions of her first book, *Luni-solar and horary tables* (or *Lunar Tables*) alone, appearing between 1833 and 1854. Her *Principles of Navigation Simplified: with Luni-Solar and Horary Tables*, first published in 1834 had three editions, while *An Epitome of Navigation and Nautical Astronomy* went to twelve editions between 1842 and 1859. Her *Planisphere of the Fixed Stars with Book of Directions*, with its beautiful plates of star charts, was first published in 1846 and reached its sixth edition in 1863. And then there were the *Diurnal Register for Barometer, Sympiesometer, Thermometer and Hygrometer* (seven editions); *A Guide Book to Lt Maury's Wind and Current Charts* (seven editions); and the *Handbook to the Local Marine Board Examinations for Officers of the British Mercantile Marine Board* (twenty-seven editions). Each was dedicated to a member of the royal family – William IV (whose mother had sponsored Janet's scholarship as a child), Queen Adelaide and the Duchess of Kent (then Princess Victoria's mother). This was a remarkable volume of work and, almost without exception, was highly praised and well received.

Her nautical tables – a vital tool in the determination of longitude at sea – were widely recognised as an invaluable aid to merchant seamen. She also discussed the use of the marine chronometer, developed by John Harrison. In September 1834, Janet obtained a British patent for 'A Mariner's Calculator', claiming 'improvements in instruments for measuring angles and distances, applicable to nautical and other purposes'. Between 1617 and 1852, only seventy-nine patents were awarded in the category 'Compasses and Nautical

Instruments'. They were awarded to renowned leaders in the field, and during this 235-year period Janet was the only woman.

Her Mariner's Calculator was an ingeniously clever concept, combining several nautical instruments in one. Janet delivered a prototype of her new device to the Admiralty for assessment, and it was given to their hydrographer, Captain Francis Beaufort. However, Beaufort's report in May 1834 was not favourable. It was not that he thought it would not work but, in the 'clumsy fingers of seamen', he thought not. He also felt it would encourage slovenliness (perhaps because it would do too much of the hard work).

By 1845, the Taylors had opened business premises at 104 Minories where she sold a wide range of nautical and mathematical instruments, illustrated by an advertisement in the Mercantile Marine Magazine, October 1854, declaring that the firm manufactured 'every description of nautical and mathematical instruments'. It was in Minories that Janet also ran her 'Nautical Academy', which would continue for over thirty years.

Perhaps the most distinctive of Janet's instruments was the exquisite Prince of Wales 'quintant', entered for the Great Exhibition of 1851. It was far more than a working nautical instrument. Magnificent and flamboyantly ornate, it combined the principal aspirations of the exhibition in one, arts and manufacturing, and of the Royal Society that had championed the Great Exhibition. At the centre of the instrument was the Prince of Wales crest, '*Ich Dien*' (I serve), which was also on the box. The elegance of Janet's creation, however, was lost on the jury. In the section relating to 'Instruments made in the United Kingdom', it was simply reported that Janet had exhibited a sextant 'intended for show rather than use' and there was no mention of any award. The Prince himself was only eleven years old, but in time the instrument came to be presented to the Royal Family, and, in turn, it came to the National Maritime Museum in May 1936.

Janet's contributions in the maritime world also extended to compass adjusting. She ran large compass adjusting business, based on the application of techniques for correcting magnets to counteract the effects of iron on ships' compasses advocated by the Astronomer Royal, Professor George Airy.

Janet Taylor passed away on 26 January 1870, at the Vicarage, St Helen Auckland, the home of her sister and brother-in-law, Canon Matthew Chester, the incumbent of the parish. Her death certificate recorded her occupation simply as 'Teacher of Navigation'. But she was far more than this. Her story is a remarkable one, especially for a woman in the nineteenth century.

Ruth Lichterman Teitelbaum

Born: 1 January 1924, New York, USA
Died: 9 August 1986 (aged 62)
Field: Technology
Award: Women in Technology International
Hall of Fame (1997)

Ruth Lichterman was born in New York City, New York, USA, on 1 January 1924, to Sarah and Simon Lichterman. She attended Hunter College and graduated with a bachelor's degree, majoring in mathematics. Soon after, she was hired by the Moore School of Electrical Engineering at the University of Pennsylvania, a school being funded by the US Army during the Second World War. It was there that a group of about eighty women, known as 'computers', worked doing manual calculations of ballistic trajectories through complex differential calculations with the aid of desktop calculators, an analogue technology of the period. In 1945, the Army decided to fund an experimental project involving the first all-electronic digital computer. Six of the women 'computers' were selected to be its first programmers, including Ruth.

The computer on which the group worked was called ENIAC (Electronic Numerical Integrator and Computer), a machine with about 18,000 vacuum tubes and forty black 8-foot (2.43 m) panels. It was 8 feet tall and 80 feet (24.4 m) long. The total cost was about $487,000 USD, equivalent to $7,000,000 USD in 2019. The construction contract was signed on 5 June 1943 and work on the computer began in secret at the University of Pennsylvania's Moore School of Electrical Engineering the following month, under the code name 'Project PX'.

As the ENIAC project was classified, the programmers were denied access to the machine until they received their security clearances. As the first programmers, there were no programming manuals or courses, only the logical diagrams to help them determine how to make ENIAC work. This involved them physically programming the ballistics programme by using the 3,000 switches and dozens of cables and digit trays to route the data and programme pulses through the machine. Ruth can be seen in the picture opposite, wiring the right side of the ENIAC with a new programme.

After the war, Ruth relocated with ENIAC to Aberdeen, Maryland, where she taught the next generation of ENIAC programmers. She worked closely with another programmer, Marlyn Meltzer, and the pair performed a vital function for the 15 February 1946 demonstration to the world's public and press. This was to carry out a manual calculation to prove the machine was working properly. Their work helped debug the machine as the engineers and programmers prepared for the public unveiling. On the day, ENIAC ran the ballistics trajectory programmed by the six women and, with perfect results, captivated the world's imagination. The following year, ENIAC was turned into the world's first 'stored program' computer and these six programmers were the only generation of programmers to programme it at the machine level.

The women all left the programme and most of them, including Ruth, did so to marry. Ruth remained for two more years to train future ENIAC programmers. She was the last of the original six to remain employed at Aberdeen, retiring on 10 September 1948 when she married Adolph Teitelbaum one week later. Although the male engineers who built the machine soon became famous, little is known of the women programmers. To remedy this, the ENIAC Programmer Project was founded by lawyer and historian Kathy Kleiman, who devoted years to researching the University of Pennsylvania Archives and Library of Congress, as well as recording extensive broadcast-quality oral histories with four of the original six ENIAC programmers in the late 1990s with senior PBS producer David Roland. As she undertook the research, Kathy said that people kept coming forward 'with stories of grandmothers, their great aunts, their mothers – there were so many women involved in early technology that we don't know about'. Kathy says that the women not only programmed the first computer, but 'dedicated years after the war to making programming easier and more accessible for all of us who followed'. In 2013, she teamed up with award-winning documentary producers Jon Palfreman and Kate McMahon of the Palfreman Film Group to tell this incredible story in the documentary short *The Computers*.

On 9 August 1986, Ruth Teitelbaum passed away in Dallas, Texas, aged sixty-two, having received little credit for what she had achieved during her lifetime, despite playing a pivotal role. In 1997, more than a decade after her death, Ruth was inducted into the Women in Technology International Hall of Fame, along with the other original ENIAC programmers. These five women were Marlyn Meltzer, Frances Spence, Kathleen Antonelli, Jean Bartik and Betty Holberton. Ruth's husband accepted the award in her memory.

Valentina Vladimirovna Tereshkova

Born: 6 March 1937 in Bolshoye Maslennikovo, Tutayevsky District, Yaroslavl Oblast, Russian SFSR, Soviet Union
Field: Engineering and Space travel
Awards: Hero of the Soviet Union (1963); Order of Lenin (1963 and 1981); Gold Medal of Peace Joliot-Curie (France, 1964); Order of Honour (2003); Order of Friendship (2011); Order of Alexander Nevsky (2013) and many others

Valentina Tereshkova was born on 6 March 1937 in the village of Maslennikovo in Tutayevsky District, Yaroslavl Oblast, in central Russia. Her father, Vladimir Tereshkov (who was later killed in the Second World War), was a tractor driver and her mother worked in a cotton mill. There were two other children, Vladimir and Ludmila. Their parents had migrated from Belarus and Valentina started school in 1945 at the age of eight, leaving in 1953 aged sixteen. She then continued her education by correspondence courses.

At a young age, Valentina became interested in parachuting and trained in skydiving at the local Aeroclub. She made her first jump aged twenty-two on 21 May 1959 while employed as a textile worker in a local factory. It was her expertise in skydiving that led to her selection into the cosmonaut programme. In 1961, she became Secretary of the local Komsomol (Young Communist League) and later joined the Communist Party of the Soviet Union.

After the flight of Yuri Gagarin in 1961, Valentina was chosen on 16 February 1962 to join the female cosmonaut corps, being one of five women from a field of 400 selected. Qualifications included that they be parachutists under thirty years of age, under 5 feet 9 inches (170 cm) tall and below 154 lbs (70 kg) in weight.

Valentina was also selected due to her 'proletarian' background and because her father was a war hero. Her training included weightless flights, isolation tests, centrifuge tests, rocket theory, spacecraft engineering, 120 parachute jumps and pilot training in MiG-15UTI jet fighters. The women spent several months in intensive training, concluding with examinations in November 1962. Valentina was one of the three leading candidates, and a joint mission profile was developed that would see two women launched into space, on solo Vostok flights on consecutive days in March or April 1963.

Originally it was intended that Valentina would launch first in Vostok 5, while another woman, Ponomaryova, would follow her into orbit in Vostok 6. However, this flight plan was altered in March 1963, when it was decided that Vostok 5 would carry a male cosmonaut, Valery Bykovsky, flying the joint mission, with a solo woman aboard Vostok 6 in June 1963. The State Space Commission selected Valentina to pilot Vostok 6 at their meeting on 21 May 1963 and this was personally confirmed by the Premier of the Soviet Union, Nikita Khrushchev. At the age of twenty-six, Valentina was exactly ten years younger than Gordon Cooper, who was the youngest astronaut of the seven aboard Project Mercury, the first manned-space effort by the United States in 1959.

After the successful launch of Vostok 5 on 14 June 1963, Valentina began final preparations for her own flight. On the morning of 16 June 1963, she and her backup, Irina Solovyova, were both dressed in spacesuits and taken to the launch pad by bus. After completing her communication and life support checks, the door to Vostok was sealed and she became the first woman in space. Her call sign in this flight was *Chaika* (Seagull) and it was commemorated with the name of an asteroid, 1671 Chaika. Although she was an honorary inductee into the Soviet Air Force in order to join the Cosmonaut Corps, she was the first civilian to fly into space.

Experiencing nausea and physical discomfort for much of the flight, Valentina orbited the earth forty-eight times and spent almost three days in space. With a single flight, she logged more flight time than the combined times of all the American astronauts who had flown before that date. She also kept a flight log and took photographs of the horizon, which were later used to identify aerosol layers within the atmosphere. Her total time in space was two days, twenty-two hours, fifty minutes. After her flight, Valentina enrolled at the Zhukovsky Air Force Academy and graduated with distinction as a cosmonaut engineer.

In September 1963, Valentina donated a silver cup at the women's 1963 European Rowing Championships held in Khimki near Moscow for the most successful nation, which went to the team from the Soviet Union as they won gold in all five boat classes. On 3 November 1963, Valentina married cosmonaut Andriyan Nikolayev at the Moscow Wedding Palace, supposedly under pressure from Soviet leader Nikita Khrushchev, seeing a propaganda advantage in the pairing of the two single cosmonauts. Khrushchev presided at the wedding party, with leading government and space programme leaders. On 8 June 1964, Valentina gave birth to the couple's daughter, Elena Andrianovna Nikolaeva-Tereshkova, who became a doctor. In October 1969, the pioneering female cosmonaut group was dissolved. Although it was planned to have further flights by women, it took nineteen years until the second woman, Svetlana Savitskaya, flew into space.

In 1977, Valentina earned a PhD in engineering. Valentina and Nikolayev divorced in 1982 and, in the same year, she remarried the orthopaedist Yuliy G. Shaposhnikov. Valentina won many awards, including the Hero of the Soviet Union medal, the USSR's highest award. She was also awarded the Order of Lenin twice, as well as the Order of the October Revolution, numerous other medals, and foreign orders including the Karl Marx Order, United Nations Gold Medal of Peace and the Simba International Women's Movement Award. There were other honours in which she was considered a 'hero', including the Hero of Socialist Labour of Czechoslovakia, Hero of Labour of Vietnam and Hero of Mongolia. In 1990, she received an honorary doctorate from the University of Edinburgh, Scotland. The Tereshkova Crater on the far side of the Moon was named after her.

Because of her public prominence, Valentina was selected for several high political positions between 1966 and 1991. She lost her political office after the collapse of the Soviet Union in 1991, but in 2011 she was elected to the State Duma, the lower house of the Russian legislature, as a member of United Russia.

Valentina Tereshkova retired in 1997 with the rank of Major General, Soviet Air Force. She was invited to Russian Prime Minister Vladimir Putin's residence in Novo-Ogaryovo for the celebration of her seventieth birthday. During her visit she said that she would like to fly to Mars, even if it meant that it was a one-way trip. Her husband, Yuliy, died in 1999. On 5 April 2008, she was a torchbearer of the 2008 Summer Olympics torch relay in Saint Petersburg, Russia.

The band Public Service Broadcasting has a song entitled 'Valentina' on their 2015 album *The Race for Space* in her honour. The BBC drama *Call the Midwife* (Season 7, Episode 4; broadcast 11 February 2018) used her space flight as an example of heroism by a woman. She is revered as a hero in Russian space history.

Mildred Trotter

Born: 3 February 1899 in Monaca,
Pennsylvania, USA
Died: 23 August 1991 (aged 92)
Field: Anatomy and anthropology
Awards: Woman of Achievement in Science
(*St Louis Globe-Democrat*, 1955); First woman
to receive the Viking Fund Medal in Physical
Anthropology (Wenner-Gren Foundation for
Anthropological Research, 1956)

Mildred Trotter was born on 3 February 1899 in Monaca, Pennsylvania,
to schoolteacher Jennie (Zimmerley) and farmer and community school
director James R. Trotter. Of German and Irish extraction, her parents were
active Democrats and Presbyterians. There were three other children – Sarah
Isabella, Jeannette Rebecca and Robert – all of whom were raised on the
family farm.

Mildred attended a single-room grammar school until 1913, completing
her high school education in nearby Beaver Falls. Majoring in zoology, she
matriculated to Mount Holyoke College in 1916. It was here that her interest
in science and zoology was generated by a number of female professors,
notably Ann Morgan, Elizabeth Adams and Christianna Smith. As a result of
their influence as role models, in a later interview Mildred said, 'I never even
thought, let alone worried, about being a woman in science'.

After Mildred completed her bachelor's degree in 1920, she attracted
the attention of Professor Smith, who specialised in comparative anatomy
and histology, and had established the school's first course in medical
zoology. She recommended her for the position of assistant to Dr Charles
Haskell Danforth, an associate professor in the Department of Anatomy
at Washington School of Medicine. Smith soon arranged for her to go
to Washington as a 'Fellow in Hypertrichosis', where she assisted in a
programme involving the growth of excessive hair. Upon discovering that
this work could form part of a master's degree in anatomy, Mildred accepted
the role, completing her master's degree in 1921. The following year, she was
appointed as an assistant in the Department of Anatomy where she joined
five other women on the medical staff. In 1924 she completed her doctorate
in anatomy, this leading to a promotion to an instructor of anatomy.

Mildred was awarded a National Research Council Fellowship for the 1925–26 academic year, which took her to England to the University of Oxford. Here she studied bones, particularly museum specimens from ancient Egypt and the Roman era in Britain. However, she decided to leave after only twelve months, despite being offered a second year, so that she could return to Washington University to accept a promotion to assistant professor promised by the Head of the Anatomy Department, Robert J. Terry. She served at this rank for four years before being elevated to a tenured associate professor position. From 1941 to 1948 and from 1949 to 1967, Mildred also served as president of the St Louis Anatomical Board.

After sixteen years, Mildred had become frustrated at not being able to gain promotion and, in 1948, took unpaid leave from Washington to be a volunteer director of the Graves Registration Service of the US Army, based at Schofield Barracks. For the following fourteen months, she obtained permission from the US Army to carry out allometric studies (the relationship between body size and the shape of the anatomy) of the identification of the remains of US servicemen and servicewomen in the Pacific Theatre, using the height of a person based on the relative length of their long bones. This study was critical, as the identities of the deceased had been lost. The only records available for the possible victims were their sex and their height at the time of induction, with Trotter's formulas being later used by the Federal Bureau of Investigation in the US. Mildred and her team assisted in the identification of the remains of US servicemen and servicewomen, identifying 94 per cent of those analysed.

In the autumn of 1951, Mildred undertook further work for the US at Fort McKinley in the Philippines for three months and during the 1950s, she played a leading role in convincing Missouri lawmakers to make it possible for people to leave their bodies to medical schools for research and teaching purposes.

Working with the psychologist Goldine C. Gleser in 1952, Mildred derived statistical regression equations for the calculation of the height of a person based on the length of certain human long bones with the data being drawn from American casualties from the Korean War.

In 1955, Mildred was one of the founders of the American Association of Physical Anthropologists and served as president for three years between 1955 and 1957, being the first woman to do so. Then she assumed the role of president of the Missouri State Anatomical Board, a position she held for ten years. In 1958, at the age of fifty-nine, Mildred became the first woman to hold the rank of professor at Washington University, her field being Gross Anatomy, with the word 'Gross' being dropped shortly after.

In late 1963, Mildred was instrumental in starting the teaching programme at the Department of Anatomy at Makerere in Kampala, Uganda. Under the Rockefeller Foundation's programme, she was a visiting professor at Makerere College.

Back in Washington, having reached the compulsory retirement age of sixty-eight in 1967, Mildred was conferred the title of emeritus professor and continued to be an active author, researcher and teacher. Throughout her career, Mildred taught over 4,000 students, including two Nobel Laureates in pharmacology, the biochemist Dr Earl Wilbur Sutherland, Jr (1971) and microbiologist Dr Daniel Nathans (1978). In 1975, Washington University School of Medicine Alumni Association endowed a lectureship in her name, aiming to bring a distinguished female scientist to the university every year, and in 1980 conferred on her an honorary DSc. In her honour, the American Association of Physical Anthropologists created the Mildred Trotter Prize for exceptional work in the field of physical anthropology.

Mildred Trotter continued to travel extensively after her formal retirement and attended lectures in cultural anthropology, art and music at Washington University's College of Arts and Science. She remained active until suffering an incapacitating stroke the day before her eighty-sixth birthday in 1975. On 23 August 1991, she passed away at the age of ninety-two, her body donated to the Washington University School of Medicine.

Anna Johnson Pell Wheeler

Born: 5 May 1883 in
Hawarden, Iowa, USA
Died: 26 March 1966
(aged 82)
Field: Mathematics
Award: Alice Freeman
Palmer Fellowship (Wellesley
College, 1906)

Anna Johnson was born on 5 May 1883 in Calliope (now Hawarden), Iowa,
USA, the second daughter of Swedish immigrants Andrew Gustav Johnson
and Amelia (Friberg) Johnson who came from Lyrestad in Skaraborg län,
Västergötland, Sweden. In 1872, about ten years before Anna was born,
they had arrived in the USA where they settled in Union Creek in Dakota
Territory. They lived in a dugout hollowed from the side of a small hill while
Andrew made a living as a farmer. In 1882, the family moved to Calliope,
where Anna was born.

Anna was the youngest of her parents' three surviving children (a fourth
child died in infancy), including a sister, Esther (four years older), and a
brother, Elmer (two years older). Esther and Elmer had been born in Union
Creek. Andrew eventually abandoned farming and became a furniture
dealer and later an undertaker. When Anna was nine years old, the family
moved to Akron, Iowa, where she attended a public high school. As her
mother was disappointed that she had received an inadequate education,
she was determined that her daughters would not suffer the same fate.
Anna was assisted by her older sister, Esther, who recognised Anna's gift for
mathematics and was an inspiration to her.

In 1899, aged sixteen, Anna enrolled at the University of South Dakota.
Her outstanding mathematical talent was evident, but she also studied
English, physical culture, history, German, Latin, French, physics and
chemistry. The professor of mathematics there, Alexander Pell (a Russian
emigrant and former Russian double agent), recognised her mathematical
ability and persuaded her to undertake a career in that field. Anna and her
sister Esther were at university together, boarding in the home of Pell and
his wife, Emma. In 1899–1900 Anna studied algebra and trigonometry, and

in 1900–01 modern geometry, the theory of equations, and solid analytical geometry. The next years included calculus, analytical mechanics and plane analytical geometry in 1901–02; and the theory of substitutions and potential, partial differential equations and Fourier series, and differential equations in 1902–03. Esther also enrolled in many of these classes.

In 1903, Anna completed her bachelor's degree, but she had ambitions to continue her study of mathematics. She wrote to Esther in January 1900, 'Sure, I would like to go to Europe. I could easily stay in Germany to study mathematics since I know so much German already.' After winning a scholarship to study for her master's degree at the University of Iowa, in 1904 she was awarded the degree for her thesis, 'The extension of Galois's theory to linear differential equations'. During this year, in addition to taking five mathematics courses and a philosophy course, she also taught a first-year calculus subject.

Anna obtained a second master's degree the following year, from Radcliffe College in Massachusetts, also with a scholarship. After a further year of study at Radcliffe, she won an Alice Freeman Palmer Fellowship from Wellesley College to study at Göttingen University during the academic year 1906/07. Alexander Pell, who was twenty-five years her senior and whose wife had died in 1904, continued to correspond with Anna during her years of graduate studies. After the end of her fellowship in July 1907, they were married in Göttingen. The couple returned to the University of South Dakota, where Anna lectured two courses in the mathematics department, returning to Göttingen in 1908 with the intention of undertaking a PhD with the eminent mathematician David Hilbert. However, due to conflicts with him, Anna dropped out and returned to the USA the following January, enrolling for a PhD at the University of Chicago under the supervision of Eliakim Hastings Moore. Anna graduated in 1909 with a thesis, 'Biorthogonal Systems of Functions with Applications to the Theory of Integral Equations', which was part of the emerging area of functional analysis. The thesis was published in two parts in the *Transactions of the American Mathematical Society* in 1911.

In the spring of 1911, Anna's husband Alexander suffered a stroke and she took over his mathematics classes at the Armour Institute in Chicago. The following year she accepted a teaching position at Mount Holyoke College, lecturing there for seven years before moving to Bryn Mawr College in 1918 as an associate professor. In 1917, her last year at Mount Holyoke College, Anna co-authored (with R. L. Gordon) a paper regarding Sturm's theorem, solving a problem that had eluded the prominent mathematicians James Joseph Sylvester (1853) and Edward Burr Van Vleck (1899). That paper, with her theorem, was forgotten for almost 100 years and was revived in 2015 by Akritas, Malaschonok and Vigklas in their paper, 'On the Remainders Obtained in Finding the Greatest Common Divisor of Two Polynomials', published in the *Serdica Journal of Computing*, Vol 9(2), 123–138.

Alexander passed away in 1920. In 1924, Anna became head of the mathematics department and was promoted to professor in 1925. That same year she married Arthur Wheeler, a colleague in classics at Bryn Mawr, who moved to Princeton as a professor of Latin. Now with the surname 'Pell Wheeler', Anna also moved to Princeton, commuting to Bryn Mawr to teach on a part-time basis. This also allowed her to participate in the mathematical activities at Princeton. Arthur died in 1932 and Anna returned to Bryn Mawr as a full-time faculty member, Emmy Noether (*see entry*) joining her in 1933. She was departmental chair until her retirement in 1948.

Anna received many honours during her life. In 1927, she became the first woman to give the Colloquium Lectures at the American Mathematical Society meetings. Her topic was 'Theory of quadratic forms in infinitely many variables and applications'. She was an active member of both the American Mathematical Society (AMS) and the Mathematical Association of America.

She served on the Board of Trustees and the Council (1924–26) of the AMS. She was an editor of the *Annals of Mathematics* for eighteen years and received honorary doctorates from the New Jersey College for Women (1932) and Mount Holyoke College (1937). The 1940 Women's Centennial Congress named her as one of the 100 American women to have succeeded in careers not open to women a century before. In 1960, an anonymous donor gave money to establish the Anna Pell Wheeler Prize at Bryn Mawr.

Anna Pell Wheeler died of complications from a stroke on 26 March 1966, aged sixty-two. In accordance with her wishes, she was buried beside her first husband, Alexander Pell, in Lower Merion Baptist Church Cemetery in Bryn Mawr. As well as being an outstanding mathematician, she was an influential and dedicated teacher.

Sheila Marie Evans Widnall

Born: 13 July 1938 in Tacoma,
Washington, USA
Field: Aeronautical Engineering
Awards: Lawrence Sperry Award
(1972); National Academy of
Engineering (1985); New Englander
of the Year (1996); Women in Aviation
Pioneer Hall of Fame (1996)

Sheila Marie Evans was born on 13 July 1938 in Tacoma, Washington, USA, to Rolland John and Genevieve Alice (Krause) Evans. Her father worked as a rodeo cowboy, later becoming a production planner for Boeing Aircraft Company and, after that, a teacher. Her mother was a juvenile probation officer.

When she was seventeen, Sheila's uncle, who worked for a mining company, gave her a piece of uranium. She made it the centrepiece of a project that won first prize at her local science fair in Tacoma, attracting the attention of one of the judges who was graduate MIT. He suggested she go there too and offered to organise a scholarship for her. She replied 'Okay, where's that?'

In September 1956, Sheila enrolled at MIT, one of only twenty-one women in a class of 900, graduating in 1960 with a Bachelor of Science degree in aeronautics and astronautics. In June that year, Sheila married William Soule Widnall, also an aeronautical engineer. She then entered graduate school at MIT, earning a Master of Science degree in 1961 and a Doctor of Science in 1964, both in aeronautics and astronautics.

After graduation, Sheila took a position as an assistant professor in mathematics and aeronautics at MIT and was the first alumna to serve on the faculty in the school of engineering. She was promoted to associate professor in 1970, gaining full professorship four years later. During her time at MIT, she served as head of the Division of Fluid Mechanics from 1975 to 1979, and as director of the Fluid Dynamics Laboratory between 1979 and 1990.

Sheila's field of specialisation was fluid dynamics, especially issues associated with air turbulence created by rotating helicopter blades. Her research focused on the vortices or eddies of air created at the ends and at the trailing edge of helicopter blades as they whirl through the air. These vortices caused noise, instability, and vibrations that affect the integrity of the blades and the stability of the aircraft. She also researched similar problems relating to aircraft that make vertical short take-offs and landings (V/STOL aircraft) and the associated noise.

As part of her investigation, she established the anechoic wind tunnel at MIT, where researchers study the phenomenon of noise and V/STOL aircraft. She soon established a reputation as an expert in her field and lectured widely on her research on vortices and their relation to aerodynamics. Sheila authored seventy papers on fluid dynamics, as well as other areas of science and engineering. She also served as associate editor of the scientific publications *Journal of Aircraft*, *Physics of Fluids* and the *Journal of Applied Mechanics*.

Sheila was actively involved in administration, public policy and consulting. In 1972 she won the Lawrence Sperry Award from the American Institute of Aeronautics and Astronautics, and in 1974 she became the first director of university research of the US Department of Transportation. The following year she won the Outstanding Achievement Award from the Society of Women Engineers, and in 1979 MIT nominated her to be the first woman to chair its 936-member faculty. Her association with the Air Force developed through her appointment by US President Jimmy Carter to two three-year terms on the Air Force Academy's board of visitors, chairing it from 1980 to 1982. She was elected to the National Academy of Engineering in 1985.

In recognition of Sheila's efforts on behalf of women in science and engineering, in 1986 MIT awarded her the Abby Rockefeller Mauzé Chair, an endowed professorship awarded to those who promote the advancement of women in industry and in the arts and professions. A year later she won the Washburn Award from the Boston Museum of Science, its highest honour, awarded to an individual 'who has made an outstanding contribution toward public understanding and appreciation of science and the vital role it plays in our lives'. In 1988, as newly elected president of the American Association for the Advancement of Science (AAAS), Sheila addressed the association on her longstanding interest in seeing more women become scientists and engineers and the problems they faced in attaining higher degrees and achieving professional goals. She also served on advisory committees to the Military Airlift Command and to Wright-Patterson Air Force Base in Dayton, Ohio.

Commencing in 1991, Sheila chaired MIT's Committee on Academic Responsibility for twelve months, after which she became associate provost at the university. In addition to her term as president of the AAAS, she has served on the board of directors for the American Institute of Aeronautics and Astronautics, as a member of the Carnegie Commission on Science, Technology and Government. She also acted as a consultant to businesses and colleges, including American Can Corporation, Kimberly-Clark, McDonnell Douglas Aircraft, and Princeton University.

Sheila has written extensively about aerodynamics, publishing articles and presenting seminars about the changing attitudes and trends in education for prospective engineers and scientists. One of her most notable works is on the elliptical instability mechanism, with the physicist Raymond Pierrehumbert. In her role as Secretary of the Air Force, her responsibilities included all administrative, training, recruiting, logistical support and personnel matters, as well as research and development operations. In 1996, she was named New Englander of the Year by the New England Council and was admitted into the Women in Aviation Pioneer Hall of Fame.

Sheila Widnall and her husband, William, have two adult children, William and Ann Marie. In their spare time, the couple enjoy bicycling, wind surfing and hiking in the Cascade Mountains in her native Washington.

Rosalyn Sussman Yalow

Born: 19 July 1921 in New York City, New York, USA
Died: 30 May 2011 (aged 89)
Field: Medical physicist
Awards: William S. Middleton Award for Excellence in Research (the highest honour of the Veterans' Affairs Medical Center) (1972); AMA Scientific Achievement Award (with Solomon A. Berson) (1975); Nobel Prize in Physiology or Medicine (shared with Roger Guillemin and Andrew V. Schally) (1977)

Rosalyn Sussman was born on 19 July 1921 in New York City to Clara (Zipper) Sussman who emigrated to the USA from Germany at the age of four, and Simon Sussman, born on the Lower East Side of New York, which she later described as 'the melting pot for Eastern European immigrants', who conducted his own small business selling cardboard and packing twine. Neither parent went to high school. Rosalyn and her older brother, Alexander, were avid readers from a young age and made weekly trips to the public library. She was a determined child, and first discovered a passion for mathematics during seventh grade, while attending Walton High School. After completing secondary school at the age of fifteen, Rosalyn attended Hunter, a free city university for women in New York City's college system (now the City University of New York, CUNY). Her interest was drawn to physics, which she found exciting, recognising that almost every major experiment seemed to attract a Nobel Prize. She was also inspired by the work of Marie Curie (*see entry*).

In January 1939, in her third year at college, Rosalyn was captivated by a colloquium presented by the eminent physicist Enrico Fermi – 'hanging from the rafters', she later said – on the newly discovered nuclear fission and the availability of radioisotopes for medical investigation. It was then Rosalyn decided to embark on a career in physics, although her parents would have preferred her to be an elementary school teacher. In September 1940, during her senior year, she was offered a position as secretary to the biochemist Rudolf Schoenheimer. She said that this was 'supposed to provide an entrée for me into graduate courses, via the backdoor', but first she had to learn shorthand. She graduated from Hunter in January 1941 *magna cum laude* (with great distinction) and as a member of America's oldest honour society *Phi Beta Kappa*, with a bachelor's degree in chemistry and physics. Although an outstanding academic, she attended a business school for a short time to improve her prospects of earning an income. In mid-February, Rosalyn received an offer of a teaching assistantship in physics at the University of Illinois, the most prestigious of the schools to which she had applied. She stayed on as a secretary until June and, during the summer, took two tuition-free physics courses under government auspices at New York University.

In September 1941, Rosalyn enrolled in the graduate physics programme at the University of Illinois at Champaign-Urbana. At the first meeting of the Faculty of the College of Engineering, she found herself the only woman among its 400 faculty and teaching assistants. In fact, the dean of the faculty announced that she was the first woman there since 1917. On the first day of graduate school, Rosalyn met Aaron Yalow, who was also beginning graduate study in physics there. On 6 June 1943 they were married, later having two children, Benjamin (b. 1952) and Alanna (b. 1954).

As a result of the attack on Pearl Harbour on 7 December 1941, the campus was filled with young Army and Navy students who had been sent to the campus by their respective services for training. Rosalyn's research was in nuclear physics, and she became skilled in making and using apparatus for the measurement of radioactive substances. In 1945, Rosalyn graduated with a PhD in nuclear physics under the supervision of Dr Maurice Goldhaber, whose wife was Trude Goldhaber (*see entry*), and who later became Director of Brookhaven National Laboratories. Rosalyn then accepted a position as assistant engineer at the Federal Telecommunications Laboratory, a research laboratory for International Telephone and Telegraphic (ITT), as the only female engineer. When her research group left New York in 1946, she returned to Hunter College to teach physics to returning veterans in a pre-engineering programme.

In December 1947, Rosalyn joined the Bronx Veterans' Affairs Hospital (VA) as a part-time consultant, while keeping her position at Hunter until the Spring Semester of 1950. While teaching full-time, she, with Dr Bernard Roswit, equipped and developed the Radioisotope Service. They started research projects with other physicians in the hospital in a number of clinical fields. By January 1950, Rosalyn left teaching and joined the VA full-time and, in July that year, was joined by Dr Solomon A. Berson, who was completing his residency in internal medicine there. This would be the beginning of a twenty-two-year partnership.

Soon after meeting Berson, Rosalyn gave up collaborative work with others and concentrated on their joint research, firstly investigating the application of radioisotopes in blood volume determination, clinical diagnosis of thyroid diseases and the kinetics of iodine metabolism. They extended these techniques to the study of the distribution of globin, which had been suggested for use as a plasma expander, and of serum proteins. These methods were then applied to smaller peptides such as hormones, with insulin being the hormone most readily available in a highly purified form. They soon deduced from the retarded rate of disappearance of insulin from the circulation of insulin-treated subjects, that all these patients develop antibodies to the animal insulins. In studying the reaction of insulin with antibodies, they developed a tool with the potential for measuring circulating insulin. It took several more years of work to transform the concept into the reality of its practical application to the measurement of plasma insulin in humans. It was in 1959 that the era of radioimmunoassay (RIA) began, and RIA is now used to measure hundreds of substances of biological interest in thousands of laboratories worldwide, even in developing countries.

In 1961, Rosalyn won the Eli Lilly Award of the American Diabetes Association, an organisation that provides scholarships for up to 100 scholars to attend Scientific Sessions, the world's largest scientific and medical conference focused on diabetes and its complications. A year later she was awarded the Gairdner Foundation International Award, which recognises the world's most creative and accomplished biomedical scientists who are advancing humanity, along with the American College of Physicians Award, which recognises excellence and distinguished contributions by individuals to internal medicine.

In 1972, Berson died prematurely at age fifty-three of a massive heart attack and, at Rosalyn's request, the laboratory which they shared was designated the 'Solomon A. Berson Research Laboratory', so that his name would continue to be on her papers as long as she published and so that his contributions would not be forgotten. The same year, she was awarded the William S. Middleton Award for Excellence in Research. This was the highest honour awarded annually by the Biomedical Laboratory Research and Development Service to senior biomedical research scientists in recognition of their outstanding scientific contributions and achievements, pertaining to the healthcare of Veterans. Also in 1972, Rosalyn was given the Koch Award of the Endocrine Society, in recognition of individuals for their dedication to excellence in research, education and clinical practice in the field of endocrinology.

In 1975, Rosalyn and Berson (posthumously) were awarded the American Medical Association Scientific Achievement Award, comprising a gold medallion presented to individuals on special occasions in recognition of their outstanding work in scientific achievement. The year after, she became the first female and first nuclear physicist to receive the Albert Lasker Award for Basic Medical Research. As with Christiane 'Janni' Nüsslein-Volhard in later years (*see entry*), Rosalyn's Lasker was followed by a Nobel when, in 1977, she became the sixth woman, and the first American-born woman, to win the Nobel Prize (her field was Physiology or Medicine), with Roger Guillemin and Andrew V. Schally, for her role in devising the RIA technique. RIA can be used to measure many substances found in tiny quantities in fluids within and outside of organisms (such as viruses, drugs and hormones). There is a long list of current possible uses but, in one example, RIA allows blood-donations to be screened for various types of hepatitis. The technique can also be used to identify hormone-related health problems. In addition, RIA can be used to detect in the blood many foreign substances, including some cancers, and can be used to measure the effectiveness of dose levels of antibiotics and drugs.

Rosalyn Sussman Yalow

In 1986, Rosalyn was awarded the A. Cressy Morrison Award in Natural Sciences of the New York Academy of Sciences, offered by Abraham Cressy Morrison to individuals with superlative papers on a scientific subject within the field of the New York Academy of Sciences and its Affiliated Societies. In 1988, she received the National Medal of Science, which is given to American individuals who deserve the highest honour in science and technology. She also received five honorary doctorates.

At the age of eighty-two, Rosalyn Yalow suffered a serious fall, causing her to be confined to a wheelchair and making further activity impossible. She passed away at the age of eighty-nine at the Bronx, New York, on 30 May 2011. Her husband had died in 1992 and she was survived by their son Benjamin, daughter Elanna and two grandchildren.

Ada E. Lifshitz Yonath

Born: 22 June 1939 in Geula, Jerusalem, British Mandate of Palestine (now Israel)
Field: Crystallography
Awards: Israel Prize in Chemistry (2002); Harvey Prize in Science and Technology (2002); Massry Prize (2004); Louisa Gross Horwitz Prize (2005); Wolf Prize in Chemistry (2007); Paul Ehrlich and Ludwig Darmstaedter Prize (2007); L'Oréal-UNESCO Award for Women in Science (2008); Albert Einstein World Award of Science (2008); Nobel Prize in Chemistry (2009)

Ada Lifshitz was born on 22 June 1939 in Geula, Jerusalem, Israel, to Esther Lifshitz and Hillel Lifshitz, a rabbi. Her parents were Zionist Jews who migrated from Zduńska Wola in Poland in 1933, before Israel was established. They ran a grocery store in Jerusalem, all the while enduring financial difficulties. They shared a four-room apartment with two other families. Ada was a keen reader from an early age and had a curiosity about the world around her. Despite the family's impoverished circumstances, her parents were determined that she receive the best possible education. As a result of their efforts, she enrolled in a prestigious secular grammar school in the upmarket Beit Hakerem district. Her father passed away when she was aged eleven and, as her mother had difficulty coping financially, Ada took on all manner of jobs, including cleaning, babysitting and providing tutoring to younger children. But it was not enough, and a year later Ada and her mother moved to Tel Aviv, to be close to her mother's sisters, where Ada completed her high school education at Tichon Hadash. She paid the fees herself, with money earned from tutoring mathematics to younger students and even some of her own age.

After finishing secondary school, Ada spent her compulsory army service in the 'top secret office' of the Medical Forces, where she was exposed to clinical and medical issues. When this ended, she enrolled in the Hebrew University of Jerusalem where she graduated with a bachelor's degree in chemistry in 1962 and a master's degree in biochemistry in 1964. She then enrolled for a PhD at the Weizmann Institute of Science under the supervision of Wolfie Traub from the Department of Structural Chemistry, where she examined the high-resolution structure of collagen. Obtaining her doctorate in 1968, Ada then undertook a postdoctoral year at the Mellon Institute in Pittsburgh, Pennsylvania, before moving to MIT to learn high resolution structural methods for protein three-dimensional analysis. After completing her postdoctoral research at the end of 1970, Ada returned to the Weizmann Institute where she established the first biological crystallography laboratory in Israel, one that was the only laboratory for such studies for almost a decade.

At this stage, Ada had an interest in one of the major outstanding questions concerning living cells: the process of protein biosynthesis and she aimed to determine the three-dimensional structure of the ribosome (the cells' factory for translating the instructions written in the genetic code into proteins) and, in doing so, reveal the mechanics guiding the process. The ribosome is a complex of proteins and ribonucleic acid (RNA) chains. Its structure is extraordinarily intricate and it is unusually flexible, unstable and lacking internal symmetry – all making crystallisation an extremely difficult task. She later compared her task to 'climbing Mt Everest only to discover that a higher Everest stood in front of [her]' and said she was 'met with reactions of disbelief and even ridicule in the international scientific community'.

This research occupied her for the next twenty years. Ada began these studies in collaboration with Professor Heinz-Günter Wittmann of the Max Planck Institute for Molecular Genetics in Berlin, who supported her academically and financially. At the same time, she maintained her laboratory at the Weizmann Institute.

During this period, a centre for macromolecular assemblies was established by Helen Kimmel at the Weizmann Institute, and Ada led a large team of international researchers, commencing with an attempt to comprehend one of the fundamental components of life. This led to a detailed understanding of the actions of some of the most widely prescribed antibiotics, her results aiming to assist in the development of more efficient antibacterial drugs, as well as providing scientists with new weapons in the fight against antibiotic resistant bacteria. This problem has been called 'one of the most important medical challenges of the twenty-first century'.

In the early 1980s, Ada and her team created the first ribosome micro crystals and, by mid-decade, had visualised a tunnel spanning the large ribosomal subunit. They assumed, based on previous biochemical works, that this is the path through which the nascent protein progresses as it is being formed – until it emerges out of the ribosome. During the course of their research, they developed a number of new techniques that are today widely used in structural biology labs worldwide.

In 1993, Ada envisaged the path taken by the nascent proteins, namely the ribosomal tunnel, and later revealed the dynamic elements enabling its involvement in elongation arrest, gating, intra-cellular regulation and nascent chain trafficking into their folding space. In 2000 and 2001, she determined the complete high-resolution structures of both ribosomal subunits and discovered, within the otherwise asymmetric ribosome, the universal symmetrical region that provides the framework and navigates the process of polypeptide polymerisation. She subsequently showed that the ribosome is a ribozyme that places its substrates in stereochemistry suitable for peptide bond formation and for substrate-mediated catalysis. She was able to elucidate the modes of action of over twenty different antibiotics targeting the ribosome, clarifying mechanisms of drug resistance and synergism, deciphering the structural basis for antibiotic selectivity and demonstrating how it plays a key role in clinical usefulness and therapeutic effectiveness. Ada described this discovery as a high point in her research.

In 2009, Ada was awarded the Nobel Prize in Chemistry 'for studies of the structure and function of the ribosome', shared with Venkatraman Ramakrishnan and Thomas A. Steitz. With this achievement she became the first Israeli woman to win the Nobel Prize, the first woman from the Middle East to win a Nobel Prize in the sciences, and the first woman in forty-five years to win the Nobel Prize for Chemistry, after Dorothy Crowfoot Hodgkin (*see entry*) in 1964, and the third in a century, including Marie Curie (*see entry*) in 1911. In addition to the Nobel Prize, Ada received numerous other awards, including, in 2005, the Louisa Gross Horwitz Prize for Biology and Chemistry, awarded by Columbia University to a researcher or group of researchers who have made an outstanding contribution in basic research in the fields of biology or biochemistry. In 2007, she was awarded the Paul Ehrlich and Ludwig Darmstaedter Prize. One of the highest endowed and internationally most distinguished awards in medicine in Germany, the award carries prize money of 100,000 Euro. The following year she won the Albert Einstein World Award of Science, the recipient of which is chosen by an interdisciplinary committee composed of world-renowned scientists, among them twenty-five Nobel laureates, to recognise and encourage scientific and technological research and development.

Ada was made a member of many prestigious organisations, including the US National Academy of Sciences, the American Academy of Arts and Sciences, the Israel Academy of Sciences and Humanities, the European Molecular Biology Organisation (EMBO), the European Academy of Sciences and Art, the German National Academy of Sciences Leopoldina, the Korean Academy for Science and Technology, the Royal Society of Chemistry (UK), the International Academy of Microbiology and the International Academy of Astronautics. In addition, she holds honorary doctorates from almost all Israeli Universities, along with those from High Energy Accelerator Research Organization (Japan), Oslo University (Norway), Fujian University (China), New York University and Mount Sinai Universities (New York City), Hamburg University, Patras University (Greece), Oxford and Cambridge Universities (UK) and the Baptist University (Hong Kong).

In 2019, Ada Yonath was serving as the director of the Helen and Milton A. Kimmelman Center for Biomolecular Structure and Assembly of the

Weizmann Institute of Science. In a 2014 interview with *The Conversation*, Ada was asked to provide a tip for early career scientists. Her response was, 'Passion. It's not enough to be curious – one has to really love what one does. For men and women, science is demanding and there are many, many dark periods, low periods.' She has one child, a daughter, Hagit Yonath, who is a physician at Sheba Medical Center, Tel Hashomer, and a granddaughter, Noa, who, at the age of five, invited Ada to her kindergarten to talk about the ribosome.

In 2018, Ada was awarded an Honorary Doctorate from Carnegie Mellon University, and in 2020 was elected a Foreign Member of the Royal Society. In 2023 she received an Honorary Doctorate from the Jagiellonian University, a public research university in Kraków, Poland.

Bibliography

General

Alic, Margaret, *Hypatia's Heritage: A History of Women in Science from Antiquity to the Late Nineteenth Century* (USA: Beacon Press, 2001).

Biographies of Women Mathematicians <https://www.agnesscott.edu/lriddle/women/women.htm>. (An excellent website with links to many other resources).

Contributions of 20th Century Women to Physics <http://cwp.library.ucla.edu/>. (An excellent website with links to many other resources).

Ignotofsky, Rachel, *Women in Science* (Berkeley: Ten Speed Press, 2016).

Latta, Susan. M., *Bold Women of Medicine 21 Stories of Astounding Discoveries, Daring Surgeries, and Healing Breakthroughs* (Chicago Review Press, 2017).

McGrayne, Sharon Bertsch, *Nobel Prize Women in Science: Their Lives, Struggles, and Momentous Discoveries* (USA: National Academies Press, 2001).

Ogilvie, Marilyn Bailey, *Women in Science: Antiquity through the Nineteenth Century, A Biographical Dictionary with an Annotated Bibliography* (The MIT Press, 1990).

Rossiter, Margaret W., *Women Scientists in America: Before Affirmative Action, 1940–1972* (Johns Hopkins University Press, 1998).

Saini, Angela, *Inferior: How Science Got Women Wrong – and the New Research That's Rewriting the Story* (Beacon Press, 2017).

Swaby, Rachel, *Headstrong: 52 Women Who Changed Science and the World* (New York: Broadway Books, 2015).

Wasserman, Elga R., *The Door in the Dream: Conversations with Eminent Women in Science* (Joseph Henry Press, 2000).

Revolvy <https://www.revolvy.com/>.

Frances 'Fran' Elizabeth Allen

Abbate, Janet, *Oral-History: Frances 'Fran' Allen*, Archived at the ETHW, New Brunswick, New Jersey (Interview #573 for the IEEE History Center, 2 Aug 2001).

Bergstein, Brian, 'First Woman Honored with Turing Award' (*The Globe and Mail*, 21 Feb 2007).

Lohr, Steve, 'Scientist at Work: Frances Allen, Would-Be Math Teacher Ended Up Educating a Computer Revolution' (*The New York Times*, 6 August 2002).

Perelman, Deb, 'Turing Award Anoints First Female Recipient' (*eWEEK*, 27 Feb 2007).

Steele, Guy L. Jr., 'An Interview with Frances E. Allen, *Communications of the ACM*, 54(1) (2011), 39–45.

Thomas, Jeffrey, 'Turing Award Winner Sees New Day for Women Scientists, Engineers', *USINFO* (Bureau of International Information Programs, U.S. Department of State, 16 March 2007).

'Frances Allen – 2004 Women of Vision ABIE Award Winner for Technical Leadership', Anita B.org Institute (1 October 2004) <https://anitab.org/profiles/abie-award-winners/frances-allen>.

Virginia Apgar

Amschler, Denise, 'Apgar, Virginia (1909–1974)', in Anne Commire (ed.), *Women in World History: A Biographical Encyclopedia* (Waterford CT: Yorkin Publications, 1999), pp. 415–418.

Apgar, Virginia and Joan Wagner Beck, *Is My Baby All Right? A Guide to Birth Defect* (New York: Trident Press, 1972).

Apgar, Virginia, 'A Proposal for a New Method of Evaluation of the Newborn Infant' *Current Researches in Anesthesia & Analgesia*, 32(4) (1953), 260–267.

Blakemore, Erin, 'The Doctor who Saved Countless Newborn Babies' (*Time: History*, 29 August 2016) <http://time.com/4460720/virginia-apgar>.

'The Virginia Apgar Papers: Biographical Information', *Profiles in Science*, US National Library of Medicine <https://profiles.nlm.nih.gov/CP>.

Phoebe Sarah 'Hertha' Marks Ayrton

Appleyard, Rollo, *The History of the Institution of Electrical Engineers* (London, 1939).

Ayrton, Hertha, *The Electric Arc* (New York: D. Van Nostrand Co., 1902).

'Ayrton, Hertha (Sarah) Marks', *Contributions of 20th Century Women to Physics* <http://cwp.library.ucla.edu/>.

Crawford, Elizabeth, *The Women's Suffrage Movement: A Reference Guide*, 1866–1928 (London: Routledge, 1999).

Hirsch, Pam, *Barbara Leigh Smith Bodichon: Feminist, Artist and Rebel, 1869–1946* (London: Girton College Register, 1998).

Mason, Joan, 'Hertha Ayrton (1854–1923) and the Admission of Women to the Royal Society of London', *Notes and Records of the Royal Society*, 45(2) (1991), 201–220.

Mason, Joan, 'Hertha Ayrton' in Nina Byers and Gary Williams (eds), *Out of the Shadows: Contributions of 20th Century Women to Physics* (Cambridge University Press, 2006).

Mason, Joan, 'Ayrton [*née* Marks], (Phoebe) Sarah [Hertha] (1854)–1923)' *Oxford Dictionary of National Biography* (online ed.) (23 September 2004, 23 September 2010).

Moore, Glenis, 'Hertha Ayrton – First Lady of the IEE', *Electronics and Power*, 32(8) (1986) 583.

Riddle, Larry, 'Hertha Marks Ayrton', *Biographies of Women Mathematicians* <https://www.agnesscott.edu/lriddle/women/ayrton.htm>.

Sharp, Evelyn, *Hertha Ayrton: A Memoir* (London: E. Arnold, 1926).

Massimilla 'Milla' Baldo-Ceolin

'Baldo-Ceolin, Milla', *Contributions of 20th Century Women to Physics* <http://cwp. library.ucla.edu/>.

Bonolis, Louisa and Michael Friedlander, 'Massimilla Baldo-Ceolin', *Physics Today*, 65(8) (2012) 60 <https://doi.org/10.1063/PT.3.1686>.

'Faces and Places: Obit. Milla Baldo Ceolin (1924–2011)', *CERN Scientific Information Service*, 52(2) (2012) 38.

'Workshop: Neutrino Telescopes', *CERN Courier*, 30(4) (1990), 15–17.

Peruzzi, Giulio, 'Milla Baldo Ceolin, La Signora Dei Neutrini' (*ScienzainRete*, 8 December 2011) <https://www.scienzainrete.it/contenuto/articolo/milla-baldo-ceolin-signora-dei-neutrini>.

Dorigo, Tommaso, 'Padova Loses Milla' (*Science2.0*, 25 November 2011) <https://www.science20.com/quantum_diaries_survivor/blog/padova_loses_milla-84978>.

Alice Augusta Ball

Brown, Jeannette, *African American Women Chemists* (New York: Oxford University Press, 2012), pp. 19–24.

Cederlind, Erika, 'A Tribute to Alice Ball: A Scientist whose Work with Leprosy was Overshadowed by a White Successor' (*The Daily of the University of Washington*, 29 February 2008).

Mendheim, Beverly, 'Lost and Found: Alice Augusta Ball, an Extraordinary Woman of Hawai'i Nei' (*Northwest Hawaii Times*, September 2007).

Wermager, Paul and Carl Heltzel, 'Alice A. Augusta Ball', *ChemMatters*, 25(1) (2007) 16–19.

'Alice Ball Biography', *Biography* <https://www.biography.com/people/alice-ball>.

'Alice Augusta Ball (1892-1916)', *Blackpast* <http://www.blackpast.org/aaw/ball-alice-augusta-1892-1916>.

Nora Stanton Blatch Barney

Litoff, Judy Barrett and Judith McDonnell, *European Immigrant Women in the United States: A Biographical Dictionary* (United Kingdom: Taylor & Francis, 1994), p. 14.

Walpole, Ben, 'ASCE Recognizes Stanton Blatch Barney: Pioneering Civil Engineer, Suffragist' (*ASCE News*, 28 August 2015).

Bois, Danuta, 'Nora Stanton Blatch Barney (1883-1971)', *Distinguished Women of Past and Present* (1999) <http://www.distinguishedwomen.com/bio.php?womanid=1014>.

'Mrs. Nora S. Barney, Architect, 87, Dies' (*New York Times*, 20 January 1971) <https://www.nytimes.com/1971/01/20/archives/mrsras-barny.html>.

'Nora Stanton Blatch Barney: American Civil Engineer and Architect', *IEEE Global History Network* <https://www.britannica.com/biography/Nora-Stanton-Blatch-Barney>.

Florence Bascom

Bascom, Florence, *The Ancient Volcanic Rocks of South Mountain, Pennsylvania* (Washington, D.C.: US Government Printing Office, 1986).

Bascom, Florence, 'The University in 1874–1887', *The Wisconsin Magazine of History*, VIII(3) (1925), 300–308.

Bibliography

Clary, R. M. and J. H. Wandersee, 'Great Expectations: Florence Bascom (1842–1945) and the Education of Early US Women Geologists', in C. V. Burek and B. Higgs (eds), *The Role of Women in the History of Geology* (Geological Society, London, Special Publications, 2007), vol. 281, 123–136.

Schneidermann, Jill, 'A Life of Firsts: Florence Bascom' (*GSA Today*, Geological Society of America, July 1997), 8–9.

'Florence Bascom Biography', *Browse Biography* (3 May 2011) <http://www.browsebiography.com/bio-florence_bascom.html>.

Laura Bassi

Elena, Alberto, 'In Lode Della Filosofessa di Bologna: An Introduction to Laura Bassi', *Isis*, 82(3) (1991), 510–518.

Frize, Monique, *Laura Bassi and Science in 18th Century Europe: The Extraordinary Life and Role of Italy's Pioneering Female Professor* (Springer, 2013).

Findlen, Paula, 'Science as a Career in Enlightenment Italy: The Strategies of Laura Bassi', *The History of Science Society*, 84(3) (1993) 441–469.

DeBakcsy, Dale, 'The Life of Laura Bassi (1711–1778): The World's First Female Full Professor of Science', *Women You Should Know* (6 June 2018) <https://womenyoushouldknow.net/laura-bassi-worlds-first-female-professor-science/>.

Laura Bassi and the Bassi-Veratti Collection, Stanford University, Stanford, California <https://bv.stanford.edu/en/about/laura_bassi>.

O'Connor, John, J. and Edmund F. Robertson, 'Laura Maria Catarina Bassi' (December 2008) <http://www-groups.dcs.st-and.ac.uk/history/Biographies/Bassi.html>.

Ulrike Beisiegel

Ausgezeichnete Wissenschaftlerin (in German) (Hamburg, Germany: Hamburger Aabendblatt, 12 October 1996).

Ehren-Medaille für Uni-Präsidentin Beisiegel (in German) (Kassel, Germany: Hessische/Niedersächsische Allgemeine (HNA), 24 May 2014).

Liste der Heinz Maier-Leibnitz-Preisträger 1978 bis 2015 (PDF) (in German) (Bonn Germany: Deutsche Forschungsgemeinschaft, 2015), p. 3.

Prof. Dr. Ulrike Beisiegel wird künftige Präsidentin der Universität Göttingen (in German) (Informationsdienst Wissenschaft Online, 10 March 2010).

Prof. Ulrike Beisiegel (Vienna, Austria: Österreichische Agentur für Wissenschaftliche Integrität, 2011).

Abbott, Alison, 'German University Head Lauds Progress of Women Scientists', *Nature International Weekly Journal of Science* (10 April 2011) <https://www.nature.com/news/2011/110410/full/news.2011.223.html>.

Elizabeth Helen Blackburn

Brady, Catherine, *Elizabeth Blackburn and the Story of Telomeres* (Cambridge, Massachusetts: The MIT Press, 2007).

Corbyn, Zoë, 'Elizabeth Blackburn on the Telomere Effect: "It's About Keeping Healthier for Longer"' (*The Guardian*, 29 January 2017) <https://www.theguardian.com/science/2017/jan/29/telomere-effect-elizabeth-blackburn-nobel-prize-medicine-chromosomes>.

'Elizabeth Blackburn', *Famous Female Scientists* <http://famousfemalescientists.com/ elizabeth-blackburn>.

Park, Alice, 'The Time 100: Elizabeth Blackburn' (*Time Magazine*, 3 May 2007) <http://content.time.com/time/specials/2007/time100/ article/0,28804,1595326_1595329_1616029,00.html>.

Rogers, Kara, 'Elizabeth H. Blackburn: American Molecular Biologist and Biochemist', *Encyclopædia Britannica* (22 November 2018) <https://www.britannica.com/ biography/Elizabeth-Blackburn>.

Elizabeth Blackwell

Baker, Rachel, *The First Woman Doctor: The Story of Elizabeth Blackwell* (New York: M.D. J. Messner Inc., OCLC 848388, 1994).

Boyd, Julia, *The Excellent Doctor Blackwell: The Life of the First Woman Physician* (The History Press Ltd, Thistle Publishing, 2005).

Kline, Nancy, *Elizabeth Blackwell: First Woman MD* (Conari Press, 1997).

Morantz, Regina, M., 'Feminism, Professionalism and Germs: The Thought of Mary Putnam Jacobi and Elizabeth Blackwell', *American* Quarterly, 34(5) (1982), 461–478.

Sahli, Nancy Ann, *Elizabeth Blackwell, M.D., (1871–1910): A Biography* (New York: Arno Press, 1982).

Wilson, Dorothy Clarke, *Lone Woman: The Story of Elizabeth Blackwell, the First Woman Doctor* (Boston: Little Brown, OCLC 56257, 1970).

Mary Adela Blagg

Whitaker, Ewen A., *Mapping and Naming the Moon: A History of Lunar Cartography and Nomenclature* (United Kingdom: Cambridge University Press, 1999).

Bartels, Meghan, 'The Woman who Named the Moon and Clocked Variable Stars', *Astronomy*, 14 November 2016) <http://www.astronomy.com/news/2016/11/ mary-adela-blagg>.

Christoforou, Peter, *English Astronomy Pioneer Mary Adela Blagg (1858–1944)*, www.astronomytrek.com (17 December 20016) <https://www.astronomytrek. com/english-astronomy-pioneer-mary-adela-blagg-1858-1944>.

Morgan, Barbara, 'Blagg, Mary Adela (1858–1944)', Women in World History: A Biographical Encyclopedia (2002) <https://www.encyclopedia.com/women/ encyclopedias-almanacs-transcripts-and-maps/blagg-mary-adela-1858-1944>.

Shears, Jeremy, 'Selenography and Variable Stars: Women & The RAS: Mary Blagg', Oxford Journals, Oxford University Press (1 October 2016), *Astronomy & Geophysics*, 57(5) (2016), 5.17–5.18 <https://academic.oup.com/astrogeo/ article/57/5/5.17/2738839>.

Katharine 'Katie' Burr Blodgett

'Blodgett, Katharine Burr', *Contributions of 20th Century Women to Physics* <http:// cwp.library.ucla.edu/>.

Byers, Nina and Gary A. Williams, *Out of the Shadows: Contributions of Twentieth-century Women to Physics* (Print book, English ed.) (Cambridge, UK; New York: Cambridge University Press, 2006).

Bibliography

Rossiter, Margaret W., *Women Scientists in America: Struggles and Strategies to 1940* (Baltimore: Johns Hopkins University Press, 1982).

Siegel, Patricia J. and Kay T. Finley, *Women in the Scientific Search : An American Bio-bibliography, 1724–1979* (Metuchen N.J.: Scarecrow Press, 1985).

Stanley, Autumn, 'Blodgett, Katharine Burr', in Edward T. James (ed.), *Notable American Women* (Cambridge, MA: Belknap Press of Harvard University Press, 2004), pp. 66–67.

Shearer, Benjamin F. and Barbara Smith, *Notable Women in the Physical Sciences: A Biographical Dictionary* (ebook ed.) (Westport, Conn.: Greenwood Press, 1997).

Rachel Littler Bodley

Elliott, Clark A., *Biographical Dictionary of American Science: The Seventeenth Through the Nineteenth Centuries* (Westport and London: Greenwood Press, 1979), p. 33.

Grinstein, Louise S., Carol A. Biermann and Rose K. Rose (eds), *Women in the Biological Sciences: A Biobibliographic Sourcebook* (Westport, Conn: Greenwood Press, 1997).

Oakes, Elizabeth, *International Encyclopedia of Women Scientists* (New York: Facts on File, 2002), pp. 36–37.

Ogilvie, Marilyn Bailey, *Women in Science: Antiquity Through the Nineteenth Century: A Biographical Dictionary with Annotated Bibliography* (MIT Press, 1986), p. 42.

Shearer, Benjamin F. and Barbara S. Shearer, *Notable Women in the Life Sciences: A Biographical Dictionary* (1. publ. ed.) (Westport, Conn.: Greenwood Press, 1996).

'Bodley, Rachel (1831–1888)', *Women in World History: A Biographical Encyclopedia* (2002) <https://www.encyclopedia.com/women/encyclopedias-almanacs-transcripts-and-maps/bodley-rachel-1831-1888>.

Alice Middleton Boring

Adler, Kraig, *Contributions to the History of Herpetology, Society for the Study of Amphibians and Peptiles* (1989), p. 202.

Ogilvie, Marilyn Bailey and Clifford J. Choquette, *A Dame Full of Vim and Vigor: A Biography of Alice Middleton Boring; Biologist in China* (Harwood, 1999).

Zheng, Shu, 'Alice M. Boring: A Pioneer in the Study of Chinese Amphibians and Reptiles' *Protein Cell*, 6(9) (2015), 625–627.

Morgan, Barbara, 'Boring, Alice Middleton (1883–1955)', *Women in World History: A Biographical Encyclopedia* <https://www.encyclopedia.com/women/encyclopedias-almanacs-transcripts-and-maps/boring-alice-middleton-1883-1955>.

Elizabeth Brown

Brück, Mary T., *Women in Early British and Irish Astronomy: Stars and Satellites* (Dordrecht: Springer, 2009), pp. 151–156.

Creese, Mary R. S., 'Brown, Elizabeth (1830–1899)', *Oxford Dictionary of National Biography* (Oxford: Oxford University Press, 2004).

Creese, Mary R. S., 'Elizabeth Brown (1830–1899), Solar Astronomer', *Journal of the British Astronomical Association*, 108(4) (1998), 193–197.

'In Memoriam, Elizabeth Brown, FRMetSoc' *Journal of the British Astronomical Association*, 9(5) (1899), 214–215.

The Biographical Dictionary of Women in Science: Pioneering Lives from Ancient Times to mid-20th Century, Marilyn Ogilvie, Joy Harvey and Margaret Rossiter (eds) (Abingdon: Routledge, 2000), p. 189.

Linda Brown Buck

'Buck, Linda B.', *Who's Who* (online Oxford University Press ed.), ukwhoswho.com. (A & C Black, an imprint of Bloomsbury Publishing plc., 2016).

Wayne, Tiffany K., and Linda B. Buck, *American Women of Science since 1900* (Santa Barbara, CA, USA, 2010).

'Linda B. Buck: Biographical', The Nobel Prize in Physiology or Medicine (2004) <https://www.nobelprize.org/prizes/medicine/2004/buck/biographical>.

'Linda Buck PhD', *Academy of Achievement* (26 February 2010) <http://www.achievement.org/achiever/linda-buck>.

(Susan) Jocelyn Bell Burnell

Allan, Vicky, 'Face to Face: Science Star Who Went under the Radar of Nobel Prize Judges' (*The Herald*, Glasgow, 5 January 2015).

'Burnell, Jocelyn Bell', *Contributions of 20 Century Women to Physics* <http://cwp.library.ucla.edu/>.

Drake, Nadia, 'Meet the Woman Who Found the Most Useful Stars in the Universe', *National Geographic* (6 Sept 2018) <https://www.nationalgeographic.com/science/2018/09/news-jocelyn-bell-burnell-breakthrough-prize-pulsars-astronomy>.

'Jocelyn Bell Burnell, 1943–Present', *Famous Scientists the Art of Genius* <https://www.famousscientists.org/jocelyn-bell-burnell>.

Weatherall, Kate Marsh, 'The woman who discovered pulsars: An Interview with Jocelyn Bell-Burnell' (26 October 1995) <http://weatheralltech.com/bell/index.html>.

Wu, Katherine J., 'Decades After Being Passed Over for a Nobel, Jocelyn Bell Burnell Gets Her Due', *Smithsonian* (10 September 2018) <https://www.smithsonianmag.com/smart-news/decades-after-being-passed-over-nobel-jocelyn-bell-burnell-gets-her-due-180970248>.

Nina Byers

Byers, Nina and Gary Williams (eds), *Out of the Shadows: Contributions of Twentieth-Century Women to Physics* (Cambridge: Cambridge University Press, 2010).

'Byers, Nina', *Contributions of 20 Century Women to Physics* <http://cwp.library.ucla.edu/>.

Rosenzweig, James, 'Nina Byers', *UCLA Physics and Astronomy* <http://www.pa.ucla.edu/content/nina-byers>.

Winslow, Lindley, Roberto Peccei and Steven Moskowski, 'Nina Byers', *Physics Today* 68(1) (2015) <https://doi.org/10.1063/PT.3.2663>.

Bibliography

Annie Jump Cannon

'Annie Jump Cannon – American Astronomer', *Encyclopædia Britannica* (7 December 2018) <https://www.britannica.com/biography/Annie-Jump-Cannon>.

Cannon, *Annie Jump, 1863–1941. Papers of Annie Jump Cannon: An Inventory*, Harvard University Library (2014) <https://www.revolvy.com/page/Annie-Jump-Cannon>.

Hennessey, Logan, 'Annie Jump Cannon (1863–1941)' *Wellesley Women in Science* (23 Jul 2006) <http://academics.wellesley.edu/Astronomy/Annie>.

'Annie Cannon', Astronomical Society of the Pacific, University of California at Berkeley <https://howlingpixel.com/i-en/Annie_Jump_Cannon>.

Shteynberg, Catherine, 'Pickering's Women', *Smithsonian Institution Archives* (8 May 2009) <https://siarchives.si.edu/blog/pickering-women>.

Sobel, Dava, 'The Glass Universe: How the Ladies of the Harvard Observatory Took the Measure of the Stars', Narrated by Cassandra Campbell, Penguin Audio (2016) <https://www.britannica.com/biography/Annie-Jump-Cannon>.

Rachel Louise Carson

Carson, Rachel, *Silent Spring* (Houghton Mifflin Company, Anniversary edition, 22 October 2002) (Originally Published: 27 September 1962).

Carson, Rachel, *Under the Sea-Wind* (United Kingdom: Penguin Classics, New edition, 3 April 2007) (Originally Published: 1941).

Gottlieb, Robert, *Forcing the Spring: The Transformation of the American Environmental Movement* (Washington DC: Island Press, 2005).

Jezer, Marty, *Rachel Carson: Biologist and Author. American Women of Achievement* (Chelsea House Publications, 1988).

Lear, Linda, *Rachel Carson: Witness for Nature* (Houghton Mifflin Harcourt, 2009).

Matthiessen, Peter, *Courage for the Earth: Writers, Scientists, and Activists Celebrate the Life and Writing of Rachel Carson* (New York: Mariner Books, 2007).

Moore, Kathleen Dean and Lisa H. Sideris, *Rachel Carson: Legacy and Challenge* (Albany, New York: SUNY Press, 2008).

Paull, John, *The Rachel Carson Letters and the Making of Silent Spring* (United States: Sage Open, 3 July 2013).

Quaratiello, Arlene, *Rachel Carson: A Biography* (Amherst, New York: Prometheus, 2010).

'Rachel Carson (1907–1964)', Debra Michals (ed.) *National Women's History Museum* <https://www.womenshistory.org/education-resources/biographies/rachel-carson>.

Sideris, Lisa H., 'Fact and Fiction, Fear and Wonder: The Legacy of Rachel Carson', *Soundings*, 91(3–4) (2009) 335–69.

Souder, William, *On a Farther Shore: The Life and Legacy of Rachel Carson* (New York: Crown Publishers, 2012).

'Legacy of Rachel Carson's Silent Spring', *American Chemical Society National Historic Chemical Landmarks* (26 October 2012) <https://www.acs.org/content/acs/en/education/whatischemistry/landmarks/rachel-carson-silent-spring.html#top>.

Lear, Linda, *The Life and Legacy of Rachel Carson*, Houghton Mifflin Harcourt (2009) <http://www.rachelcarson.org/Bio.aspx>.

Mary Lucy Cartwright

Byers, H. A., 'Cartwright, Mary Lucy', *Contributions of 20 Century Women to Physics* <http://cwp.library.ucla.edu/>.

Dyson, Freeman, 'Nature's Numbers by Ian Stewart', *Mathematical Intelligencer*, 19(2) (1997), 65-67.

Freeman, J. Dyson, 'Mary Lucy Cartwright (1900–1998): Chaos Theory', in Nina Byers and Gary Williams (eds), *Out of the Shadows: Contributions of Twentieth-Century Women to Physics* (Cambridge University Press, 2006), pp. 169–177.

Girton College, *Girton College Register: 1944-1969, Volume 2* (Cambridge: Girton College, 1991).

Hayman, Walter K., 'Dame Mary (Lucy) Cartwright, D.B.E. 17 December 1900 – 3 April 1998: Elected F.R.S. 1947', *Biographical Memoirs of Fellows of the Royal Society*, 46 (2000), 19.

Jardine, Lisa, 'A Point of View: Mary, Queen of Maths' (*BBC News, Magazine*, 8 March 2013) <https://www.bbc.com/news/magazine-21713163>.

McMurran, Shawnee L. and James J. Tattersall, 'The Mathematical Collaboration of M. L. Cartwright and J. E. Littlewood', *American Mathematical Monthly*, 103(10) (1996), 833–845.

O'Connor, John J. and Edmund F. Robertson, 'Mary Cartwright' <http://www-history.mcs.st-and.ac.uk/Biographies/Cartwright.html>.

Williams, E. M., 'Presidential Address: The Changing Role of Mathematics in Education', *The Mathematical Gazette*, 50(373) (1966), 243–254.

'Mary-Cartwright', Revolvy < https://www.revolvy.com/page/Mary-Cartwright>.

'Obituary: Mary Cartwright' (*The Times*, 7 April 1998) <http://www-groups.dcs.st-and.ac.uk/~history/Obits/Cartwright.html>.

'Obituary: Dame Mary Cartwright – Mistress of Girton whose mathematical work formed the basis of chaos theory' (*Electronic Telegraph*, 11 April 1998) <http://cwp.library.ucla.edu/articles/ebcart11.html>.

Yvette Cauchois

Apotheker, Jan and Livia Simon Sarkadi, *European Women in Chemistry* (Weinheim: Wiley-VCH Verlag, 2011).

Bonnelle, Christiane, 'Obituary: Yvette Cauchois', *Physics Today*, 54(4) (2001), 88–89.

Bonnelle, Christiane and Betty Anderson, 'Cauchois, Yvette', *Contributions of 20 Century Women to Physics* <http://cwp.library.ucla.edu/>.

Byers, Nina and Gary Williams, 'Yvette Cauchois (1908-1999)', Ch 20 in *Out of the Shadows: Contributions of Twentieth-century Women to Physics* (Reprinted ed.) (Cambridge: Cambridge University Press, 2006).

Wuilleumier, François J., 'Yvette Cauchois and Her Contribution to X-ray and Inner-shell Ionization Processes', *AIP Conference Proceedings*, 30 (2003) 652.

Edith Clarke

Brittain, James, 'Scanning the Past: Edith Clarke and Power System Stability', *Proceedings of the IEEE*, The Institute of Electrical and Electronics Engineers, Jan. 1996.

Durbin, John, 'In Memoriam: Edith Clarke', *Index of Memorial Resolutions and Biographical Sketches* (Texas: University of Texas, 2012).

Layne, Margaret E., *Women in Engineering. Pioneers and Trailblazers* (Reston, Va.: ASCE Press, 2009).

Carey, Charles Jr., 'Edith Clarke', *American National Biography Online* <https://doi.org/10.1093/anb/9780198606697.article.1300295>.

'Edith Clarke', Revolvy <https://www.revolvy.com/page/Edith-Clarke>.

Riddle, Larry, 'Edith Clarke', *Biographies of Women Mathematicians* <https://www.agnesscott.edu/lriddle/women/clarke.htm>.

Anna Botsford Comstock

Brandt, William, *Interpretive Wood-Engraving: The Story of the Society of American Wood-Engravers* (New Castle, Delaware: Oak Knoll Press, 2009), p. 108.

Comstock, John Henry and Anna Botsford, *How to Know the Butterflies* (New York: D. Appleton and Co. Comstock, 1904).

Golemba, Beverly E., *Lesser-Known Women: A Biographical Dictionary* (Boulder u.a.: Rienner, 1992), p. 157.

'Anna Botsford Comstock: American Illustrator and Writer', *Encyclopædia Britannica* <https://www.britannica.com/biography/Anna-Botsford-Comstock>.

Esther Marly Conwell

Colburn, Robert, 'How Four Pioneering Women in Technology Got Their Big Break' (16 June 2017) <http://sites.ieee.org/houston/four-pioneering-women-technology/>.

Freile, Victoria, 'UR Professor Esther Conwell Remembered as a Trailblazer' (*Democrat & Chronicle*, 18 November 2014).

Rothberg, Lewis, Charles B. and Mildred Dresselhaus, 'Esther Marly Conwell' (2015) 68 (5), *Physics Today*, 63–63.

'Esther Conwell: Biography', *Engineering and Technology History Wiki* <https://ethw.org/Esther_Conwell>.

Iglinski, Peter, 'Esther Conwell, Pioneering Professor of Chemistry, Dies at 92', University of Rochester (17 November 2014) <http://www.rochester.edu/newscenter/esther-conwell-pioneering-professor-of-chemistry-dies-at-92>.

Iglinski, Peter, 'University of Rochester's Esther Conwell, a Pioneering Woman Scientist to Receive the National Medal of Science', University of Rochester (15 October 2010) <http://www.rochester.edu/news/show.php?id=3707>.

Kruss, Todd, 'Esther Conwell '44 (MS): "Lived and Breathed Science"', *Rochester Review* (January–February 2015) 61 <https://www.rochester.edu/pr/Review/V77N3/pdf/0704_conwell.pdf>.

Gerty Theresa Radnitz Cori

'Carl and Gerty Cori and Carbohydrate Metabolism', Australian Chemical Society (2004) <https://www.acs.org/content/acs/en/education/whatischemistry/landmarks/carbohydratemetabolism.html>.

Ginsberg, Judah, *Carl and Gerty Cori and Carbohydrate Metabolism*, A National Chemical Landmark, American Chemical Society, 21 September 2004 <https://www.acs.org/content/dam/acsorg/education/whatischemistry/landmarks/carbohydratemetabolism/carl-and-gerty-cori-and-carbohydrate-metabolism-commemorative-booklet.pdf>.

'Changing the Face of Medicine: Dr Gerty Theresa Radnitz Cori', *Celebrating America's Women Physicians* (14 October 2003) <https://cfmedicine.nlm.nih.gov/physicians/biography_69.html>.

'Gerti Cori: Biographical', The Nobel Prize in Physiology or Medicine (1947) <https://www.nobelprize.org/prizes/medicine/1947/cori-gt/biographical/>.

'Gerty Theresa Cori (1896-1957), Renowned Biochemist, America's First Woman Nobel Laureate', Australian Chemical Society <https://www.acs.org/content/acs/en/education/whatischemistry/women-scientists/gerty-theresa-cori.html>.

Larner, Joseph, 'Gerty Theresa Cori', *Biographical Memoirs: National Academy of Sciences*, vol. 61 (Washington, D.C.: The National Academies Press, 1992), pp. 110–135.

'Women in Medicine at Washington University School of Medicine – Gerty Theresa Cori', Washington University School of Medicine, St Louis, Missouri <http://beckerexhibits.wustl.edu/women/cori.htm>.

Marie Skłodowska Curie

Barbara, Goldsmith, *Obsessive Genius: The Inner World of Marie Curie (Great Discoveries)* (W. W. Norton, Reprint edition, 2005).

Byers, N. 'Curie, Marie Sklodowska'. *Contributions of 20th Century Women to Physics* <http://cwp.library.ucla.edu/>.

Curie, Eve, *Madame Curie: A Biography* (Da Capo Press, 1936, Reprint edition, 2001).

'Marie Curie', *Encyclopædia Britannica* <https://www.britannica.com/biography/Marie-Curie>.

'Marie Curie: Facts', The Nobel Prize in Chemistry (1911) <https://www.nobelprize.org/prizes/chemistry/1911/marie-curie/facts/>.

'Marie Curie the Scientist', Care and Support Through Terminal Illness <https://www.mariecurie.org.uk/who/our-history/marie-curie-the-scientist>.

Quinn, Susan, *Marie Curie* (Da Capo Press, Revised edition, 10 April 1996).

Ingrid Daubechies

Hall, Frankie Grace, 'Duke Professor Integrates Biology' (*Mathematics Technician*, 26 September 2016)

'Ingrid Daubechies', *Biographies of Women Mathematicians* <https://www.agnesscott.edu/lriddle/women/daub.htm>.

Jackson, Allyn, 'Ingrid Daubechies Receives NAS Award in Mathematics' *Notices of the AMS, American Mathematical Society*, 47(5) (2000), 571.

'Ingrid Daubechies, James B. Duke Professor of Mathematics and Professor of Electrical and Computer Engineering, Department of Mathematics, Duke University' (2016) <https://web.archive.org/web/20160202012751/http://fds.duke.edu/db/aas/math/ingrid>.

'Ingrid Daubechies', The Franklin Institute Awards (April 2011) <https://www.fi.edu/laureates/ingrid-daubechies>.

O'Connor, John J. and Edmund F. Robertson, 'Ingrid Daubechies' <http://www-groups.dcs.st-and.ac.uk/history/Biographies/Daubechies.html>.

Riddle, Larry, 'Ingrid Daubechies', *Biographies of Women Mathematicians* (27 December 2018) <https://www.agnesscott.edu/lriddle/women/daub.htm>.

Svitil, Kathy A., 'The 50 Most Important Women in Science' (*Discover,* November 2002) <http://discovermagazine.com/2002/nov/feat50/>.

Olive Wetzel Dennis

Giaimo, Cara, 'The "Lady Engineer" Who Took the Pain Out of the Train' (*Atlas Obscura,* 9 April 2018).

Hatch, Sybil E., *Changing Our World: True Stories of Women Engineers* (Reston, VA: American Society of Civil Engineers, 2006), p. 93.

Ogilvie, Maryiln Bailey and Dorothy Harvey (eds), *The Biographical Dictionary of Women in Science: Pioneering Lives from Ancient Times to the mid-20th Century* (New York: Routledge, 2000), vol. 1.

'Olive Dennis', *The Famous People* <https://www.thefamouspeople.com/profiles/olive-dennis-7092.php>.

Mildred Spiewak Dresselhaus

Angier, Natalie, 'Queen of Carbon, Dies at 86' (*The New York Times,* 23 Feb 2017).

Chung, D. D. L., 'Mildred S. Dresselhaus (1930–2017)', *Nature,* 543, 16 March 2017, 316.

Morinobu, Endo, 'Mildred S. Dresselhaus' (2017) 70 (6), *Physics Today,* p.73.

Weil, Martin, 'Mildred Dresselhaus, Physicist Dubbed "Queen of Carbon Science", Dies at 86' (*The Washington Post,* Health and Science, 22 February 2017).

'Mildred "Millie" Dresselhaus', *Physics Central* (2018) <http://www.physicscentral.com/explore/people/dresselhaus.cfm>.

Dresselhaus, S. D. and M. Keyes, 'Dresselhaus, Mildred Spiewak', *Contributions of 20th Century Women to Physics* <http://cwp.library.ucla.edu/>.

Émilie du Châtelet

Mandic, Sasha, 'Émilie du Châtelet', *Biographies of Women Mathematicians* (April 1995) <https://www.agnesscott.edu/lriddle/women/chatelet.htm>.

O'Connor, J. and Edmund F. Robertson, 'Gabrielle Émilie Le Tonnelier de Breteuil Marquise du Châtelet' <http://www-history.mcs.st-and.ac.uk/Biographies/Chatelet.html>.

Pearce, Elizabeth, 'Émilie du Châtelet, Women Who Persist', *Project Continua* (23 March 2015) <http://www.projectcontinua.org/emilie-du-chatelet>.

Wills, Matthew, 'Émilie du Châtelet: Heroine of the Enlightenment', *JSTOR Daily,* 22 July 2016 <https://daily.jstor.org/emilie-du-chatelet/>.

Zinsser, Judith, 'Mentors, the Marquise du Châtelet and Historical Memory', *The Royal Society,* 27 March 2007 <http://rsnr.royalsocietypublishing.org/content/61/2/89>.

Alice Eastwood

Abrams, Leroy, 'Alice Eastwood: Western Botanist' (2017) 70 (6), *Pacific Discovery,* 14–17.

'Bibliography of the Writings of Alice Eastwood', *Proceedings of the California Academy of Sciences, Fourth Series,* Vol. 25 (1943–1949), pp. xv–xxiv.

Howell, John Thomas. 'Alice Eastwood: 1859-1953' (1953) 3 (4), *Taxon,* 98–100.

Lindon, Hearher, Lauren Gardiner, Abigail Brady and Maria Vorontsova, 'Fewer than Three Percent of Land Plant Species Named by Women: Author Gender over 260 years' (2015) 64 (2), *Taxon*, 209–215.

MacFarland, F. M., R. C. Miller and J. T. Howell, 'Biographical Sketch of Alice Eastwood', *Proceedings of the California Academy of Sciences*, Fourth series, Vol. 25 (1943–1949), 25, ix–xiv.

'A Dictionary of the Fushcia, Fushcias in the City' (2017) <http://www.fuchsiasinthecity.com/about-fuchsias/dictionary/dictionary-e-f/dictionary-e-f.html>.

'Eastwood, Alice, 1859–1953', *JSTOR Global Plants* <https://plants.jstor.org/stable/10.5555/al.ap.person.bm000002304>.

Morin, Rebecca, 'Celebrating Women's History Month: Alice Eastwood', *Biodiversity Heritage Library* (29 March 2012) <https://blog.biodiversitylibrary.org/2012/03/celebrating-womens-history-month-alice-eastwood.html>.

Elsie Eaves

Layne, Margaret E., *Women in Engineering: Pioneers and Trailblazers* (Reston, Virginia: American Society of Civil Engineers, Editor 2009), p. 173.

Bois, Danuta, 'Distinguished Women of Past and Present Elsie Eaves' (1996) <http://www.distinguishedwomen.com/biographies/eaves.html>.

'Engineer Girl – Elsie Eaves Profile', *Engineer Girls*, *National Academy of Engineering* <https://www.engineergirl.org/125315/Elsie-Eves>.

'Obituary for Elsie Eaves' (*New York Times*, 2 April 1983) <https://www.nytimes.com/1983/04/02/obituaries/elsie-eaves.html>.

'Women's History Month: Elsie Eaves' (2 March 2018) <https://alltogether.swe.org/2018/03/womens-history-month-elsie-eaves/>.

Gertrude Belle Elion

Holloway, M., 'Profile: Gertrude Belle Elion – The Satisfaction of Delayed Gratification', *Scientific American*, 265(4) (1991), 40–44.

McGrayne, Sharon Bertsch, *Gertrude Belle Elion. Nobel Prize Women in Science* (Washington, D.C.: Joseph Henry Press, 1998), pp. 279–302.

McGrayne, Sharon Bertsch, *Gertrude Elion* (Nobel Prize Women in Science, Carol Publishing Group), pp. 280–303.

Rayner-Canham, Marelene and Geoffrey Rayner, *Women in Biochemistry. Women in Chemistry: Their Changing Roles from Alchemical Times to the Mid-twentieth Century* (Danvers, MA: American Chemical Society and Chemical Heritage Foundation, 1998), pp. 152–156.

Trescott, Martha M., *Women in the Intellectual Development of Engineering: A study in Persistence and Systems Thought. Women of science* (Bloomington and Indianapolis, IN: Indiana University Press, 1990), pp. 313–314.

'Gertrude Elion (1918–1999)', *ACS Chemistry for Life* <https://www.acs.org/content/acs/en/education/whatischemistry/women-scientists/gertrude-elion.html>.

'Gertrude B. Elion: Biographical', The Nobel Prize in Physiology or Medicine 1988 <https://www.nobelprize.org/prizes/medicine/1988/elion/biographical>.

Bibliography

Thelma Austern Estrin

Estrin, Thelma, 'Women's Studies and Computer Science: Their Intersection', *IEEE Annals of the History of Computing*, 18(3) (1996), 43–46.

Gurer, Denise, 'Pioneering Women in Computer Science', *SIGCSE Bulletin*, 34(2) (2002), 175–180.

Nebeker, Frederik, *Sparks of Genius: Portraits of Electrical Engineering Excellence* (New York: Institute of Electrical and Electronics Engineers, 1994).

'Thelma Estrin Biography', *Engineering and Technology History Wiki* (1 May 2018) <https://ethw.org/Thelma_Estrin> (This website contains links to a series of other biographical resources).

'Women in Technology Hall of Fame – Dr Thelma Estrin' <https://www.witi.com/halloffame/102608/Dr.-Thelma-Estrin-Professor-Emerita-University-of-California,-Los-Angeles>.

Margaret Clay Ferguson

Carey, Charles W, Jr, 'Ferguson, Margaret Clay', *American National Biography Online*, Oxford University Press, February 2000).

Ferguson, Margaret Clay, *A Preliminary Study of the Germination of the Spores of Agaricus Campestris and Other Basidiomycetous Fungi* (Classic Reprint, Forgotten Books, 2018).

Ferguson, Margaret Clay, *Contributions to the Knowledge of the Life History of Pinus: With Special Reference to Sporogenesis, the Development of the Gametophytes and Fertilization* (USA: Kessinger Publishing, 1904).

Proffitt, Pamela, *Notable Women Scientists* (Detroit: Gale Group, ed. 1999).

Shearer, Benjamin F. and Barbara S. Shearer (eds), *Notable Women in the Life Sciences: A Biographical Dictionary* (Westport, Conn.: Greenwood Press, 1996), pp. 128–131.

Lydia Folger Fowler

Blake, John B., 'Lydia Folger Fowler', *Notable American Women 1607–1950: A Biographical Dictionary* (Cambridge: Radcliffe College, 1971), vol. 2, pp. 654–655.

Lovejoy, Esther Poh, *Women Doctors of the World* (1957), pp. 8–21.

McHenry, Robert (ed.), *Lydia Folger Fowler – Famous American Women: A Biographical Dictionary from Colonial Times to the Present*, Vol. 2 (Springfield, MA: G. & C. Merriam Co., 1980), p. 139.

Silverthorne, Elizabeth, *Lydia Folger Fowler* (Women Pioneers in Texas Medicine, 1997), p. xxii

Stern, Madeleine B., 'Lydia Folger Fowler, M.D.: First American Woman Professor of Medicine', *New York State Journal of Medicine*, 77(7) (1977), 1137–1140.

Waite, Frederick Clayton, *Dr Lydia Folger Fowler: The Second Woman to Receive the Degree of Doctor of Medicine in the United States, Annals of American History* (New York, NY: Hoeber, 1932), pp. 290–297.

'Lydia Folger Fowler', *History of American Women* <http://www.womenhistoryblog.com/2014/09/lydia-folger-fowler.html>.

Rosalind Elsie Franklin

'Clue to Chemistry of Heredity Found' (*New York Times*, 13 June 1953).

Elkin, Lynne Osman, 'Rosalind Franklin and the Double Helix', *Physics Today*, 56 (2003), 42–48.

Piper, Anne, 'Light on a Dark Lady', *Trends in Biochemical Sciences*, 23(4) (1998), 151–154.

The Papers of Rosalind Franklin, Janus, Documents from the Churchill Archives Centre, Cambridge.

Thomas, T. Dennis, 'The Role of Activated Charcoal in Plant Tissue Culture', *Biotechnology Advances*, 26(6) (2008), 618–631.

'The DNA Molecule is Shaped like a Twisted Ladder – Rosalind Franklin', *DNA from the Beginning* <http://www.dnaftb.org/19/bio-3.html>.

'Rosalind Franklin British Scientist', *Encyclopaedia Britannica* <https://www.britannica.com/biography/Rosalind-Franklin>.

Trueblood, K., 'Franklin, Rosalind', *Contributions of 20th Century Women to Physics* <http://cwp.library.ucla.edu/>.

Sophie Germain

Cipra, Barry A., 'A Woman Who Counted', *Science*, 319(5865) (2008), 899.

Del Centina, Andrea 'Letters of Sophie Germain Preserved in Florence', *Historia Mathematica*, 32(1) (2005), 60–75.

Del Centina, Andrea, 'Unpublished Manuscripts of Sophie Germain and a Revaluation of Her Work on Fermat's Last Theorem', *Archive for History of Exact Sciences*, 62(4) (2008), 349–392.

Gray, Mary W., 'Sophie Germain', in Bettye Anne Case and Anne M. Leggett (eds), *Complexities: Women in Mathematics* (Princeton University Press, 2005), pp. 68–75.

Gray, Mary W., 'Sophie Germain (1776–1831)', in Louise S. Grinstein and Paul Campbell, *Women of Mathematics: A Bibliographic Sourcebook* (Greenwood, 1978), pp. 47–55.

Mackinnon, Nick, 'Sophie Germain, or, Was Gauss a Feminist?', *The Mathematical Gazette*, 74(470) (1990), 346–351.

Moncrief, J. William, 'Germain, Sophie', in Barry Max Brandenberger, *Mathematics*, vol. 2 (Macmillan Science Library, Macmillan Reference USA, 2002).

Swift, Amanda, 'Sophie Germain', *Biographies of Women Mathematicians* <https://www.agnesscott.edu/lriddle/women/women.htm>.

Catherine Anselm 'Kate' Gleason

Bailey, Margaret B., 'Kate Gleason: The Ideal Business Woman', *The Rochester Engineer*, 86(6) (2008), *Rochester Engineering Society*, 8–9.

Gleason, Janis F., *The Life and Letters of Kate Gleason* (Rochester, New York: RIT Press, 2010).

Karwatka, Dennis, 'Kate Gleason – First Female Engineering Student' (*Tech Directions*, 10 October 2010), 12.

Layne, Margaret E., *Women in Engineering. Pioneers and Trailblazers* (Reston, Va.: ASCE Press, 2009).

Weingart, Richard, *Engineering Legends: Great American Civil Engineers* (Engineering Legends: Great American Civil Engineers, 2005), p. 108.

'Woman Bank Head in East Rochester' (Rochester, New York: *The Post Express*, 19 August 1918), p. 3
'Kate Gleason American Businesswoman', *Encyclopaedia Britannica* (5 January 2019) <https://www.britannica.com/biography/Kate-Gleason>.

Gertrude 'Trude' Scharff Goldhaber

Bond, Peter, and Henley M. Ernest, 'Gertrude Scharff Goldhaber 1911–1998', *Biographical Memoirs: National Academy of Sciences*, vol. 77 (Washington, D.C.: The National Academies Press, 1992) <https://www.nap.edu/read/9681/chapter/9>.
Goldhaber, Maurice, 'Gertrude Scharff Goldhaber', *Contributions of 20th Century Women to Physics* <http://cwp.library.ucla.edu/>.
Goldhaber, Michael H., 'Gertrude Scharff-Goldhaber, 1911–1908: Nuclear Physicist Against the Odds' *Physics in Perspective*, 18(2) (2016) 182
'Gertrude S. Goldhaber', American Institute of Physics <https://history.aip.org/phn/11511019.html>.
'Gertrude Scharff Goldhaber', *Contributions of 20th Century Women to Physics* <http://cwp.library.ucla.edu/Phase2/Goldhaber,_Gertrude_Scharff@812345678.html>.
Henley, Ernest M. and Peter Bond, 'Gertrude Scharff Goldhaber', *Jewish Women: Jewish Women's Archive* (1 March 2009) <https://jwa.org/encyclopedia/article/goldhaber-gertrude-scharff>.
Saxon, Wolfgang, 'Gertrude Scharff Goldhaber, 86, Crucial Scientist in Nuclear Fission' (*The New York Times*, 6 February 1998).

Anna Jane Harrison

'Anna Jane Harrison', *Science History Institute, Historical Biographies* <https://www.sciencehistory.org/historical-profile/anna-jane-harrison>.
'Anna Harrison fills ACS board vacancy', *Chemical & Engineering News*, 54(4) (1976), 6.
'Biographies: Anna Jane Harrison (1912–1998)', Women in Health Sciences, Bernard Becker Medical Library Digital Collection, Washington University School of Medicine, St Louis, Missouri.
Harrison, Anna J., Papers 1854–1999, Manuscript Collection: MS 0763, Mount Holyoke College, Archives and Special Collections.
Long, Janice, 'Anna Harrison Dies at Age 85', *Chemical & Engineering News*, 76(33) (1998), 9.
Rogers, Kara, 'Anna Jane Harrison', *Encyclopaedia Britannica* (19 December 2018) <https://www.britannica.com/biography/Anna-Jane-Harrison>.
Saxon, Wolfgang, 'Anna J. Harrison, 85, Led US Chemical Society' (*The New York Times,* 16 August 1998).
Shearer, Benjamin F. and Barbara Smith Shearer, *Notable Women in Physical Sciences: A Biographical Dictionary* (Westport, Connecticut: Greenwood Press, 1997).

Caroline Lucretia Herschel

Brock, Claire, 'Public Experiments', *History Workshop Journal* (2004), 306–312.
Brock, Claire, *The Comet Sweeper: Caroline Herschel's Astronomical Ambition* (Icon Books, 2004).

Fernie, J. Donald, 'The Inimitable Caroline', *American Scientist,* November–December (2007), 486–488.

Herschel, Mrs John, *Memoir and Correspondence of Caroline Herschel* (London: John Murray, Albemarle Street, 1876).

Hoskin, Michael, 'Caroline Lucretia Herschel', *Oxford Dictionary of National Biography Online* <https://www-oxforddnb-com.simsrad.net.ocs.mq.edu.au/view/10.1093/ref:odnb/9780198614128.001.0001/odnb-9780198614128-e-13100?rskey=mHk3Gn&result=3>.

Hoskin, Michael, *Discoverers of the Universe William and Caroline Herschel* (Princeton NJ: Princeton University Press, 2011).

Hoskin, Michael, *William and Caroline Herschel: Pioneers in Late 18th-Century Astronomy* (New York: Springer, 2014).

Nysewander, Melissa, 'Caroline Herschel', *Biographies of Women Mathematicians* (Atlanta: Agnes Scott College, 1998).

Ogilvie, Marilyn B., *Searching the Stars: The Story of Caroline Herschel* (History Press, 2011).

Beatrice Alice Hicks

'Beatrice A. Hicks', *National Women's Hall of Fame* <https://www.womenofthehall.org/inductee/beatrice-a-hicks>.

Busignies, Henri G., *Beatrice Hicks*, The National Academies Press, <https://www.nap.edu/read/565/chapter/24>.

Cummings, Charles F., *Hicks, Beatrice A. Jan. 2, 1919 – Engineering executive*, Current Biography (The H. W. Wilson Company, 1957).

James, Edward T., Janet Wilson James and Paul S.Boyer, *Notable American Women: A Biographical Dictionary, Volume 5: Completing the Twentieth Century* (Cambridge, MA: Belknap Press, 2004).

National Academy of Engineering of the United States of America, Memorial Tributes (Washington, D.C.: National Academy Press, 1984), p. 118.

Shompole, Lydia, 'Spotlight on Women in Electrical Engineering: Beatrice Hicks, First President of the Society of Women Engineers', PowerStudies Inc. (May 8, 2017) <https://www.powerstudies.com/blog/spotlight-women-electrical-engineering-beatrice-hicks>.

Dorothy Crowfoot Hodgkin

Byers, N., 'Hodgkin, Dorothy Crowfoot', *Contributions of 20th Century Women to Physics* <http://cwp.library.ucla.edu/>.

Dodson, Guy, Jenny P. Glusker and David Sayre (eds.), *Structural Studies on Molecules of Biological Interest: A Volume in Honour of Professor Dorothy Hodgkin* (Oxford: Clarendon Press, 1981).

Ferry, G., 'Hodgkin, Dorothy Mary Crowfoot', *Oxford Dictionary of National Biography online* (21 May 2009).

Glusker, J. P. and M. J. Adams, 'Dorothy Crowfoot Hodgkin', *Physics Today*, 48(5) (1995), 80.

Hudson, Gill, 'Unfathering the Thinkable: Gender, Science and Pacificism in the 1930s', in Marina Benjamin (ed.) *Science and Sensibility: Gender and Scientific Enquiry, 1780–1945* (Oxford: Basil Blackwell, 1991), pp. 264–86.

Johnson, L. N. and D. Phillips, 'Professor Dorothy Hodgkin, OM, FRS', *Nature Structural Biology*, 1(9) (1994), 573–76.

Moncrief, J. W., 'Hodgkin, Dorothy Mary Crowcroft' <https://www.encyclopedia.com/people/medicine/biochemistry-biographies/dorothy-mary-crowfoot-hodgkin>.

Opfell, Olga S., *Lady Laureate: Women Who Have Won the Nobel Prize* (Metuchen, NJ & London: Scarecrow Press, 1978), pp. 209–23.

Perutz, M., 'Professor Dorothy Hodgkin', *Quarterly Reviews of Biophysics*, 27(4) (2009), pp. 333–37.

Erna Schneider Hoover

'Erna Hoover – Biography', *World of Computer Science* (2012) <https://www.women.cs.cmu.edu/ada/Resources/Women>.

'Erna Schneider Hoover – Computerized Telephone Switching System', Lemelson MIT <http://lemelson.mit.edu/resources/erna-schneider-hoover>.

'Erna Schneider Hoover', *Engineering and Technology Wiki* <https://ethw.org/Erna_Schneider_Hoover>.

Nutt, Amy Ellis, 'Fame calls on 2 titans of telephony in NJ' (*The Star-Ledger*, 18 June 2008), https://www.nj.com/news/index.ssf/2008/06/fame_calls_on_2_titans_of_tele.html?.

'Queens of Tech', *YouTube* <https://www.youtube.com/watch?v=mqIfLPqE3iU>.

Grace Brewster Murray Hopper

Beyer, Kurt W., *Grace Hopper and the Invention of the Information Age* (Cambridge, Massachusetts: MIT Press, 2009).

'Grace Hopper Biography: Mathematician, Military Leader, Computer Programmer', *Biography* <https://www.biography.com/people/grace-hopper-21406809>.

'Grace Hopper: United States Naval Officer and Mathematician', *Encyclopaedia Britannica* <https://www.britannica.com/biography/Grace-Hopper>.

Marx, Christy, *Grace Hopper: The First Woman to Program the First Computer in the United States, Women Hall of Famers in Mathematics and Science* (New York City: Rosen Publishing Group, August 2003).

Norman, Rebecca, *Biographies of Women Mathematicians: Grace Murray Hopper* (Agnes Scott College, June 1997).

Williams, Kathleen Broome, *Grace Hopper: Admiral of the Cyber Sea* (Annapolis, Maryland: Naval Institute Press, 15 November 2004).

Margaret Lindsay Huggins

Becker, Barbara J., *Unravelling Starlight: William and Margaret Huggins and the Rise of the New Astronomy* (Cambridge, England: Cambridge University Press, 2011).

Becker, Barbara J., 'Margaret Huggins and Tulse Hill Observatory', *Astronomy & Geophysics*, 57(2) (2016), 2.13.

Brück, M.T. and I. Elliott, 'The Family Background of Lady Huggins', *Irish Astronomical Journal*, 20(3) (1992), 210.

Brück, Mary T., 'Companions in Astronomy: Margaret Lindsay Huggins and Agnes Mary Clerke', *Irish Astronomical Journal*, 20(2) (1991) 70.

Brück, Mary T., 'Huggins, Margaret Lindsay Murray', in Thomas Hockey, Virginia Trimble and Thomas R. Williams, *The Biographical Encyclopedia of Astronomers* (New York: Springer Publishing, 2009).

Chant, Clarence Augustus, 'Death of Lady Huggins', *Journal of the Royal Astronomical Society of Canada*, 9(4) (1915), 149–150.

'Dame Margaret Lindsay Huggins', *Royal Astronomical Society Monthly Notices*, 76(4) (1916), 278–282.

Hearnshaw, J. B., *The Analysis of Starlight* (New York: Cambridge University Press, 1986).

'Lady Huggins', *The Observatory*, 38(488) (1915), 254–256.

Frances Betty Sarnat Hugle

'Grandma Got Stem: Frances Hugle' (2 April 2015) <https://ggstem.wordpress.com/2015/04/02/frances-hugle/>.

'Frances Hugle', DBpedia <http://dbpedia.org/page/Frances_Hugle>.

'Frances Hugle', MyHeritage <https://www.myheritage.com/names/frances_hugle>.

Shirley Ann Jackson

Camp, Carole Ann, *American Women Inventors* (Berkeley Heights, NJ: Enslow Publishers, 2004).

June, Audrey, 'Shirley Ann Jackson Sticks to the Plan' (*The Chronicle of Higher Education*, 5 June 2007).

'Presidential Perspectives: Biography of Shirley Ann Jackson, PhD' (*The New York Times*, 21 July 2003).

'Rensselaer President Leads List of Highest-Paid Private College Leaders' (*The New York Times*, 8 December 2014).

'How Dr Shirley Ann Jackson Defied Gender and Race Barriers to Become the Ultimate Role Model in Science', *Joyce Riha Linik Education* (2 November 2016) <https://iq.intel.com/dr-shirley-ann-jackson-ultimate-role-model-science/>.

Taylor, Mildred Europa, 'Face to Face Africa' (2 March 2018) <https://face2faceafrica.com/article/black-woman-made-caller-id-call-waiting-possible>.

Sophia Louise Jex-Blake

'Jex-Blake, Sophia', *Who's Who*, Vol. 59 (1907), pp. 938–939.

Knox, William, *Lives of Scottish Women* (Edinburgh: Edinburgh University Press, 2006), p. 71.

Lutzker, Edythe, *Women Gain a Place in Medicine* (New York: McGraw Hill, 1969), p. 149.

Roberts, Shirley, *Sophia Jex-Blake: A Woman Pioneer in Nineteenth-century Medical Reform* (London: Routledge, 1993).

Roberts, Shirley, 'Blake, Sophia Louisa Jex- (1840–1912)', Oxford *Dictionary of National Biography* (online ed.) (September 2004).

'Sophia Jex-Blake', *Spartacus Educational* <https://spartacus-educational.com/Wjex.htm>.

Todd, Margaret, *The Life of Sophia Jex-Blake* (London: MacMillan & Co., 1918).

'Sophia Jex-Blake', *Undiscovered Scotland* <https://www.undiscoveredscotland.co.uk/usbiography/j/sophiajexblake.html>.

'Sophia Jex-Blake (1840–1912)', Historic Alumni, College of Medicine & Veterinary Medicine, University of Edinburgh <https://www.ed.ac.uk/medicine-vet-medicine/about/history/historic-alumni/sophia-jex-blake>.

Nalini Joshi

Arvind, Usha Ramanujam, 'Women of Influence' (*Indian Link*, 8 October 2015) <https://www.indianlink.com.au/women-of-influence/>.

Borrello, Eliza, 'Professor Nalini Joshi, mistaken for wait staff at functions, highlights gender bias in Australian science' (*ABC News online*, 31 Mar 2016) <https://www.abc.net.au/news/2016-03-30/women-scientists-highlight-gender-bias-in-australian-stem/7285312>.

'Nalini Joshi' <http://wp.maths.usyd.edu.au/nalini/intro/>.

'One Plus One: Nalini Joshi', interview, 6 December 2018 <https://www.abc.net.au/news/programs/one-plus-one/2018-12-06/one-plus-one:-nalini-joshi/10591426>

Ong, Pristine, 'Interview with Nalini Joshi December 6, 2012, University of Sydney', *Asia Pacific Mathematics Newsletter*, 3(1) (2013), 41–45 <https://www.asiapacific-mathnews.com/03/0301/0041_0045.pdf>.

'Professor Nalini Joshi', Australian Academy of Science <https://www.science.org.au/fellowship/fellows/professor-nalini-joshi>.

'Professor Nalini Joshi: 2012 Australian Laureate Recipient', Australian Research Council <https://web.archive.org/web/20121004235232/http://arc.gov.au/pdf/FL12/NALINI%20JOSHI.pdf>.

Helen Dean King

Clause, Bonnie Tocher, 'The Wistar Rat as a Right Choice: Establishing Mammalian Standards and the Ideal of a Standardized Mammal', *Journal of the History of Biology*, 26(2) (1993), 329–349.

Clause, Bonnie Tocher, 'The Wistar Institute Archives: Rats (Not Mice) and History' (*Mendel Newsletter*, February 1998).

Ogilvie, Marilyn Bailey, 'Inbreeding, Eugenics, and Helen Dean King (1869–1955)', *Journal of the History of Biology*, 40(3) (2007), 467–507.

Kaufman, Dawn M., Donald W. Kaufman and Glennis A. Kaufman, 'Women in the Early Years of the American Society of Mammalogists (1919–1949)', *Journal of Mammalogy*, 77(3) (1996), 642.

Herr, Mickey, 'King of The Rats: How One Female Scientist Colonized the Modern Lab', *Hidden City Philadelphia* (16 July 2018) <https://hiddencityphila.org/2018/07/king-of-the-rats-how-one-female-scientist-colonized-the-modern-lab/>.

Maria Margaretha Winckelman Kirch

Alic, Margaret, *Hypatia's Heritage* (British Library: The Women's Press Limited, 1986), p. 121.

'Maria Winckelmann Kirch Facts', *Your Dictionary* <https://biography.yourdictionary.com/maria-winckelmann-kirch>.

Schiebinger, Londa, *Maria Winkelmann at the Berlin Academy*, Vol. LXXVIII (Isis., 1987), pp. 174–200.

Schiebinger, Londa, *The Mind Has No Sex? Women in the Origins of Modern Science* (Harvard University Press, 1989), pp. 21–38.

Wielen, Roland, 'Kirch, Maria Margaretha Winkelman', in Thomas Hockey, Virginia Trimble, Thomas R. Williams, Katherine Bracher, Richard A. Jarrell, Jordan D. Marché, JoAnn Palmeri and Daniel W. E. Green (eds), Biographical Encylopedia of Astronomers (online ed. 24 December 2016).

Yount, Lisa, *A to Z of Women in Science and Math* (Infobase Publishing, 2007), p. 157.

Margaret Galland Kivelson

'How Do You Find an Alien Ocean? Margaret Kivelson Figured It Out' (*New York Times*, 8 October 2018).

Kivelson, M. G., 'The Rest of the Solar System', *Annual Review of Earth and Planetary Sciences*, 36 (2008), 1–32.

'Kivelson, Maragert Galland', *Contributions of 20th Century Women to Physics* <http://cwp.library.ucla.edu/>.

'Margaret Kivelson – 2005 John Adam Fleming Medal Winner', AGU100 Advancing Earth and Space Science (2000) <https://honors.agu.org/winners/margaret-kivelson>.

'Margaret Kivelson', Google Scholar Citations <https://scholar.google.com/citations?user=T70E5X8AAAAJ&hl=en>.

'Margaret Kivelson, Institute of Geophysics & Planetary Physics, University of California, Los Angeles (UCLA)' <http://www.igpp.ucla.edu/people/mkivelson.html>.

Oakes, Elizabeth, *Encyclopedia of World Scientists* (Infobase Publishing, 2007), pp. 404–405.

Sofia Vasilyevna Korvin-Krukovskaya Kovalevskaya

Belits, Yuriy, 'Brief Biography of Sofia Kovalevskaya' (17 March 2005), <https://web.archive.org/web/20060920215141/http://www-math.cudenver.edu/~wcherowi/courses/m4010/s05/belitspdf.pdf>.

Cooke, Roger, *The Mathematics of Sonya Kovalevskaya* (Springer-Verlag, 1984).

Kennedy, Don H., *Little Sparrow, A Portrait of Sofia Kovalevsky* (Athens: Ohio University Press, 1983).

Koblitz, Ann Hibner, *A Convergence of Lives – Sofia Kovalevskaia: Scientist, Writer, Revolutionary Lives of Women in Science*, 99-2518221-2 (2nd printing, revised) (New Brunswick, N.J.: Rutgers University Press, 1993).

Koblitz, Ann Hibner, 'Sofia Vasilevna Kovalevskaia', in Louise S. Grinstein and Paul J. Campbell (eds), *Women of Mathematics: A Bio-Bibliographic Sourcebook* (New York: Greenwood Press, 1987).

O'Connor, John J. and Edmund F. Robertson, 'Sofia Kovalevskaya' (MacTutor History of Mathematics Archive, University of St Andrews).

The Legacy of Sonya Kovalevskaya: Proceedings of a Symposium Sponsored by the Association for Women in Mathematics and the Mary Ingraham Bunting Institute, held October 25–28, 1985. Contemporary Mathematics, 0271-4132; 64 (Providence, R.I.: American Mathematical Society, 1987).

Bibliography

Doris Kuhlmann-Wilsdorf

'Barrier-breaking Professor Doris Wilsdorf Dies at 88' (*The Daily Progress*, 29 March 2010).

'Doris Kuhlmann-Wilsdorf', *Contributions of 20th Century Women to Physics* <http://cwp.library.ucla.edu/Phase2/Kuhlmann-Wilsdorf,_Doris@900123456.html>.

Oakes, Elizabeth H., *Encyclopedia of World Scientists* (Infobase Publishing, 2007), p. 417.

'Patents by Inventor Doris Kuhlmann-Wilsdorf', *Justia Patents* <https://patents.justia.com/inventor/doris-kuhlmann-wilsdorf>.

Pipkin, Josie, 'Pioneering Professor Doris Kuhlmann-Wilsdorf Dies' (*UVAToday*, 30 March 2010) <https://news.virginia.edu/content/pioneering-professor-doris-kuhlmann-wilsdorf-dies-service-thursday>.

Ratha, Bhatka B. and Edgar A. Starke, Jr., 'Doris Kuhlmann-Wilsdorf', *Memorial Tributes*, Vol. 21 (National Academy of Engineering, 2017), p. 192.

Wayne, Tiffany K., *American Women of Science Since 1900: Essays A-H*, Vol.1. (ABC-CLIO, 2011), pp. 590–591.

Stephanie Louise Kwolek

'Citation Conferring an Honorary Doctor of Science Degree on Stephanie Louise Kwolek' (University of Delaware, *UDaily*, 31 May 2008).

'Dupont Scientists Honored with Lavoisier Medals for Technical Achievement' (*(PRNewswire*, 27 April 1995).

Bensaude-Vincent, Bernadette, 'Stephanie L. Kwolek', Transcript of an Interview Conducted by Bernadette Bensaude-Vincent at Wilmington, Delaware, Philadelphia: Chemical Heritage Foundation (21 March 1998) <https://oh.sciencehistory.org/sites/default/files/kwolek_sl_0168_suppl.pdf>.

Ferguson, Raymond C., 'Stephanie Louise Kwolek', Transcript of an Interview Conducted by Raymond C. Ferguson in Sharpley, Delaware, Philadelphia: Beckman Center for the History of Chemistry (4 May 1986) <https://oh.sciencehistory.org/sites/default/files/kwolek_sl_0028_suppl.pdf>.

'Stephanie Kwolek', *Famous Scientists* (21 June 2014) <https://www.famousscientists.org/stephanie-kwolek>.

'Stephanie Kwolek Obituary' (*The Guardian*, 28 June 2014) <https://www.theguardian.com/science/2014/jun/26/stephanie-kwolek>.

Hedwig Eva Maria Kiesler 'Hedy Lamarr'

Barton, Ruth, *Hedy Lamarr: The Most Beautiful Woman in Film* (Lexington: University of Kentucky Press, 2010).

'Hedy Lamarr: Inventor of More than the 1st Theatrical-film Orgasm' (*Los Angeles Times*, 28 November 2010).

'Hollywood Star whose Invention Paved the Way for Wi-Fi' (*New Scientist*, 8 December 2011).

Lamarr, Hedy, *Ecstasy and Me: My Life as a Woman* (New York: Bartholomew House, 1966).

Rhodes, Richard, *Hedy's Folly: The Life and Breakthrough Inventions of Hedy Lamarr* (New York: Doubleday, 2012).

Shearer, Stephen Michael, *Beautiful: The Life of Hedy Lamarr* (New York: St. Martin's Press, 2010).

Young, Christopher, *The Films of Hedy Lamarr* (New York: Citadel Press, 1979).

Inge Lehmann

Bolt, Bruce A., 'Inge Lehmann', *Physics Today*, 47(1) (1994), 61.

Hjortenberg, Erik, 'Inge Lehmann's Work Materials and Seismological Apistolary Archive', *Annals of Geophysics*, 52(6) (2009), 679–698.

Jacobsen, A. Lif Lund, *Inge Lehmann and the Rise of International Seismology 1925–1970* (Carlsberg Foundation).

Lehmann, Inge, 'Seismology in the Days of Old', *Eos, Transactions American Geophysical Union*, 68(3) (1987), 33–35.

Maiken, Lolck, 'Lehmann, Inge, Complete Dictionary of Scientific Biography', *Gale Virtual Reference Library*, 22 (2008), 232–236.

Swirles, Lady Jeffreys, Bertha, 'Inge Lehmann: Reminiscences', *Quarterly Journal of the Royal Astronomical Society*, 35(2) (1994), 231.

'Inge Lehmann', *Famous Scientists* (29 May 2015) <https://www.famousscientists.org/inge-lehmann>.

Rita Levi-Montalcini

Aloe, Luigi, 'Rita Levi-Montalcini: The Discovery of Nerve Growth Factor and Modern Neurobiology', *Trends in Cell Biology*, 14 (2004), 395–99.

Levi-Montalcini, Rita, 'The Nerve Growth Factor 35 Years Later', *Science*, 237 (1987), 1154–62.

Levi-Montalcini, Rita, *In Praise of Imperfection: My Life and Work* (New York: Basic Books, 1988).

Odelberg, Wilhelm (ed.), *Les Prix Nobel: The Nobel Prizes, 1986* (Stockholm: Nobel Foundation, 1987).

Purves, Dale and Jeff W. Lichtman, *Principles of Neural Development* (Sunderland, MA: Sinauer Associates, 1985).

'Biographies – Rita Levi-Montalcini (b. 1909)', Women in Health Sciences, Washington University in St Louis (2004) <http://beckerexhibits.wustl.edu/mowihsp/bios/levi_montalcini.htm>.

'Rita Levi-Montalcini: Facts', The Nobel Prize in Physiology or Medicine 1986 <http://nobelprize.org/nobel_prizes/medicine/laureates/1986/levi-montalcini-autobio.html>.

Barbara Jane Huberman Liskov

'Barbara Liskov Wins Turing Award' (*MIT News*, 10 March 2009) <http://news.mit.edu/2009/turing-liskov-0310>.

Huberman (Liskov), Barbara Jane, *A Program to Play Chess End Games* (Stanford University Department of Computer Science, Technical Report CS 106, Stanford Artificial Intelligence Project Memo AI-65, 1968).

Liskov, Barbara, 'Distributed programming in Argus', *Communications of the ACM*, 31(3) (1988), 300–312.

Liskov, Barbara, H., 'The design of the Venus operating system', *Communications of the ACM*, 15(3) (1972), 144–149.

'MIT's Magnificent Seven: Women Faculty Members Cited as Top Scientists' (*MIT News Office*, 5 Nov 2002).

'Oral-History: Barbara Liskov', *Engineering and Technology Wiki* (12 September 2018) <https://ethw.org/Oral-History:Barbara_Liskov_(1991)>.

Van Vleck, Tom, 'A. M. Turing Award Barbara Liskov' (2012) <https://amturing.acm.org/award_winners/liskov_1108679.cfm>.

(Augusta) Ada Byron Lovelace

'Ada Lovelace Symposium – Celebrating 200 Years of a Computer Visionary', *Podcasts* (UK: University of Oxford).

Baum, Joan, *The Calculating Passion of Ada Byron* (Archon, 1986).

Hooper, Rowan, 'Ada Lovelace: My Brain is More than Merely Mortal', *New Scientist* (15 October 2012) <https://www.newscientist.com/article/dn22385-ada-lovelace-my-brain-is-more-than-merely-mortal/>.

Phillips, Ana Lena, 'Crowdsourcing Gender Equity: Ada Lovelace Day, and Its Companion Website, Aims to Raise the Profile of Women in Science and Technology', *American Scientist*, 99(6) (2011), 463.

Woolley, Benjamin, *The Bride of Science: Romance, Reason, and Byron's Daughter* (McGraw-Hill, 2002).

Elizabeth 'Elsie' Muriel Gregory MacGill

Bourgeois-Doyle, Richard I., *Her Daughter the Engineer: The Life of Elsie Gregory MacGill* (Ottawa: NRC Research Press, 2008).

Bourgeois-Doyle, Richard I., 'Six decades later, SWE pioneer Elsie MacGill continues to inspire', *SWE Magazine*, 57(2) (2011), 28–32.

Hatch, Sybil, *Changing Our World: True Stories of Women Engineers* (Reston, Virginia: American Society of Civil Engineers, 2006).

MacGill, Elizabeth M. G., 'Factors Affecting Mass Production of Aeroplanes', *Flight*, 38 (1656) (1940), 228–231.

MacGill, Elizabeth M. G., *My Mother, the Judge: A Biography of Judge Helen Gregory MacGill* (Toronto: Ryerson Press, 1955; reprinted in 1981 by Toronto: PMA Books).

Sissons, Crystal, *Queen of the Hurricanes: The Fearless Elsie MacGill* (Toronto: Second Story Press, 2014).

Margaret Eliza Maltby

Gill, Raymond, 'Genetics & Genealogy – Miss Maltby and Her Ward: Using DNA to Investigate a Family Mystery', *American Ancestors*, 17(2) (2016), 49–52.

'Maltby, Margaret Eliza'. *Contributions of 20th Century Women to Physics* <http://cwp.library.ucla.edu/>.

Shearer, Benjamin F. and Barbara S. Shearer (eds), *Notable Women in the Physical Science: A Biographical Dictionary* (Westport, Conn: Greenwood Press, 1997).

'Maltby, Margaret E. (1860–1944)' <https://www.encyclopedia.com/women/encyclopedias-almanacs-transcripts-and-maps/maltby-margaret-e-1860-1944>.

Lynn Petra Alexander Margulis

Lake, James A., 'Lynn Margulis (1938–2011)', *Nature*, 480 (7378) (2011), 458.

Mann, C., 'Lynn Margulis: Science's Unruly Earth Mother', *Science*, 252 (5004) (1991), 378–381.

Margulis, Lynn, *Gaia is a Tough Bitch. Chapter 7 in the Third Culture: Beyond the Scientific Revolution by John Brockman* (Simon & Schuster, 1995).

Schaechter, M., 'Lynn Margulis (1938–2011)', *Science*, 335 (6066) (2012), 302.

Teresi, Dick, 'Discover Interview: Lynn Margulis Says She's Not Controversial, She's Right' (*Discover Magazine*, April 2011).

Weber, Bruce, 'Lynn Margulis, Evolution Theorist, Dies at 73' (*The New York Times*, 24 November 2011).

Lynn Margulis, Curriculum Vitae <http://www.bib.ub.edu/fileadmin/bibs/biologia/Lynn_margulis/curriculum_margulis.pdf>.

Mileva Marić-Einstein

Benedict, Maria, *The Other Einstein* (Sourcebooks Landmark, 2016).

Calaprice, A. and T. Lipscombe, *Albert Einstein: A Biography* (Westport and London: Greenwood Press, 2005).

Popović, M., *In Albert's Shadow: The Life and Letters of Mileva Marić, Einstein's First Wife* (Johns Hopkins University Press, 2003).

Troemel-Ploetz, Senta, 'The Woman Who Did Einstein's Mathematics', *Women's Studies International Forum*, 13(5) (1990), 415–32.

'Mileva Marić-Einstein', *History of Scientific Women* <https://scientificwomen.net/women/maric-einstein-mileva-62>.

Antonia Caetana de Paiva Pereira Maury

Bailey, Brooke, *The Remarkable Lives of 100 Women Healers and Scientists* (Holbrook, Mass: B. Adams, Inc., 1994), pp. 138–139.

Gingerich, Owen, *Maury, Antonia Caetana de Paiva Pereira*, Dictionary of Scientific Biography, Vol. 9 (New York: C. Scribner's Sons, 1974), pp. 194–195.

Hoffleit, Dorrit, 'Antonia C. Maury', *Sky and Telescope*, 11 (1952), 106.

Larsen, Kristine M., Benjamin F. Shearer and Barbara S. Shearer, 'Antonia Maury (1866–1952), Astronomer', *Notable Women in the Physical Sciences: A Biographical Dictionary* (Westport, Conn.: Greenwood Press, 1997), pp. 255–259.

Maury, Antonia C., 'The Spectral Changes of *Beta Lyrae*', *Annals of the Astronomical Observatory of Harvard College*, 84(8) (1933), 207–255.

'Antonia Maury', *Vassar Encyclopedia* (2008) <http://vcencyclopedia.vassar.edu/alumni/antonia-maury.html>.

Trimble, Virginia, 'Maury, Antonia Caetana de Paiva Pereira', in Thomas Hockey, Virginia Trimble and Thomas R. Williams, *The Biographical Encyclopedia of Astronomers* (New York: Springer Publishing, 2009), p. 1422.

Maria Göppert Mayer

Ferry, Joseph, *Maria Goeppert Mayer* (Philadelphia: Chelsea House Publishers, 2003).

Haber, Louis, *Women Pioneers of Science* (1st ed.) (New York: Harcourt Brace Jovanovich, 1979).

Bibliography

'Maria Goeppert Mayer: Facts', The Nobel Prize in Physics 1963 <https://www.nobelprize.org/prizes/physics/1963/mayer/facts/>.

Nobel Lectures, Physics 1963–1970 (Amsterdam, Elsevier Publishing Company, 1972).

Opfell, Olga S., *The Lady Laureates: Women Who Have Won the Nobel Prize* (Metuchen, N.J.: Scarecrow Press, 1978), pp. 194–208.

Sachs, Robert, 'Maria Goeppert Mayer 1906–1972: A Biographical Memoir' (*Biographical Memoirs*, National Academy of Sciences, 1979).

Wuensch, Daniela, *Der Letzte Physiknobelpreis Für Eine Frau? Maria Goeppert Mayer: Eine Göttingerin Erobert Eie Atomkerne. Nobelpreis 1963. Zum 50. Jubiläum* (Göttingen, Germany: Termessos Verlag, 2013).

(Eleanor) Barbara McClintock

Comfort, Nathaniel C., 'Barbara McClintock's Long Postdoc Years'. *Science*, 295 (5554) (2002), 440.

Comfort, Nathaniel C., 'The Real Point is Control: The Reception of Barbara McClintock's Controlling Elements', *Journal of the History of Biology*, 32(1) (1999), 133–162.

Kass, Lee B. and Christophe Bonneuil, *Mapping and Seeing: Barbara McClintock and the Linking of Genetics and Cytology in Maize Genetics, 1928–1935*, in Hans-Jörg Rheinberger and Jean-Paul Gaudilliere, *Classical Genetic Research and its Legacy: The Mapping Cultures of 20th Century Genetics* (London: Routledge, 2004), pp. 91–118.

Lamberts, William J., 'McClintock, Barbara', *American National Biography Online* (Oxford University Press, February 2000).

Pray, Leslie and Kira Zhaurova, 'Barbara McClintock and the Discovery of Jumping Genes (Transposons)', *Nature Education*, 1(1) (2008), 169.

The Barbara McClintock Papers. Controlling Elements: Cold Spring Harbor, 1942-1967, US National Library of Medicine <https://profiles.nlm.nih.gov/ps/retrieve/Narrative/LL/p-nid/49>

Elise 'Lise' Meitner

Bentzen, Søren M., 'Lise Meitner and Niels Bohr – A Historical Note', *Acta Oncologica*, 39(8) (2000), 1002–1003.

Haber, Louis, *Women Pioneers of Science* (New York: Harcourt Brace Jovanovich, 1979).

'Meitner, Lise', *Contributions of 20th Century Women to Physics* <http://cwp.library.ucla.edu/Phase2/Meitner,_Lise@844904033.html>.

Rife, Patricia, *Lise Meitner and the Dawn of the Nuclear Age* (Basel: Birkhäuser, 1999).

Rife, Patricia, 'Lise Meitner, 1878–1968', *The Encyclopedia of Jewish Women* <https://jwa.org/encyclopedia/article/meitner-lise>.

Sime, Ruth Lewin, 'Lise Meitner', in Nina Byers and Gary Williams, *Out of the Shadows: Contributions of 20th-Century Women to Physics* (Cambridge University Press, 2006).

Maud Leonora Menten

Menten, Maud, 'A Study of the Oxidase Reaction with Alpha-naphthol and Paraphenylenediamine', *The Journal of Medical Research*, 40(3) (1919), 433–458.3.

Skloot, Rebecca, 'Some Called Her Miss Menten' (*PITTMED,* University of Pittsburgh School of Medicine Magazine, October 2000), 18–21, https://www.pittmed.health.pitt.edu/oct_2000/miss_menten.pdf>.

Stock, Aaron and Anna-Mary Carpenter, 'Professor Maud Menten', *Nature*, 189(4769) (1961), 965.

'Dr. Maud L. Menten – She Had Vision, Enthusiasm and Compassion', *The Canadian Medical Hall of Fame* (1998) <http://www.cdnmedhall.org/inductees/dr-maud-l-menten>.

'Scarlet Fever Deaths Avoided in City' (*The Pittsburgh Press*, 19 May 1942) <https://www.newspapers.com/newspage/143360593>.

'The Mystery of Maud Menten' (*Sandwalk*, 11 September 2014), <https://sandwalk.blogspot.com/2014/09/the-mystery-of-maud-menten.html>.

Maria Sibylla Merian

Andréolle, Donna Spalding and Veronique Molinari (eds), *Women and Science, 17th Century to Present: Pioneers, Activists and Protagonists* (Cambridge Scholars Publishing, 2011), p. 42.

Etheridge, Kay, 'Maria Sibylla Merian and the Metamorphosis of Natural History', *Endeavour*, 35(1) (2011), 16–22.

Pieters, Florence F. J. M. and D. Winthagen, 'Maria Sibylla Merian, Naturalist and Artist (1647-1717): A Commemoration on the Occasion of the 350th Anniversary of Her Birth', *Archives of Natural History*, 26(1) (1999), 1–18.

Swaby, Rachel, *Headstrong: 52 Women Who Changed Science – And the World* (New York: Broadway Books, 2015), pp. 47–50.

Todd, Kim, 'Maria Sibylla Merian (1647-1717): An Early Investigator of Parasitoids and Phenotypic Plasticity', *Terrestrial Arthropod Reviews*, 4(2) (2011), 131–144.

Rogers, Kara, 'Maria Sibylla Merian – German-born Naturalist and Artist', *Encyclopædia Britannica* <https://www.britannica.com/biography/Maria-Sibylla-Merian>.

Maryam Mirzakhani

Chang, Kenneth, 'Maryam Mirzakhani, Only Woman to Win a Fields Medal', *Dies at 40* (*The New York Times*, 16 July 2017).

Jacobson, Howard, 'The World Has Lost a Great Artist in Mathematician Maryam Mirzakhani' (*The Guardian*, 29 July 2017).

Webb, Jonathan, 'First Female Winner for Fields Maths Medal' (*BBC News*, 2014).

Joshi, Nalini, 'Maryam Mirzakhani: Remembering a Brilliant Mathematician Who Inspired a World of Possibilities' (*ABC Science,* 7 January 2018).

'Maryam Mirzakhani's Work on Riemann Surfaces Explained in Simple Terms' (*Matific*, 14 Aug 2014).

'Maryam Mirzakhani – First woman and the first Iranian to win the Fields Medal', *Wonder Women* <https://www.wndrwmn.com/maryam-mirzakhani>.

Valette, Alain, *The Fields Medalists 2014* (Neuchâtel, Switzerland: Institut de mathématiques, Université de Neuchâtel, 2014) <https://www.unine.ch/files/live/sites/math/files/shared/documents/prepublications/2014/Fields2014%20english%20light.pdf>.

Bibliography

Maria Mitchell

Among the Stars: The Life of Maria Mitchell. Astronomer, Educator, Women's Rights Activist (Nantucket, MA: Mill Hill Press, 2007).

Bergland, Renée, *Maria Mitchell and the Sexing of Science* (Boston, MA: Beacon Press, 2008), p. 57.

Gormley, Beatrice, *Maria Mitchell the Soul of an Astronomer* (William B. Eerdmans Publishing Co, Grand Rapids, MI, 1995), pp. 4–6.

Hoffleit, Dorrit, 'The Maria Mitchell Observatory – For Astronomical Research and Public Enlightenment', *The Journal of the American Association of Variable Star Observer*, 30(1) (2001), 62.

'Maria Mitchell's Gold Medal', Maria Mitchell Association, Nantucket's Science Center <https://www.mariamitchell.org/about/awards/maria-mitchells-gold-medal>.

Marcia Neugebauer

'Neugebauer, Marcia', *Contributions of 20th Century Women to Physics* <http://cwp.library.ucla.edu/Phase2/Neugebauer,_Marcia@931234567.html>.

'Marcia Neugebauer – Distinguished Visiting Scientist', *WITI Hall of Fame, Jet Propulsion Laboratory* (1997) <http://www.witi.com/center/witimuseum/halloffame/119285/Marcia-Neugebauer-Distinguised-visiting-scientist-Jet-Propulsion-Laboratory>.

Neugebauer, Marcia, 'Pioneers of Space Physics: A Career in the Solar Wind', *Journal of Geophysical Research: Space Physics*, 102(A12) (1997), 26,887–26,894.

O'Connell, Franklin, 'The Venus Mission, How Mariner 2 Led the World to the Planets', *NASA Jet Propulsion Laboratory* (14 December 1962) <https://www.jpl.nasa.gov/mariner2>.

Thompson, Samantha M., 'Oral History – Marcia Neugebauer', *American Institute of Physics* (18 July 2017) <https://www.aip.org/history-programs/niels-bohr-library/oral-histories/42831>.

Ida Tacke Noddack

Gregersen, Erik, 'Ida Noddack – German Chemist', *Encyclopædia Britannica* <https://www.britannica.com/biography/Ida-Noddack>.

Habashi, Fathi, 'Ida Noddack and the Missing Elements', *Education in Chemistry*, 46 (2009) <https://eic.rsc.org/feature/ida-noddack-and-the-missing-elements/2020167.article>.

'Ida Noddack – Scientific Woman', *History of Scientific Women* <https://scientificwomen.net/women/noddack-ida-74>.

Noddack, Ida, 'On Element 93', *Zeitschrift für Angewandte Chemie*, 47(37) (1934), 653.

Santos, Gildo Magalhäes, 'A Tale of Oblivion: Ida Noddack and the "Universal Abundance" of Matter', *Notes and Records of the Royal Society of London*, 68(4) (2014) <https://doi.org/10.1098/rsnr.2014.0009>.

'Noddack, Ida (1896–1978)', *Women in World History: A Biographical Encyclopedia* <https://www.encyclopedia.com/women/encyclopedias-almanacs-transcripts-and-maps/noddack-ida-1896-1978>.

Amalie 'Emmy' Noether

Alexandrov, Pavel S., 'In Memory of Emmy Noether', in Brewer, James, W. and Smith, Martha K. (eds), *Emmy Noether: A Tribute to Her Life and Work* (New York: Marcel Dekker, 1981), pp. 99–111.

Byers, Nina, 'Emmy Noether', in Nina Byers and Gary Williams, *Out of the Shadows: Contributions of 20th Century Women to Physics* (Cambridge: Cambridge University Press, 2006).

Conover, Emily, 'Emmy Noether Changed the Face of Physics; Noether Linked Two Important Concepts in Physics: Conservation Laws and Symmetries' (*ScienceNews*, 12 June 2018) <https://www.sciencenews.org/article/emmy-noether-theorem-legacy-physics-math>.

Einstein, Albert, 'Professor Einstein Writes in Appreciation of a Fellow-Mathematician' (*New York Times*, 5 May 1935).

Ne'eman, Yuval, *The Impact of Emmy Noether's Theorems on XXIst Century Physics* (Teicher, 1999), pp. 83–101.

'Noether, Amalie Emmy'. *Contributions of 20th Century Women to Physics* <http://cwp.library.ucla.edu/Phase2/Noether,_Amalie_Emmy@861234567.html>.

Scharlau, Winfried, *Emmy Noether's Contributions to the Theory of Algebras* (Teicher, 1999), p. 49.

Christiane 'Janni' Nüsslein-Volhard

Arias, Alfonso M., 'Drosophila Melanogaster and the Development of Biology in the 20th Century', *Methods in Molecular Biology*, 420 (2008), 1–25.

Biozentrum Lectures, University of Basel (2018) <https://www.biozentrum.unibas.ch/?id=24229>.

'Christiane Nüsslein-Volhard', *DNA from the Beginning* <http://www.dnaftb.org/37/bio-2.html>.

'Christiane Nüsslein-Volhard', *Famous Scientists, The Art of Genius* <https://www.famousscientists.org/christiane-nusslein-volhard>.

'Christiane Nüsslein-Volhard – Biographical', The Nobel Prize in Physiology or Medicine 1995 <https://www.nobelprize.org/prizes/medicine/1995/nusslein-volhard/biographical/>.

Nüsslein-Volhard Christiane and Eric Wieschaus, 'Mutations Affecting Segment Number and Polarity in Drosophila', *Nature*, 287(5785) (1980), 795–801.

Orden pour le Mérite für Wissenschaften und Künste, 'Christiane Nüsslein-Volhard' <http://www.orden-pourlemerite.de/sites/default/files/vita/nuesslein-volhard-vita.pdf>.

Weston, Kathy, 'An Interview with Nobel Laureate Janni Nüsslein-Volhard' (*Crosstalk*, 8 December 2017).

Muriel Wheldale Onslow

Creese, Mary R. S., 'Onslow, Muriel Wheldale (1880–1932)', *Oxford Dictionary of National Biography* (2004).

McDonald, I. G., 'Obituary Notice: Muriel Wheldale Onslow: 1880–1932', *Biochemical Journal*, 26(4) (1932), 915–916.

Rayner-Canham, Marelene and Geoffrey Rayner-Canham, 'Muriel Wheldale Onslow (1880–1932), Pioneer Plant Biochemist', *The Biochemist*, 'Past Times' (April 2002), 49–51 <http://www.biochemist.org/bio/02402/0049/024020049. pdf>.

Bibliography

Richmond, Marsha L., 'Muriel Wheldale Onslow and Early Biochemical Genetics', *Journal of the History of Biology*, 40(3) (2007), 389–426.

Wheldale, Muriel, *The Anthocyanin Pigments of Plants* (Cambridge: Cambridge University Press, 1916).

Ruby Violet Payne-Scott

Australian Institute of Physics, 'Ruby Payne-Scott Award' <http://aip.org.au/medals-awards-and-prizes/ruby-payne-scott-award/>.

Goss, William. M. and Richard X. McGee, *Under the Radar the First Woman in Radio Astronomy, Rudy Payne-Scott* (Heidelberg: Springer, 2009).

Goss, William M. and Claire Hooker, 'Payne-Scott, Ruby Violet (1912–1981)', *Australian Dictionary of Biography*, vol. 18 (Melbourne University Publishing, 2012), online: <http://adb.anu.edu.au/biography/payne-scott-ruby-violet-15036>.

Goss, William M., *Making Waves: The Story of Ruby Payne-Scott: Australian Pioneer Radio Astronomer* (Heidelberg: Springer, 2013).

Halleck, Rebecca, 'Overlooked No More: Ruby Payne-Scott, Who Explored Space with Radio Waves' (*The New York Times*, 29 August 2018).

'Ruby Payne-Scott Biography', *CSIROpedia* (2009) <https://csiropedia.csiro.au/payne-scott-ruby>.

'The Secret Life of Miss Ruby Payne-Scott', *National Archives of Australia* <http://www.naa.gov.au/collection/snapshots/find-of-the-month/2009-march.aspx>.

Rózsa Péter

Albers, Donald J., Gerald L. Alexanderson and Constance Reid (eds), *Rózsa Péter 1905–1977, More Mathematical People* (Harcourt Brace Jovanovich, 1990).

Andrásfai, Béla, 'Rózsa (Rosa) Péter', *Periodica Polytechnica Electrical Engineering*, 30(2–3) (1986), 139–145.

Morris, Edie and Leon Harkleroad, 'Rózsa Péter: Recursive Function Theory's Founding Mother', *The Mathematical Intelligencer*, 12(1) (1990), 59–64.

O'Connor, John, J. and Edmund F. Robertson, 'Rózsa Péter' <http://www-history.mcs.st-and.ac.uk/Biographies/Peter.html>.

'Rózsa Péter: Founder of Recursive Function Theory', *Women in Science: A Selection of 16 Contributors* (San Diego Supercomputer Center, 1997).

Tamássy, István, 'Interview with Róza Péter', *Modern Logic*, 4(3) (1994), 277–280.

Agnes Luise Wilhelmine Pockels

Derrick, Elizabeth M., 'Agnes Pockels, 1862-1935', *Journal of Chemical Education*, 59(12) (1982), 1030–1031.

Giles, Charles H. and Stanley D. Forrester, 'The Origins of the Surface Film Balance: Studies in the Early History of Surface Chemistry, Part 3' (*Chemistry and Industry*, 9 January 1971), 43–53.

Kruse, Andrea and Sonja M. Schwarzl, 'Zum Beispiel Agnes Pockels' (*Nachrichten aus der Chemie*, June 2002).

Pockels, Agnes, 'Surface Tension', *Nature*, 43 (1891), 437–439.

'Pockels, Agnes.' *Contributions of 20th Century Women to Physics* <http://cwp.library.ucla.edu/Phase2/Pockels,_Agnes@871234567.html>.

McCarthy, Stephen, 'Agnes Pockels', *175 Faces of Chemistry* (November 2014) <http://www.rsc.org/diversity/175-faces/all-faces/agnes-pockels>.

Helen Rhoda Arnold Quinn

Poggio, E. C., Helen R. Quinn and S. Weinberg, 'Smearing Method in the Quark Model', *Physical Review D.*, 13(7) (1976), 1958–1968.

'APS Members Elect Helen Quinn as Society's Next Vice President', *APS Physics* (November 2001) <https://www.aps.org/publications/apsnews/200111/quinn.cfm>.

Bardi, Jason Socrates, 'Helen Quinn Named Winner of the 2016 AIP Karl Compton Medal', *American Institute of Physics* (17 November 2015) <https://www.aip.org/news/2015/helen-quinn-named-winner-2016-aip-karl-compton-medal>.

'Helen Rhoda Quinn', *The Franklin Institute Awards* (15 March 2018), <https://www.fi.edu/laureates/helen-rhoda-quinn>.

Quinn, Helen, 'Belief and Knowledge – A Plea About Language' (January 2007) <https://ned.ipac.caltech.edu/level5/March07/Quinn/Quinn.html>

Quinn, Helen, 'You Could be a Mathematician' (*ICTP*, 16 Sept. 2013) <https://www.ictp.it/about-ictp/media-centre/news/news-archive/2013/9/helen-quinn.aspx>.

Mina Spiegel Rees

Alt, Franz L., 'Fifteen Years ACM', *Communications of the ACM – Celebrating ACM's 40th Anniversary*, 30(10) (1987), 850–857.

Green, Judy, Jeanne La Duke, Saunders Mac Lane and Uta C. Merzbach, 'Mina Spiegel Rees (1902–1997)', *Notices of the AMS*, 45(7) (1998), 866–873 <http://www.ams.org/notices/199807/memorial-rees.pdf>.

'Mina S. Rees: Award Recipient', *IEEE Computer Society* <https://www.computer.org/web/awards/pioneer-mina-rees>.

Rees, Mina, 'The mathematics program of the Office of Naval Research', *Bulletin of the American Mathematical Society*, 54 (1948), 1–5.

Shell-Gellasch, Amy, *In Service to Mathematics: The Life and Work of Mina Rees* (Boston: Docent Press, Boston, 2001).

Shell-Gellasch, Amy, 'Mina Rees and the Funding of the Mathematical Sciences', *The American Mathematical Monthly*, 109(10) (2002), 873–889.

Williams, Kathleen Broome, *Improbable Warriors: Women Scientists and the U.S. Navy in World War II* (Annapolis, Maryland: Naval Institute Press, 2001).

Ellen Henrietta Swallow Richards

Clarke, Robert, *Ellen Swallow* (Chicago: Follett Pub. Co., 1973).

'Ellen Swallow Richards', *MIT History* <https://libraries.mit.edu/mithistory/community/notable-persons/ellen-swallow-richards/>.

'Mrs. Ellen H. Richards Dead. Head of Social Economics in Massachusetts Institute of Technology' (*The New York Times*, 31 March 1911).

Richards, Ellen H., *Euthenics, the Science of Controllable Environment* (Boston: Whitcomb & Barrows, 1910).

Shearer, Benjamin F., *Notable Women in the Physical Sciences: A Biographical Dictionary* (Westport, Conn.: Greenwood Press, 1997).

Swallow, Pamela C., *The Remarkable Life and Career of Ellen Swallow Richards: Pioneer in Science and Technology* (John Wiley & Sons, July 2014).

'Tributes to Ellen Swallow Richards', *Technology Review*, 13 (1911), 365–373.

Vare, Ethlie Ann and Jennifer Hangerman, *Adventurous Spirit: A Story about Ellen Swallow Richards* (Minneapolis: Carolrhoda Books, 1992).

Mary Ellen Rudin

Albers, Donald J. and Constance Reid, 'An Interview with Mary Ellen Rudin', *The College of Mathematics Journal*, 19(2) (1988), 114–137.

Albers, Donald J., Gerald L. Alexanderson and Constance Reid (eds), 'Mary Ellen Rudin', in *More Mathematical People: Contemporary Conversations* (Harcourt Brace Jovanovich, 1990), pp. 282–302.

Benkart, Georgia, Mirna Džamonja and Judith Roitman, 'Memories of Mary Ellen Rudin', *Notices of the American Mathematical Society*, 62(6) (2015), 617–629.

Jones, F. B., 'Some glimpses of the early years', *The Work of Mary Ellen Rudin*, Franklin D. Tall (ed.), *Annals of the New York Academy of Sciences*, 705, 1993, pp. xi–xii.

Henrin, Claudia, 'Mary Ellen Rudin', *Celebratio Mathematica* (1997) <https://celebratio.org/Rudin_Mary/article/552/>.

O'Connor, John, J. and Edmund F. Robertson, 'Mary Ellen Rudin' <http://www-history.mcs.st-and.ac.uk/Biographies/Rudin.html>.

Rudin, Mary Ellen, 'A Normal Space X for Which X × I Is Not Normal', *Fundamental Mathematics*, 73(2) (1971), 179–186.

Watson, W. S., 'Mary Ellen Rudin's Early Work on Suslin Spaces', in *The Work of Mary Ellen Rudin* (Madison, WI, 1991) (Ann. New York Academy Sciences, 705, New York Academy Sciences, New York, 1993), pp. 168–182.

Hazel Marguerite Schmoll

'Hazel Schmoll', Ward Colorado Resident <https://sites.google.com/site/wardcolorado/hazelschmoll-wardresident1890-1990>.

Pettem, Silvia, 'Hazel Schmoll Made a Name for Herself as a Female Scientist' (*Daily Camera*, 15 March 2013).

Varnell, Jeanne, 'Hazel Marguerite Schmoll', in *Women of Consequence* (Boulder: Johnson Book, 1999), pp. 110-114.

Beatrice ('Tilly') Shilling

Blake-Coleman, Barrie, 'The Fabulous "Tilly" Shilling!' <http://www.inventricity.com/tilly-shilling>.

Freudenberg, Matthew, 'Negative Gravity: A Life of Beatrice Shilling' (Speedreaders. info., 2013).

Haines, Catharine M. C., *International Women in Science: A Biographical Dictionary to 1950* (ABC-CLIO, 2001), p. 288.

Hill, Jonathan, 'Beatrice Shilling' (*The Classic Motorcycle*, 20 March 2018) <https://www.classicmotorcycle.co.uk/beatrice-shilling/>.

'Beatrice Shilling', *Thrust Vector* <https://thrustvector.wordpress.com/2010/03/24/beatrice-shilling>.

Michelle Yvonne Simmons

Guillat, Richard, 'Star of the Sub-atomic' (*The Weekend Australian Magazine* 15–16 April 2017).

'Quantum Physicist: Michelle Simmons', *ABC Science: Meet a Scientist*, Australian Broadcasting Corporation (28 February 2012) <http://www.abc.net.au/science/articles/2012/02/28/3441521.htm>.

Ross, John, 'Christopher Pyne Launches Nature Partner in Quantum Computing' (*The Australian*, 5 November 2014).

Storm, Mark, 'Australia Day Address Orator Michelle Simmons Horrified at "Feminised" Physics Curriculum' (*Sydney Morning Herald*, 24 January 2017).

'Your Stories: 2018 Australian of the Year, Professor Michelle Simmons', *UniSuper (Interview)* <https://www.unisuper.com.au/about-us/super-informed-enews/august-2018/meet-professor-michelle-simmons>.

Mary Fairfax Greig Somerville

Cleese, Mary R. S., 'Somerville [*née* Fairfax; *other married name* Greig], Mary', *Oxford Dictionary of National Biography* (21 May 2009) <https://www-oxforddnb-com.simsrad.net.ocs.mq.edu.au/view/10.1093/ref:odnb/9780198614128.001.0001/odnb-9780198614128-e-26024?rskey=P6IOYO&result=1>.

Fara, Patricia, 'Mary Somerville: a scientist and her ship', *Endeavour*, 32(3) (2008), 83–85.

Neeley, Kathryn A., *Mary Somerville: Science, Illumination, and the Female Mind* (Cambridge & New York: Cambridge University Press, 2001).

O'Connor, John J. and Edmund F. Robertson, *Mary Fairfax Greig Somerville*, MacTutor History of Mathematics (University of St Andrews, Scotland, November 1999).

Somerville, Martha, *Personal Recollections, from Early Life to Old Age, of Mary Somerville: With Selections from Her Correspondence* (Roberts Brothers, digitised 2007, original in Harvard University, 1874).

Somerville, Martha, *Personal Recollections, From Early Life to Old Age, of Mary Somerville* (Boston: Roberts Brothers, 1874, written by her daughter, reprinted by AMS Press, January 1996).

'Somerville, Mary Fairfax Greig', *Dictionary of Scientific Biography*, 11 & 12 (New York: Charles Scribner's Sons), pp. 521–522.

'Janet' Jane Ann Ionn Taylor

Alger, Ken R., 'Mrs Janet Taylor – Authoress and Instructress in Navigation and Nautical Astronomy' (1804-1870)', *Fawcett Library Papers*, LLRS Publications 6 (1982).

Croucher, John S. and Rosalind F. Croucher, *Mistress of Science* (Stroud: Amberley Publishing, 2016).

Croucher, Rosalind F. and John S. Croucher, 'Compasses and sinking ships: Mrs Janet Taylor's contribution in the compass adjusting controversy of mid-19th Century England', *International Journal of Maritime History*, 30(2) (2018), 234–251.

Bibliography

Croucher, Rosalind F. and John S. Croucher, 'Mrs Janet Taylor and the Civil List Pension', *Women's History Review*, 21(2) (2013), 253–280.

Croucher, John S. and Rosalind F. Croucher, 'Mrs Janet Taylor's "Mariner's Calculator": Assessment and Reassessment', *British Journal for the History of Science*, 44(4) (2011), 493–507.

Croucher, John S., 'An Exceptional Woman of Science', *Bulletin of the Scientific Instrument Society* 84 (2005), 22–27.

Ruth Teitelbaum

Haigh, Thomas, Peter Mark and Crispin Rope, *Eniac in Action: Making and Remaking the Modern Computer* (Cambridge, MA: MIT Press, 2016), p. 26.

Martin, Gay, *Recent Advances and Issues in Computers* (Phoenix, Arizona: The Oryx Press, 2000), pp. 106-107.

Minoff, Annie and Jared Goyette, 'Finding the forgotten women who programmed the world's first electronic computer', *Science Friday,* Public Radio International (30 March 2015) <https://www.pri.org/stories/2015-03-30/finding-forgotten-women-who-programmed-world-s-first-electronic-computer>.

'Ruth Teitelbaum', *Engineering and Technology History Wiki* (25 February 2016) <https://ethw.org/Ruth_Teitelbaum>.

'Ruth Teitelbaum', Revolvy <https://www.revolvy.com/page/Ruth-Teitelbaum>.

Valentina Vladimirovna Tereshkova

Eidelman, Tamara, 'The Extraordinary Destiny of an 'Ordinary' Woman', *Russian,* 46(3) (2003), 19.

Lothian, Antonella, *Valentina: The First Woman in Space* (The Pentland Press, 1993).

O'Neil, Bill, 'Whatever became of Valentina Tereshkova?', *New Scientist,* 139(1886) (1993), 21.

Woodmansee, Laira, 'Two Who Dared', *Ad Astra*, 17(2) (2005), 48.

'First Woman in Space', *This Day in History* (21 August 2018) <https://www.history.com/this-day-in-history/first-woman-in-space>.

'Valentina Tereshkova', *Encylopedia Britannica* (1 February 2019) <https://www.britannica.com/biography/Valentina-Tereshkova>.

Mildred Trotter

Byers, Steven N., *Introduction to Forensic Anthropology* (4th ed.) (Harlow: Pearson Education, 2011), p. 490.

Goldine C. Gleser Papers, 1950s–1970s, OhioLINK Finding Aid Repository (1952) <http://www.sciencestories.io/Q15130715#1>.

Trotter, Mildred and Goldine C. Gleser, 'A Re-evaluation of Estimation of Stature Based on Measurements of Stature Taken During Life and of Long Bones After Death', *American Journal of Physical Anthropology,* 16(1) (1958), 79–123.

Trotter, Mildred and Goldine C. Gleser, 'Estimation of Stature from Long Bones of American Whites and Negroes', *American Journal of Physical Anthropology*, 10(4) (1952), 463–514.

Williams, Robert C., *The forensic Historian: Using Science to Re-examine the Past* (Armonk, NY: M. E. Sharpe, 2013), Preface: 'Mildred Trotter and the Boneyard of War', pp. xi–xiii.

'Mildred Trotter (1899–1991)', Women in Health Sciences – Biographies (Bernard Becker Medical Library Digital Collection, 2009) <http://beckerexhibits.wustl.edu/mowihsp/bios/trotter.htm>.

Anna Johnson Pell Wheeler

Akritas, Alkiviadis. G., Gernadi I. Malaschonok and P. S. Vigklas, 'On a Theorem by Van Vleck Regarding Sturm Sequences', *Serdica Journal of Computing*, 7(4) (2013), 101–134.

Akritas, Alkiviadis G., 'Anna Johnson and Her Seminal Theorem of 1917' (2016) <https://faculty.e-ce.uth.gr/akritas/publications/100.pdf>

Green, Judy and Jeanne LaDuke, 'Pioneering Women in American Mathematics – The Pre-1940 PhDs', *History of Mathematics*, vol. 34 (American Mathematical Society and the London Mathematical Society, 2008), pp. 633–638.

Grinstein, Louise S. and Paul J. Campbell, 'Anna Johnson Pell Wheeler: Her life and work', *Historia Mathematica*, 9(1) (1982), 37–53.

O'Connor, John, J. and Edmund F. Robertson, 'Anna Johnson Pell Wheeler' (The MacTutor History of Mathematics archive, January 1997) <http://www-history.mcs.st-andrews.ac.uk/Biographies/Wheeler.html>.

Pell, Anna K. and Richard L. Gordon, 'The Modified Remainders Obtained in Finding the Highest Common Factor of Two Polynomials', *Annals of Mathematics, Second Series,* 18(4) (1917), 188–193.

Sheila Marie Evans Widnall

Biography, Dr. Sheila E. Widnall (Office of the Secretary of the Air Force/Public Affairs, November 1993).

Jehl, Douglas, 'MIT Professor is First Woman Chosen as Secretary of Air Force' (*New York Times*, 4 July 1993), p. 20.

Sears, William R., 'Sheila E. Widnall: President-Elect of AAAS' (*Association Affairs*, 6 June 1986), 1119–1200.

Stone, Steve, 'Air Force Secretary Salutes Female Aviators' (*Norfolk Virginian-Pilot*, 10 October 1993) p. B3.

'USAF Head Approved, in Aviation Week & Space Technology' (*Physics Today*, 9 August 1993), p. 26.

'Dr. Sheila E. Widnall – 23 July 1997' (*Air Force Times*, 2 August 1993), p. 4, <http: //www.af.mil:80/news/biographies/widnallse.html>.

Rosalyn Sussman Yalow

'Festschrift for Rosalyn S. Yalow: Hormones, Metabolism, and Society', *Mt Sinai Journal Medicine* 59(2) (1992), 95–185.

Goldsmith, S. J., 'Georg de Hevesy Nuclear Medicine Pioneer Award Citation–1986. Rosalyn S. Yalow and Solomon A. Berson', *Journal Nuclear Medicine*, 28(10) (1987), 1637–39.

Kyle, Robert A. and Marc A. Shampo, 'Rosalyn Yalow – Pioneer in Nuclear Medicine', *Mayo Clin. Proc.*, 77(1) (2002), 4.

Bibliography

Opfell, Olga S., *The Lady Laureates: Women Who Have Won the Nobel Prize* (Metuchen N.J. & London: Scarecrow Press, Inc., 1978), pp. 224–233.

Schwartz, I. L., 'Solomon A. Berson and Rosalyn S. Yalow: A Scientific Appreciation', *Mt Sinai Journal Medicine*, 40(3) (1973), 284–94.

Yalow, Rosalyn S., 'The Nobel Lectures in Immunology. The Nobel Prize for Physiology or Medicine, 1977 Awarded to Rosalyn S. Yalow', *Scand. J. Immunol*, 35(1) (1992), 1–23.

Ada Yonath

Klenke, Karin, *Women in Leadership: Contextual Dynamics and Boundaries* (Bingley, UK: Emerald Group Publishing, 2011), p. 191.

Lappin, Yaakov, 'Nobel Prize Winner "Happy, Shocked"' (*Jerusalem Post*, 7 October 2009).

Yonath, Ada and Wolfie Traub, 'Polymers of Tripeptides as Collagen Models. 4. Structure Analysis of Poly (l-prolyl-glycyl-l-proline)', *Journal of Molecular Biology*, 43(3) (1969), 461.

'Ada E. Yonath: Biographical', The Nobel Prize in Chemistry 2009 <https://www.nobelprize.or g/prizes/chemistry/2009/yonath/biographical>.

'Ada Yonath', *Profiles in Biophysics* <https://www.biophysics.org/profiles/ada-yonath>.

Rogers, Kara, 'Ada Yonath', *Encyclopaedia Britannica* <https://www.britannica.com/biography/Ada-Yonath>.

Photograph Credits

Frances E. Allen receiving the Erna Hamburger Distinguished Lecture Award at the EPFL. (Elizabeth Allen, 6 May 2008, Author: Rama, Source: Creative Commons Share-Alike)

Dr Virginia Apgar. (Library of Congress. Photograph from Public Information Department, The National Foundation (the March of Dimes). Forms part of New York World-Telegram and the Sun Newspaper Photograph Collection (Library of Congress); Date: 6 July 1959)

Hertha Ayrton, portrait. (Girton College, University of Cambridge, supplied by The Public Catalogue Foundation; Author Helena Arsene Darmeseter; Date: Before 1905)

Massimilla 'Milla' Baldo-Ceolin. (CERN, European Organisation for Nuclear Research, Date: Unknown)

Alice Augusta Ball (Date: 1915, Author unknown)

Nora Stanton Barney (National Photo Company Collection/ Library of Congress, Washington, D.C.; Date: c. 1921)

Florence Bascom (Camera Craft Studios, Minneapolis, Repository: Smithsonian Institution Archives Collection: Science Service Records, 1902–1965 (Record Unit 7091). Date: c. 1900.

Laura Bassi, The first woman in the world to earn a university chair in a scientific field of studies. (Creative Commons Share-alike; Author: Mailsapartbassimore. Date: Unknown)

Ulrike Beisiegel, President of the University of Göttingen, Göttingen, Germany. (Creative Commons Share-alike, Author: Stefan Floper; Date: 2014)

Elizabeth Blackburn (Salk Institute for Biological Studies; Date: 2016)

Elizabeth Blackwell (National Library of Medicine, Wikipedia Commons Photographer Unknown; Date: Unknown)

Mary Adela Blagg in her youth. (Photographer unknown; Date unknown but before 1898)

Katharine Blodgett, one of the world's few 'big shot' women scientists. (Image provided by the *Chicago Sun-Times*)

Rachel Littler Bodley (Willard, Frances Elizabeth (1893) *A woman of the Century: Fourteen Hundred-seventy Biographical sketches Accompanied by Portraits of Leading American women in All Walks of Life*; Moulton; Author unknown; Date: before 1923)

Photograph Credits

Alice Middleton Boring. (Open Access Copyright. Date: probably *c*. 1918)

Elizabeth Brown. (Royal Astronomical Society RAS Add MS 91/62)

Linda Brown Buck. (Fred Hutch News Service; Date: Unknown)

(Susan) Jocelyn Bell Burnell. (University of Dundee, *University News*, 20 Feb 2018)

Nina Byers. (*Cern Courier*. Date: 26 August 2014; Author: Maggie Michelson)

Mrs Annie Jump Cannon, head-and-shoulders portrait, left profile. (*New York World Telegram and the Sun Newspaper*; Date: 1922)

Rachel Carson, Author of *Silent Spring*. (United States Fish and Wildlife Service; Official photo as FWS employee; Date: *c*. 1940)

Mary Cartwright, famous mathematician. (Anitha Maria S., own work, CC BY-SA4.0; Date: unknown; Creative Commons Share-Alike)

Yvette Cauchois, image. (Reproduction from *Physics Today*, 1 April 2001, page 88, with the permission of the American Institute of Physics; Author: Christiane Benelle)

Edith Clarke. (Photo courtesy of the Society of Women Engineers Archives, Walter P. Reuther Library, Wayne State University)

Anna Botsford Comstock. (Bain Collection/Library of Congress. Date: between 1904 and 1924)

Esther M. Conwell, photograph. (Esther M. Conwell papers, D.494, Rare Books, Special Collections, and Preservation, River Campus Libraries, University of Rochester. Date: unknown)

Gerty Theresa Cori, American biochemist and Nobel Prize winner. (National Library of Medicine, Images from the History of Medicine, B05353. Date: 1947)

Marie Skłodowska Curie. (Wiki Commons. Author Henri Manuel. Date: *c*. 1920)

Ingrid Daubechies. (Photo courtesy of Duke Photography: Les Todd; Duke University)

Olive Wetzel Dennis. (Donnybrook Fair, Goucher College Yearbook, Goucher College; Date: 1908)

Mildred Spiewak Dresselhaus. President Barack Obama greets recipients of the 2010 Fermi Award in the Oval Office, May 7, 2012. (Official White House Photo. Author: Pete Souza; Date: 2012)

Madame du Chatelet at her desk. (Private Collection, Author: Maurice Quentin de la Tour; Date: 18th Century)

Alice Eastwood. (California Academy of Sciences; Author: unattributed: Date: *c*. 1910)

Elsie Eaves. (Society of Women Engineers, Walter P. Reuther Library, Wayne State University; Date: *c*. 1946)

Gertrude Belle Elion. (Duke University Archives, David M. Rubenstein Rare Book & Manuscript Library)

Thelma Austern Estrin. (Walter P Reuther Library, Wayne State University)

Margaret Clay Ferguson. (Smithsonian Institution Archives; Creator/Photographer: Unknown

Lydia Folger Fowler, US physician. (Wheaton College; Author: unknown. Date: *c*. 1880)

Rosalind Elsie Franklin. (Courtesy of the King's College Archives, taken from her time working there; Date; Unknown; Author: Unknown)

Marie Sophie Germain. (Author: unknown; Date: 1790)

Catherine 'Kate' Anselm Gleason Memorial, East Rochester New York.

Gertrude Scharff-Goldhaber, nuclear physicist, portrait. (AIP, Emilio Segre Visual Archives, Physics Today Collection; Date: 1998)

Anna Jane Harrison. (Mount Holyoke College Archives and Special Collections; Date: Unknown; Author: Unknown)

Caroline Lucretia Herschel. (Melchior Gommar Tieleman; Date: 1829)

Beatrice Hicks, 1963 Society of Women Engineers Achievement Award recipient and SWE past president (1950-1952). (Society of Women Engineers, Walter P. Reuther Library, Wayne State University; Author: unknown; Date: *c.* 1960–1965)

Dorothy Crowfoot Hodgkin (© National Portrait Gallery)

Erna Schneider Hoover (Reused with permission of Nokia Corporation and AT & T Archives)

Commodore Grace M. Hopper, USN (covered). (U.S. Navy, US Federal Govt, employee of US Navy and taken as part of that person's official duties. Date: 1984)

Margaret Lindsay Huggins (1848–1915). (Author: unknown; Date: unknown)

A generic semiconductor; we owe a lot of our understanding of semiconductors today to Frances Hugle. (Courtesy of Pixabay)

Shirley Ann Jackson, President Rensselaer Polytechnic Institute speaks during the 'WHAT IF the United States remains in a jobless recovery in 2011?' session at the Annual Meeting of the New Champions in Tianjin, China, September 15, 2010. (Creative Commons Attribution Share Alike licence)

Sophia Louisa Jex-Blake (*The Life of Sophia Jex-Blake*, by Margaret Todd; Portrait by Samuel Laurence; Date: 1865)

Nalini Joshi, taken on 29 April 2014. (This is a cropped/reduced version of a photograph by Ted Sealey. It is released under the Creative Commons Attribution-ShareAlike licence)

Helen Dean King with Wistar rats, the first standardised laboratory animal. (Courtesy of The Wistar Institute, Wistar Archive Collection, Philadelphia, Pa. Date: *c.* 1920)

Maria Margaretha Winckelman Kirch: Part of the astronomical calendar that the Kirches developed. (Chur-Brandenburgischer Verbesserter Calender Auff das Jahr Christi 1701, published in Berlin, by the Prussian Academy of Sciences, digitised by the Archiv des Astronomischen Rechen-Instituts Heidelberg)

Margaret Galland Kivelson. (Licensed under the Creative Commons Attribution Share Alike Licence. Date: 2007)

Sofia Vasilyevna Korvin-Krukovskaya Kovalevskaya. (Author: unknown; Date: 1880)

Doris Kuhlmann-Wilsdorf. (Walter P. Reuther Library, Archives of Labor and Urban Affairs, Wayne State University, Photo ID 1296; Author: Unknown: Date: 1989)

Stephanie Kwolek, photograph taken at Spinning Elements, Chemical Heritage Foundation, Philadelphia, PA, USA. (Science History Institute, Creative Commons Attribution Share Alike Licence. Author: Harry Kalish)

Hedy Lamarr, publicity photo for the film *The Heavenly Body*, 1944. (Author: unknown. Date: 1944)

Inge Lehmann (Inge Lehmann Biography, Editors, TheFamousPeople.com; Date: Unknown)

Rita Levi-Montalcini at 100. (Official web site of the *Presidenza della Repubblica Italiana*; Author: *Presidenza della Repubblica Italiana*; Date: 2009)

Barbara Liskov. (Jason Dorfman; Date: unknown)

Ada Lovelace aka Augusta Ada Byron. (Reproduction courtesy of Geoffrey Bond, Public Domain, By Antoine Claudet, a rare daguerreotype, picture taken in his studio near Regents Park in London; Date: 1843 or 1850)

Elsie MacGill during her CCF tenure. (National Archives of Canada; Date: 1938) ;

Margaret Eliza Maltby PhD, New York, a formal studio photograph. (Creative Commons Attribution Share Alike Licence – family member of Margaret Maltby inherited this image; Author: JaneRayGill75; Date: *c.* 1908)

Lynn Margulis, taken of her at the III Congress about Scientific Vulgarization, Spain. (Creative Commons Attribution Share Alike Licence; Author: Jpedreira: Date: 2005)

Mileva Marić-Einstein. (Date: 1896)

Photograph of Antonia Maury from her senior year at Vassar College. (Vassar College; Date: *c.* 1887)

Maria Goeppert Mayer (Nobel Foundation; Photographer: unknown; Date: 1963)

(Eleanor) Barbara McClintock (American Philosophical Society; Date: *c.* 1951)

Lise Meitner lecturing at Catholic University, Washington DC. (Date: 1946)

Maud Leonora Menten (Acc 90–105, Records 1920s–1970s, Smithsonian Institution Archives; Author: Smithsonian Institution from United States. Permission: Smithsonian Institution @ Flickr Commons)

Maria Sibylla Merian, portrait. (Author: Unknown: Date: 17th C. Public Domain)

Maryam Mirzakhani at the Seoul convention of the International Congress of Mathematicians. (Archives of the *Mathematisches Forschungsinstitut Oberwolfach*; Creative Commons Attribution Share Alike Licence; Author: Gert-Martin Greuel; Date: 2014)

Maria Mitchell, US astronomer and pioneer of women's rights. (*Sweeper in the Sky: The Life of Maria Mitchell, First Woman Astronomer in America* by Helen Wright, Macmillan Company, New York; from a portrait by H. Dassell; Date: 1851. This is a faithful photographic reproduction of a two-dimensional, public domain work of art)

Maria Marcia Neugebauer. Photo from JPL article – The Venus Mission: How Mariner 2 led the world to the planets. (National Aeronautics and Space Administration (NASA) .Author: Jet Propulsion Laboratory (JPL), Public Domain via Wikipedia Commons; Date: 1962)

Ida Tacke Noddack (*privates Familienalbum*; Permission: Dom de, Creative Commons Attribution Share Alike; Author: Unknown; Date: *c.* 1940)

Emmy Noether, portrait. (Source: Emmy Noether (1882-1935): Author: Unknown; Date: before 1910)

Christiane 'Janni' Nusslein-Volhard (momentum-photo.com/MPI *für Entwicklungsbiologie Tübingen*; Date & Author: Unknown)

Muriel Wheldale Onslow. (From the archives of the Department of Biochemistry, University of Cambridge)

Ruby Payne-Scott as a young woman, photo. (Peter Hall [Ruby Payne-Scott's son]; Date: 2009)

Rózsa Péter. (Dr Béla Andrásfai, Rosza Peter's adopted son; Date: Unknown)

Agnes Louise Wilhelmine Pockels. (Author: unknown; Date: *c.* 1892)

Helen Rhoda Arnold Quinn. (Photo provided by Helen Quinn; Date: unknown)

Mina Rees at home. (City University of New York Archives; Date: Unknown; Author: Unknown)

Scan of photograph/fascsimile found in *The Life of Ellen H. Richards* by Caroline L. Hunt, London, Whitcomb and Barrows, 1912. (Author: Unknown; Date: between *c.* 1890 and 1900)

Mary Ellen Estill Rudin. (Used by permission of American Mathematical Society: Author: Sara Nagreen; Date: Unknown)

Hazel Schmoll as a young woman, portrait. (Carnegie Library for Local History, Boulder, CO, USA, Museum of Boulder collection; Author/Donor: Jones Studio (Boulder, Colorado); Date: 1910–1920)

Beatrice ('Tilly') Shilling in a photograph from *The Woman Engineer*, volume 4, no. 1, December 1934. (Courtesy of the IET Archives and WES)

Michelle Simmons at the ARC Centre of Excellence for Quantum Computation and Communication Technology, cropped video still. (Creative Commons Attribution CC BY licence; Author: AustraliaUnlimited; Date: 2017)

Mary Fairfax, Mrs William Somerville, 1780–1782. Writer on science, painting. (Google Cultural Institute maximum zoom level, Public Domain, – held by Scottish National Gallery; Author: Thomas Phillips, English painter; Date: pre 1845)

'Janet' Jane Ann Ionn Taylor. (Photo courtesy of Professors John and Rosalind Croucher, Sydney, Australia)

Ruth Lichterman Teitelbaum (crouching) wiring the right side of the ENIAC with a new program. (U.S. Army Photo from the archives of the ARL Technical Library)

Valentina Vladimirovna Tereshkova. (The Russian Presidential Press and Information Office; Date: 2017. Creative Commons Attribution Licence)

Mildred Trotter. (Source: Acc. 90–105 Science Service, Records 1920s-1970s, Smithsonian Institution Archives; Author: Unknown; Date: Unknown)

Anna Johnson Pell Wheeler (Courtesy of the Bryn Mawr College Archives; Author: Unknown; Date: Unknown)

Sheila E. Widnall, US Secretary of Air Force, portrait. (This image is a work of a US military or Department of Defense employee, taken or made as part of that person's official duties)

Dr Rosalyn Yalow at her Bronx Veterans Administration Hospital, October 13, 1977, after learning she was one of three American doctors awarded the Nobel Prize for Physiology or Medicine that year. (*Women of Influence,* p.28; Author: US Information Agency; Date; 1977; Creative Commons Attribution Share Alike Licence)

Prof. Ada E. Yonath at Dept. of Computational Biology & Bioinformatics, University of Kerala during her visit to Kerala in 2013. (Home Page Haree Forografie, own work; Author: Hareesh N. Nampoothin; Date: 2013; Creative Commons Attribution Share Alike Licence)